How Knowledge Moves

How Knowledge Moves

Writing the Transnational History of Science and Technology

EDITED BY JOHN KRIGE

THE UNIVERSITY OF CHICAGO PRESS CHICAGO AND LONDON

The University of Chicago Press, Chicago 60637
The University of Chicago Press, Ltd., London

Published 2019
Printed in the United States of America

28 27 26 25 24 23 22 21 20 19   1 2 3 4 5

ISBN-13: 978-0-226-60585-2 (cloth)
ISBN-13: 978-0-226-60599-9 (paper)
ISBN-13: 978-0-226-60604-0 (e-book)
DOI: https://doi.org/10.7208/chicago/9780226606040.001.0001

Library of Congress Cataloging-in-Publication Data

Names: Krige, John, editor.
Title: How knowledge moves : writing the transnational history of science and technology / edited by John Krige.
Description: Chicago : The University of Chicago Press, 2019. | Includes index.
Identifiers: LCCN 2018027426 | ISBN 9780226605852 (cloth : alk. paper) | ISBN 9780226605999 (pbk. : alk. paper) | ISBN 9780226606040 (e-book)
Subjects: LCSH: Technology transfer—United States—History—20th century. | Science—United States—International cooperation—History—20th century. | Technology transfer—Cross-cultural studies. | Science—International cooperation—Cross-cultural studies.
Classification: LCC T174.3.H69 2019 | DDC 303.48/3—dc23
LC record available at https://lccn.loc.gov/2018027426

♾ This paper meets the requirements of ANSI/NISO Z39.48–1992 (Permanence of Paper).

Contents

Introduction: Writing the Transnational History of Science and Technology

John Krige

The transnational approach is now well established in multiple fields of academic history.[1] By decentering the nation-state as the unit of analysis, the transnational lens magnifies complex relationships of interdependency between disparate people and places that crisscross national borders. Science and technology would seem especially susceptible to such analyses, being the products of large and complex social institutions that do, at least in principle, transcend the boundaries of nations and nation-states. Yet even if the practice and circulation of (techno)science would seem to demand a transnational approach, it cannot be said that their historiography, especially of more recent times, has proved adequate to its object. The increasing investment by governments in science, technology, and development, particularly after World War II, has encouraged studies that situate scientific and technological activities and achievements almost exclusively in a national frame. This narrowing of perspective has been reinforced by having the primary archives and institutions of science and technology organized at local, regional, and national levels. The collection of essays in this book invites us instead to stand back and, rather than foregrounding local contexts, to study national actors and institutions first and foremost as nodes in transnational networks that tied people together through "aspiration, expertise, and affiliation of various kinds."[2]

Most of the essays published here were first presented at a workshop held in the School of History and Sociology at the Georgia Institute of Technology in 2016. On that occasion some two-dozen people from many parts of the world discussed the substance of their precirculated papers and reflected on *how* to write the transnational history of science and technology. Inspired by the possibilities of a transnational approach as a "perspective rather than a clear-cut method," but aware of the many rich seams that it opens up to inquiry, we sought to capture its specificity when scientific and technological *knowledge* in its many modes of being was the object of inquiry.[3] James Secord has suggested that the "central question" of the history of science is "How and why does knowledge circulate?"[4] This is obviously one of the central questions of the history of technology too, at least if we understand technology not simply as an artifact but as a form of knowledge, including tacit "know-how," that is embedded in material objects and practices that are designed to transform the world around us.[5] By emphasizing the transnational dimension of how and why knowledge circulates across borders, we fused a "central question" of the history of science and technology with an important new approach being espoused by many other kinds of historians without diffusing our object or unduly blurring its contribution to a transnational agenda.

The scientific and technological knowledge that flows from one node to another in a network can take many forms, is mobilized in diverse social and institutional settings, and transforms social relationships on different scales. It can be tacit or propositional, cutting-edge or mundane. It is expressed in multiple practices—experimental, educational, managerial, policy oriented—as well as in modes of control and domination. It engages universities and corporations, missionaries and philanthropic foundations, national governments as well as regional and international organizations.

Knowledge, as "information," as "expertise," as "know-how," crosses borders in many ways—in written or printed form (books, including textbooks and manuals, letters, newspapers, academic publications, technical reports, blueprints, trade journals) or embedded in devices (like an inertial guidance system) as well as in living things (like human beings and cloned animals). The international circulation of knowledge made possible today by sophisticated communications technologies gives the impression that—proprietary and classified knowledge apart—knowledge simply moves across borders without friction in a "flat" networked world. The essays in this volume take a very different position. They problematize circulation itself. They emphasize the social and material constraints that *impede* the movement of knowledge across borders. They pay attention to the material culture, to the mundane practices, and to the local and national resources that have to be mobilized to build and to maintain transnational networks that bind together social actors with a commitment to producing and sharing scientific and technological knowledge. They share Charles Bright and Michael Geyer's concern that much writing on globalization "tends to presume the (relative) openness of the world" rather than analyzing "the structured networks and webs through which interconnections are made and maintained—as well as contested and renegotiated."[6] They are a response to Chris Bayly's injunction to "grapple with the problem of modeling the element of power into the concept of circulation."[7]

This volume deals with transnational encounters during the long twentieth century, with particular emphasis on the period after 1945. The specificity of this periodization for the transnational history of knowledge circulation lies in the increasing centrality of science and technology to the economic, political, and military strength of the modern state. In 1945 scientific discovery and technological innovation moved to the heart of the political process. Beginning in the 1960s countries came to be judged by the percentage of their gross national products dedicated to research and development. This new official metric of national achievement was embedded in a geopolitical context that pitted Western democracies against communist rivals, while the global order was transformed by the process of decolonization. The United Nations was established in 1945 with 51 sovereign member states; by 1966 it had 122. The state cannot but be a central actor—whose role needs to be unraveled—in a transnational history of knowledge flows in the twentieth century. More specifically, the United States, as the world's leading scientific and technological power for the first twenty-five years after the end of World War II, if not before, and with long-standing aspirations to build a democratic world order, cannot but be implicated in the construction of transnational networks on a global scale. Michael McGerr is surely right to suggest, with a touch of irony, that "[t]ransnationalism may well be a form of imperialism; the transnational world may well emerge from such unlovely phenomena as American power and American exceptionalism."[8] Correlatively, the essays here provide a benchmark, a point of reference for studies that engage other periods of time and other regions of the world, studies that adopt space-time frames in which the relationship between scientific and technological knowledge and the state took different forms, and when the movement of such knowledge across borders was subject to different modes of control from different centers of power.

This volume specifically directs attention to the processes shaping knowledge flows *when national borders matter.* It deals with situations in which an analysis of knowledge-in-movement cannot ignore its value as a national resource. When national power and sovereignty are entangled with access to and control over knowledge in its multiple forms and when scientific and technological prowess are markers of national achievement (Michael Adas) and legibility is of cardinal importance for governance (James C. Scott)—"two key inheritances from nineteenth-century rationalism"—states will affirm their right to regulate the circulation of knowledge.[9] The ideology of globalization celebrates deregulation, privatization, and individual gratification via access to "free markets," reducing the state to the role of facilitator. A transnational history of knowledge as developed here treats knowledge not as a commodity to be bought and sold but as a resource to be nurtured and protected. It is a resource whose full value is thrown into relief when we track its trajectory beyond the national frame—more precisely, when we follow knowledge and its bearers as they chart their way through the complex apparatus put in place in the name of national sovereignty to control their passage beyond borders.

Five main ideas inform the analysis that follows: the centrality of travel, the role of the regulatory state, the meaning of "borders" and "networks," the significance of nationality and political allegiance, and the intersection between the local and the global. Their pertinence is fleshed out in the individual essays that make up this volume, each of which is briefly introduced in the second section of the introduction. Collectively these essays make a strong statement for the novelty and importance of a transnational approach to the circulation of scientific and technological knowledge, a "way of seeing" that reveals new, unexpected, and previously occluded dimensions of the social world.

THE CENTRALITY OF TRAVEL How are we to characterize the movement of knowledge across borders? Workshop participants were critical of popular metaphors like "circulation" and "flow" and of Emily Rosenberg's use of the term "currents," in analogy to the flow of electricity, to describe this movement.[10] The term "currents" plays down the role of human agency and intentionality and lacks the sense of purpose that is crucial to plotting the paths mapped out by knowledge itineraries. It may capture the movement of "disembodied" information or "knowledge" that encircles the globe with modern communications technologies, but it skips over the rough terrain and multilayered interactions that are constitutive of the transnational moments described here. "Circulation" also implies a "looping back" that often does not happen in a transnational context, where knowledge vectors can be unidirectional or zigzag haphazardly across multiple borders.

To avoid these pitfalls, this volume treats the movement of knowledge embodied in people (and things) as a *social accomplishment*. It pays special attention to *travel* across borders, putting human individuals and their motivations for going abroad and for engaging with and transforming (and being transformed by) others in a foreign place at the core of the analysis.[11] Travel requires funds, which are usually secured from institutional patrons, be they local, national, regional, or international, that in turn have their own motives for supporting the transnational movement of knowledge. It also requires documents—passports, visas, export licenses—that authorize one to move abroad with one's knowledge. Travel calls forth a vast bureaucratic apparatus and leaves a paper trail that documents every move of the subject, a trail that reveals the contingency of border crossings, of which most travelers are often not even aware. Unraveling how travel is possible exposes the internal negotiations between travelers and/or their patrons and the departments and agencies of the regulatory state that require that certain conditions be met before people can move abroad "freely" with their knowledge (from the mundane, like travel documents, to the contentious, like an acceptable political stance, to the detailed, like where you can go and whom you can visit without jeopardizing national security or economic competitiveness). It is in these piles of paperwork that states perform their sovereignty. Travel also involves engaging with and adjusting to people of different cultures working in institutional contexts that are shaped by unfamiliar norms and who oppose the arrival of disruptive knowledge from abroad. Thinking of movement as a social achievement foregrounds the contingency and the labor required for scientific and technological knowledge and "knowledgeable bodies" (to use Mario Daniels's happy phrase—see chapter 1) to cross borders.

The movement of knowledge from one site to another is sometimes theorized using a linear model of "transfer" that sees production followed by circulation, followed by selective adaption by local actors at the "receiving end." The epistemological hubris that underpins this center-periphery model, itself an echo of Cold War modernization theory, has been discredited by studies that emphasize the local specificity of knowledge production, wherever it may occur. Networks that bind together diverse sites of knowledge production, and that build communities that reciprocally share standardized knowledge, provide a richer account of the "global" production and circulation of knowledge than the "hub-and-spoke" representation of center and periphery.[12] Kapil Raj goes further in his study of the transnational sharing of knowledge between British officials and indigenous cognoscenti in eighteenth- and nineteenth-century India.[13] He stresses the importance of face-to-face contact as a site of engagement, transformation, and knowledge co-production for all parties engaged in the transaction, a transaction, it should be stressed, made possible by travel along imperial sea routes. Raj highlights "the mutable nature of knowledge-makers themselves, as much as of the knowledges and skills that they embodied, their transformations and reconfigurations in the course of their geographical and social displacements."[14] Face-to-face encounters in which transnational social actors dissolve "borders" are dynamic engagements in which both formal propositional and tacit craft knowledge are exchanged between social actors. They can also be unsettling, transformative experiences, occurring in unfamiliar places and dominated by asymmetric power relations. By going beyond the level of networks, and by digging down to the microlevel of interpersonal exchanges made possible by travel, several chapters in this volume throw light on the tensions, misunderstandings, and conflicts—as well as the productive encounters—that constitute transnational knowledge formation.

THE ROLE OF THE REGULATORY STATE The transnational approach effaces the national container as the unit of analysis. However, as the previous paragraph implies, it is one thing for a historian to break the national frame so as to allow movement across borders to come into view. It is another for a transnational actor to do so in practice. Movement across national (or regional) borders injects the *regulatory state* back into the core of transnational history. This is tragically evident from the dramatic plight of refugees fleeing war zones in Europe and from the anxiety felt by citizens of countries recently targeted in an unexpected executive order by the American president in January 2017. Cross-border movement is regulated by policies of inclusion and exclusion and is controlled using a variety of instruments, from the pen to the sword. Regulating the flow of knowledge during the Cold War, in particular, required striking a "balance" between local practitioners acquiring the information brought into the country by a foreign national and the risks to national security and economic competitiveness of the presence of a foreign national on local soil. Ideologies claiming that knowledge circulates freely in a world that is "flat" forget the restrictions placed by "free world" governments on the movement of communist and left-leaning scientists during the Cold War. They ignore the near-total prohibition of knowledge sharing with some countries (like North Korea) and with researchers from certain military-centered research institutions in China today. They also ignore the gray area of sensitive but unclassified knowledge that is controlled by an array of regulatory regimes that selectively discriminate against some countries, firms, and individuals, and whose boundaries are constantly (re)negotiated as the cutting edge of knowledge shifts within an existing field or carves open entirely new fields of inquiry.[15]

Akira Iriye has stressed the new insights that became available when international history relinquished its emphasis on formal relationships between states and explored transnational phenomena produced by nonstate actors.[16] In this volume the boundaries between state and nonstate actors are blurred. Diplomacy and high politics are indeed at one remove from the day-to-day activities of the regulatory state that permeates transnational lives. But they do impinge on those activities by establishing, at an interstate level, the regulatory framework that sets the terms for transnational movement. Within that framework the transnational actors we discuss, while usually not having explicit roles as representatives of their governments, do have a variety of relationships to the state other than being anonymous "desk officers" who authorize cross-border movement. Trained scientists and engineers are a national asset, and their advanced knowledge is a resource that readily serves as an instrument of "soft power," thus blurring their "nonstate" identity. Some capitalized on state-driven projects that valued their specific skills to promote "American" values abroad. Some exploited friendly relationships between governments to advance their national research agendas. Some took advantage of their close ties with officials of the regulatory state to expedite the travel of people they were sponsoring. The central point to bear in mind is that the association of knowledge with power—power to transform nature and the social world—unwittingly or otherwise engages nonstate actors with national governments that compete with one another across a spectrum of activities.

BORDERS AND NETWORKS What is a border, or a national frontier? Charles Maier writes, "A frontier is partially a virtual construction. It is as much a site of the demonstrative extent of power as a real barrier. . . . [T]he frontier defines authority, and those who govern lose legitimacy if their frontiers become totally permeable."[17] This conception of "frontier" refers to more than the geographically bounded limits to the territory of the nation-state. National borders need not be material nor are they policed only at physical customs and immigration posts. Border crossings can occur when foreign nationals attend a trade show, are present in a lecture theater or a laboratory, or are trained in the use of sophisticated equipment that is dual-use, civilian and military. Transnational "boundaries" can even be embedded within sensitive technological devices (like gas centrifuges for uranium enrichment) by denying foreign nationals access to some of the knowledge constitutive of them without permission.[18] Boundary work is constitutive of borders.

The legal apparatus of regulatory states produces taxonomies that discriminate between people and knowledge that can travel unhindered across borders and those that require authorization (a visa, an export license, . . .) on pain of sanction. These taxonomic grids are contingent responses to emerging political situations in which the state demonstrates its power by erecting barriers to free movement. Historically states have always taken measures to protect knowledge, and the movement of knowledgeable people, on which their power depends.[19] What sets the period covered by this volume apart is the centrality of the production and circulation of knowledge to the consolidation and projection of state power. This is the geopolitical context in which, immediately after World War II, the entire system of classification was reinforced by the apparatus of the national security state to control the free circulation of sensitive knowledge. In which visa policies evolved in the 1950s to target the cross-border movement of "knowledgeable bodies." In which export controls were expanded to regulate the circulation of sensitive but unclassified technical data and information. In which the distinction between basic and applied science was developed, in consultation with leading members of the American research community, to create a space for the transnational circulation of "basic" knowledge while legitimizing tighter controls on socially useful products and processes ("applied science")—the context in which "basic science" was internationalized, while "applied science" was nationalized.

In commenting on his magisterial new book on territoriality, Charles Maier recently wondered whether "the emancipation of information from local constraints in the digital age herald[s] a post-territorial dispensation" or whether "national authorities [will] manage to control data and communication, whether through traditional sovereign claims or new technologies." This is too quick. Let us not be deluded by two billion active Facebook users into thinking that "local constraints" on knowledge flows are being dissolved by communications technologies that circle the globe, or that states can do little more than have recourse to sovereignty or surveillance to reassert their control over the circulation of data. On the contrary. Several chapters in this volume highlight the range of instruments that national governments, notably the United States, have devised to regulate the flow of knowledge across "borders," some of them "epistemic" (fundamental research vs. proprietary or industrial research), some enshrined in legal regimes (export controls), some flowing from international agreements (Atoms for Peace), some built into the structure of patronage (classification). They remind us of the determination by states to coconstruct national systems of knowledge production *and* the regulatory regimes needed to control the circulation of knowledge within and across borders. These efforts may eventually "fail" or at least be reconfigured in response to the emergence and domination of an international system subtended by privatization and deregulation. Their contingency apart, today they "still testify to the enduring importance of territoriality," as Maier emphasizes.[20]

Transnational networks connect remote places where knowledge is produced, exchanged, and appropriated. They do not respect borders. It is common practice, but misleading, to think of them simply as links that can be represented by lines drawn on a two-dimensional "map." The world, as Frederick Cooper puts it, is "a space where economic and political relations are very uneven; it is full of lumps, places where power coalesces surrounded by those where it does not, places where social relations become dense amid others that are diffuse."[21] Global inequalities in the production and appropriation of science and technology demand that we imagine networks as lumpy, three-dimensional structures made up of hierarchical interpersonal encounters. Transnational actors do not simply travel from one place to another; their knowledge is an asset that they deploy to reconfigure existing spaces and themselves and what they know. Correlatively, networks are not rigid struts but dynamic relationships that evolve over time and that persist only as long as the networked participants reap some benefit from them. Neither the *links* nor the *nodes* in a conventionally understood network can, in the perspective developed in this volume, be taken for granted as unproblematic foundations for historical analysis.

Reciprocity, a sense of mutual advantage, is a sine qua non for ongoing and successful participation in a network. The possibility of sharing knowledge is obviously an important motivation for establishing a network linking researchers at different places. But it is not the only, or even the most important, reason for the transnational knowledge flows discussed in this volume. The reciprocity that maintains the engagement of actors in networks need not be measured only in terms of the knowledge that is shared. For governments, for example, transnational links can also serve as instruments of foreign policy and of economic and cultural hegemony, they can be intrinsic to building a regional "federated" system of states, they can help establish footholds in an indigenous community, they can provide access to resources not available at home, and so on. Though the chapters in this volume highlight the transnational movement of science and technology as the object of analysis, they recognize that stakeholders have many other—though related—motives for building transnational networks. These too are factored into stakeholders' calculations of reciprocity and into their assessments of why it is worth investing time, money, and resources in building and maintaining transnational networks.

The American-led project to "modernize" countries casting off the yoke of colonialism in the 1950s exemplifies the multiple dimensions of reciprocity built on the scaffolding of the transnational movement of knowledge. The asymmetry in knowledge/power between the industrialized North and the not-yet-"modern" South provided an opportunity for scholars, foundations, and various branches of the government to refashion the structure of societies and the identities of elites and peasants alike in line with American models of social organization, individual mobility, political process, and democratic values. Point IV of President Truman's inaugural speech in January 1949 pledged to use America's scientific advances, industrial progress, and "imponderable resources in technological knowledge" for the "improvement and growth of underdeveloped areas."[22] In a stroke Truman collapsed the complex histories and diverse social structures of a multitude of countries struggling toward sovereignty into a one-dimensional "lack," a generalized absence amenable to a single solution: American-led modernization. He also legitimated a discourse, the language of "development," that provided a policy goal that all could share. It fused the local aspirations of fractions of local, national elites with the global ambitions of liberal internationalists in the industrialized powers. The telos of "development" was not tarnished by the presumption of superiority that had inspired the "civilizing mission" embarked on by the European imperialist powers: "We don't do empire," as Donald Rumsfeld (George Bush II's secretary of defense) famously proclaimed to a reporter from the TV channel Al-Jazeera.[23] Helping "underdeveloped" countries climb the technological ladder through successive stages of economic growth was denuded of political overtones by mobilizing "objective" science and technology in multiple fields of the natural and social sciences in pursuit of the benign goal of human improvement.[24]

Modernizing knowledge flowed "out" under the auspices of the Ford and Rockefeller Foundations and the State Department and other branches of the government, transported by traveling expert consultants and teachers in economics, in agriculture, in demography, in physics and engineering, and in social planning. New knowledges were implanted by formal training (e.g., in dedicated courses in nuclear engineering in American universities, in experimental farms in India, in standardized physics textbooks in Latin America), by demonstration ("development . . . suddenly came down to a very simple proposition: one man seeing his neighbor doing better than he was doing"), by deploying economic incentives supplemented by force (e.g., in human sterilization programs in India).[25] In a reciprocal movement knowledge flowed back to central sites of power, where it was processed by teams of researchers, was published in international journals, boosted individual careers, and secured further funding for a variety of academic programs that enhanced the asymmetries of knowledge and power they were supposed to erode.[26]

Transnational actors refashion the local sites at which they put their knowledge to work and are refashioned themselves. This refashioning empowers some, disadvantages others, creates new opportunities for those who are included in their transformative projects, and marginalizes those who are excluded or who oppose those projects. In imperial or neocolonial contexts the engagement occurs in what Mary Louise Pratt calls "contact zones," "social spaces where disparate cultures meet, clash, and grapple with each other, often in highly asymmetrical relations of domination and subordination."[27] When the "external" intersects with the local in an asymmetric relationship of knowledge/power, it can disrupt existing social relations, destroy local knowledge and traditions, and transform a backwater into a center of knowledge production in the global political economy of knowledge creation and distribution. A transnational history of the movement of science and technology across borders changes our perception of where knowledge is produced, creating new maps of innovation and use that put multiple nodes in a network in sometimes unexpected relationships with one another.

NATIONALITY AND POLITICAL ALLEGIANCE Transnational history tends to celebrate the "fluid," "hybrid" identities assumed by people who are transformed by their engagements with different cultures and ways of life. States are uneasy with such ambiguity, however, and go to great lengths to define the national identity (and political allegiance, or "loyalty") of those who are engaged in the production and circulation of knowledge across borders. The successful functioning of the regulatory apparatus requires nothing less: as stressed above, knowledge is deemed to have crossed a "border" whenever it is exchanged between people of different nationalities. Such reasoning presumes that foreign nationals' primary allegiance will always be to their home countries, so that they can serve as conduits through which sensitive knowledge can be "leaked" abroad. Nationality is "fluid" only in the sense that it has no essence, no single meaning in the lifeworld of a transnational actor. Transnational subjectivities are blurred hybrids that dilute national identity; they are lived ambiguities that are incompatible with the legibility sought after, and imposed, by the bureaucratic state system. Fixed in taxonomic grids on an official document, if not in stone, nationality serves as a social category of inclusion or exclusion that can be used to gain or to deny access to the benefits that come from being formally recognized as a citizen.

The ideology of scientific internationalism brackets off nationality. Historically the members of the research community defined their identity as one that was committed to the collective pursuit of "objective truth," to the discovery of facts about the world that provided a solid foundation for convergence and agreement among different inquirers irrespective of their race, gender, and religious or political persuasions. Their loyalty to universal science transcended the concerns of nation, and they decried attempts to impede the international circulation of knowledge by state intervention (except in times of war). There are contemporary expressions of this legacy in the construction of multiple modes of international scientific and technological collaboration. These are also contemporary expressions of Robert Merton's ethos of science as expressed in the norm of universalism—that is, the view that the scientific community disregards the race, nationality, culture, and gender of a participant when assessing the validity of truth-claims.[28]

The scope of the "international" is bounded.[29] The high profile enjoyed in academia by scientific internationalism obscures the collaborations that are "controlled" for national or ideological reasons, the papers that cannot be shared, the self-censorship that places constraints on what can be said, the countries and their nationals that are not admitted to the "international conference" in the first place. It is trivial to note that for every Mertonian norm we find a counternorm in practice.[30] It is equally trivial to reduce to a "counternorm" the invocation of socially constructed criteria of identity that exclude individuals from transnational networks. Scientific internationalism is polyvalent. It helps structure a community of scholars committed to sharing knowledge across borders. It serves as a source of solidarity among them in the face of nationalistically aggressive political regimes. *And* it is a malleable ideology that concedes that the knowledge/power nexus demands that the scope of the "international" circulation of knowledge be restricted to those whose nationality or political allegiance is not deemed to threaten state interests.

The international circulation of knowledge in journals, workshops, and conferences does not mean that nationality and political allegiance are irrelevant analytical categories in a transnational study of scientific exchanges. It simply indicates that the transnational transaction in question was constructed around knowledge that was not deemed by the regulatory state to undermine national security or economic competitiveness, and that it was shared between protagonists whose loyalty to any political or ideological creed was deemed to be irrelevant.[31] Controlled movement and the free circulation of knowledge are coconstructed spheres whose limits are defined and enforced by the state apparatus. Knowledge is a national resource, and the art of statecraft lies in defining policies and instruments that help draw the line between what kind of knowledge will be shared with (or denied to) whom.[32] The willingness of the research community to deny knowledge to others on the grounds of national or political allegiance is indicative of their internalization of the values of a national regulatory regime that imposes boundaries beyond which knowledge may not flow freely. There is a price to pay for patronage.

THE LOCAL AND THE GLOBAL Transnational actors forge ties between the microlevel of personal biography, a local context that places a premium on international travel, the national level as defined by the regulatory state, and the "global" as constituted by the multiple sites where they alight as they and their knowledge travel from one country to another. These different scales at which transnational movement is "performed" are loosely coupled with one another: loosely because contingently, because the threads that tie them together to constitute a network can be snapped by unexpected and unplanned-for developments that can occur at any of the levels. These can be relatively mundane, like disputes between collaborating individuals; relatively frequent, like the loss of funding; and totally disruptive, like a major change in travel regulations or the outbreak of war. The transnational movement of knowledge flourishes in a context in which its advantages are valued and its practices embedded in institutional goals that can resist the storms that buffet the movement of ideas, people, and things across borders.

The transnational encounter is facilitated by the standardization of knowledge-producing practices in a network. Standardization is both a practical advantage and an epistemological condition for sustained transnational transactions. The coproduction of knowledge at multiple local sites requires a collective commitment to standardized equipment and the acquisition of standardized protocols, techniques, and disciplined procedures to use and manipulate it effectively. It is standardization that closes the gap between local and universal scientific and technological knowledge.[33] Understanding how reliable knowledge is produced in (transnational) networks involves "understanding movement: of people, things, 'languages' and techniques."[34] Together they define "best practice," ways of doing that are required to domesticate the capriciousness of equipment and the recalcitrance of nature and of "social facts."

The interpenetration of the local and the global has been facilitated by the use of English as the lingua franca in scientific and technological exchanges the world over. This coupling has not taken place of its own accord. It was not an "inevitable" consequence of the development of communications technologies that linked living rooms in the industrialized West with remote villages in China or command and control centers in Colorado with battlefields in Afghanistan. On the contrary, as Michael Gordin has argued, one major consequence of Cold War rivalry was the hegemony of English and its role in constructing transnational research communities—a development that has been contested, of course.[35] Once the Eisenhower administration grasped the extent of Soviet scientific and technological capabilities, it actively promoted the use of English among its allies in Europe in the hope of efficiently pooling advanced knowledge to maintain leadership. It also embarked on a major program to translate Soviet scientific and technological literature into English. Janet Martin-Nielsen estimates that the US military injected over $20 million into machine translation between the end of World War II and 1965, one of the first major uses of computers for nonnumerical tasks.[36] The National Defense Education Act of 1958 singled out three areas of strategic importance in the face of the communist threat: mathematics, science, and, third, foreign languages. In what was called "a war for men's minds," Martin-Nielsen tells us, "language and linguistics formed a critical element of the rise of American leadership in the new world order," contributing to the Anglicization not only of science and international affairs but more generally of global business practices and of culture.[37]

The promotion of English as a lingua franca has been complemented by the global expansion of local and national education and training programs that have helped standardize the "communicative practices" that are central to building transnational knowledge communities. The dissemination of nuclear science and technology in the 1950s, especially to "developing" countries, was an early exemplar of this process in a sensitive field. It helped kick-start indigenous nuclear programs envisioned by people like Homi Bhabha, who combined his immense scientific and political talent with a vision of the future that saw nuclear power too-cheap-to-meter as lifting India's standard of living up to that of America's.[38] The US Atomic Energy Commission provided nuclear materials. Fledgling nuclear engineering programs at North Carolina State, Penn State, MIT, and other universities provided demanding, hands-on training for the scientists and engineers who would run reactors provided in the framework of Eisenhower's Atoms for Peace program. Bill Leslie tells us that, by the time Bhabha died tragically in 1966, he had built the largest and best-funded laboratory in India, with a staff of 8,500, including 2,000 scientists and engineers. Glenn Seaborg, chairman of the US Atomic Energy Commission, could only marvel (with some trepidation) at the "truly remarkable peaceful nuclear power program" in India.[39] Today the unambiguous political and propagandistic dimension of these early initiatives to build a global nuclear community aligned with US interests has given way to far broader educational programs, particularly in STEM fields. Between 1991 and 2011 over 235,000 foreign nationals received PhDs in science and engineering in the United States, almost half of them from just three emerging economies: China, India, and South Korea.[40] Business and management courses are even more popular.[41]

The construction of a transnational knowledge community collides with vested local interests and their histories that resist its pressure to homogenize and to standardize. John Burnham has explained how the "extreme nationalism and provincialism" that divided research communities in psychiatry in the interwar period were gradually eroded after World War II. Beginning in 1950, with inflexion points around 1968 and 1980, a transnational community of publishing psychiatrists (and perhaps even practicing psychiatrists) was inadvertently constructed by the increasingly widespread use of English. This expansion was opposed in the name of national autonomy in Germany and, above all, in France, where it was criticized as "American colonization" or "American intellectual imperialism." Such labels were an implicit acknowledgment of the leading role played by US practitioners in defining the contours of the research frontier. The locus of power/knowledge had shifted away from Europe. Now the "linguistically handicapped Americans," as Burnham calls them, obliged others to publish in English to announce their innovations and to keep up to date in the field, so securing transnational visibility and awareness.[42] American preeminence (and limited language skills) constructed a transnational community that merged distinct local components with "universal," if anglophone, components. One had no choice but to communicate in English if one sought to transcend the limits of the local and to garner scientific credit, social capital, and international recognition, all of which were highly prized domestically. A working knowledge of English paved the highway from the local to the global.

The construction of a transnational community of scholars who are at ease in English, familiar with American paradigms and questions, and trained in the standardized best practices of research can be of mutual benefit. It enables scholars in the resource-rich research communities to tap comfortably into the global pool of knowledge. It enables researchers in emerging powers to work at the cutting edge of knowledge production, perhaps leapfrogging over decades of relative "backwardness" in their home countries. From the nation-state's point of view, the advantage of being rapidly able to exploit the newest knowledge has to be weighed against the danger of "creating one's own competitor." There is a constant tension between the open circulation of knowledge by shared communicative practices and the pressure to restrict its circulation to protect the national interest, however that may be defined.

Before the workshop was over the participants turned what they had learned back on themselves. As Gabrielle M. Spiegel observed in her 2009 American Historical Association presidential address, one of the marks of transnational history and related new fields of inquiry was that they entailed "the study of discontinuities in the experiences of, and displacements of location in, the lives of their subjects."[43] Was it a coincidence that the majority in our workshop had traveled extensively, had lived on several continents, had changed passports and nationality, sometimes more than once, or spoke several languages other than English? We realized that the increasing hegemony of the English language could elide the specificity of the local—but it did facilitate transnational conversations (including the one presented in this collection of essays, in which the majority of the contributions are from scholars whose first language is not English!). We recognized that most of us had a somewhat diffuse "cosmopolitan" identity that blurred our sense of national belonging—but it did heighten our awareness of the persistent salience of nation and state. It is often stressed that writing transnational history requires familiarity with foreign languages and archives. A transnational approach also resonates with the life experiences of a particular type of person, an individual who feels comfortable writing history from a position that is detached from any deep sense of national allegiance, an individual who lives, sometimes uneasily, in a liminal gray zone that enhances his or her critique of any form of national exceptionalism but that also exposes him or her to the fragility of being an "outsider" who is at once comfortable nowhere and everywhere.

The essays presented in this volume vary in scope and scale and cover transnational relations between many regions of the world. All of them deal with the long twentieth century, an era in which interimperial scientific collaboration (promoted as "scientific internationalism") was disrupted by the centrifugal forces of Cold War rivalry and of decolonization that together reconfigured the global space in which transnational (techno)scientific networks were built.[44] All take the United States as one node in the transnational network, though not always the dominant or central one. It is linked southward to Latin America, and it has ties to China, India, and Japan in Asia, as well as, across the Atlantic, to Italy (and into Kenya) and "French" Algeria. This is not a "hub-and-spoke" model: the lines of movement are multidirectional, have very different points of departure, and engage many different places at a scale below the continental. Physics of various kinds, mathematics, and agriculture are heavily represented, but so too is space science and its applications and the social sciences. The range of actors includes subaltern and elite individuals, philanthropic foundations, scientific organizations, governments and government agencies, and international, not to say global, institutions. These are not case studies. They are exercises in writing the transnational history of science and technology that break with the national frame in multiple registers, sometimes quite explicitly exposing the limits of nationalist historiographies.

Part I has two chapters that describe the day-to-day workings of the regulatory state in Cold War America. In the first, Mario Daniels traces the intellectual history of US policies for policing its borders in the decade or so after World War II, focusing on the role of passports and visas as instruments to control the circulation of "knowledgeable bodies." Passports and visas are among the mundane, if essential, materials required by foreign nationals to enter another country. They are visible manifestations of state control in transnational travel, the paperwork that does boundary work. Daniels's central point is that scientists and engineers have been singled out for close scrutiny by government authorities not simply because of their possible left-leaning political tendencies but because of the scientific-technological knowledge that they have and might enrich during their visits. Visa controls on scientists and engineers and export controls on knowledge are complementary instruments reinforcing each other to regulate the transnational movement of "knowledgeable bodies" that were perceived as posing a threat to national security.

Most scholars think of knowledge flows as being restricted primarily by classification or intellectual property laws. Daniels's chapter emphasizes instead the role of export controls on sensitive, unclassified knowledge as an instrument deployed by the national security state to restrict the freedom of movement across borders. My contribution, chapter 2, extends their field of action to include the regulation of both free trade and academic freedom. It highlights the nature and scope of export controls on the acquisition of tacit and formal knowledge through training and face-to-face interaction in academic and corporate settings from the 1970s to the early 2000s. Violations of these knowledge controls have led to imprisonment for individuals and to heavy fines for high-tech industries. Their importance today is commensurate with the growing concern about the transfer of intangible, tacit knowledge and know-how by firms trading with the communist bloc and by universities that train foreign nationals from countries of concern in their classrooms and laboratories. The recalibration of American power in a new age of globalization, and the rise of China as a major economic and military rival, have led to the quiet but steady expansion of the depth and breadth of the regulatory powers of the US national security state—further fueled and legitimated by the terrorist attacks of 2001.

Part II has five chapters, which deal with knowledge in colonial and postcolonial contexts. In chapter 3 Tiago Saraiva describes the brutal transformation of local patterns of agricultural activity on the fertile Mitidja plains in Algeria by the French colonial government. Large vineyards producing wine and orchards of citrus trees replaced traditional modes of production, transforming thousands of indigenous farmers, herdsmen, and fruit growers into wage laborers. After several efforts to stabilize the quality of citrus fruit sent to metropolitan markets, the head of the colonial botanical service imported and cloned Californian navel oranges produced by cooperatives near Los Angeles. These standardized fruits were successfully grown on small, five- to ten-hectare farms by white settlers (*pieds noirs*). Saraiva treats these oranges as "thick technoscientific things" that traveled across the Atlantic *along with* the social institutions, labor relations, and ideologies of cooperative farming. "Modernization" in the "contact zone" between colonial rulers and dominated *fellah*s was premised on the construction of bonds of solidarity between the French settler elite that rooted them in the soil, unlike the owners of the large wine estates. Decolonization was correspondingly violent: one of the first bombs exploded by Algerian rebels in 1954 was at the Boufarik citrus co-op. The *pieds noirs* went back to mainland France in the early 1960s when Algeria was granted its independence, going on to form the core of the xenophobic, nationalist right-wing Front National political party of Jean-Marie Le Pen and his family.

The multiple "modernities" displayed by the introduction of agricultural equipment into India before and after decolonization are the subject of Prakash Kumar's chapter. Kumar describes three moments, each employing a different approach to improving agricultural productivity: the use of an immense tractor to replace human and animal labor in preparing fields for planting; the introduction of small-scale, locally produced implements for use on family farms, along with the establishment of a US-inspired training college; and the more familiar use of implements and fertilizers on large farms that were part of what came to be called the Green Revolution. Each of these systems required training in different skills and techniques and was embedded in sets of social relationships among the indigenous farmers themselves and between them and the colonial and postcolonial state. Kumar blurs any teleological account of the much-studied Green Revolution, showing that there were alternative solutions to improving agricultural output that called for different skills and involved different relationships between growers and the state.

In chapter 5 Miriam Kingsberg Kadia deals with social science as an instrument of "soft power" in the postwar reconstruction of civil society in occupied Japan after World War II. Inspired by a widely acclaimed study by anthropologist Ruth Benedict, American scholars were persuaded that the Japanese were not "pathologically deviant"—meaning aggressive, authoritarian, rigid, and fearful of dishonor. These academics, some of them no more than ABD (all-but-dissertation) graduate students in sociology, sought not to punish but to refashion by instilling "cosmopolitan" values of democracy, capitalism, and peace in the new intellectual leaders of the defeated nation. Rather than impose these values arbitrarily from above, they nurtured tendencies already present in the tradition of Japanese social science fieldwork, which encouraged collaborative teamwork. Exhausting expeditions into the countryside, made on behalf of the American Occupation authorities, to study the impact of land reform became direct exercises in democracy, emphasizing egalitarianism and discussion and dissent in the pursuit of "objective" truths. US Occupation forces and their agents used seminars, language classes, and the distribution of 1.25 million English-language books to local libraries all over the country in an attempt to root out undemocratic, fascist tendencies that could easily spill over into communism. By the mid-1950s most senior Japanese social scientists in academia had been exposed to the transnational exercise of American "soft power," while the American scholars who returned home built new and prestigious academic careers in Japanese studies. Kadia nicely illustrates the role of American power in the construction of transnational scientific communities—a point taken up again in the afterword.[45]

The last two chapters in this section deal with space. Asif Siddiqi's study of an Italian American satellite launch base near the equator off the Kenyan coast reminds us of the long shadow that colonial and postcolonial relations cast over countries that are nominally independent. To make his case he demands that we shift the gaze of our transnational lens to the globally dispersed *sites* at which knowledge is produced and circulated. Geographical or physical sites are also spaces in which a multitude of social relations between a diverse array of actors intersect, each contributing in its own way to the "success" of the transnational project. If we concentrate only on the launcher, the satellite, and its scientific results, the "project" in usual space history parlance, we ignore the role of local governments and indigenous peoples as essential social actors, and we deftly efface the violence that structured non-project-related phenomena. The Italian authorities capitalized on colonial power relations to establish their base off Kenya, siting the base in international waters to elude the regulatory gaze of the state. Postcolonial indifference to the plight of indigenous tribes enabled them to continue mostly unhindered after independence. The town of Malindi was transformed by the foreign invasion into a squalid tourist resort, a haven for mafiosi escaping punishment in Italy, and a hotbed of trafficking in underage girls. The high-tech knowledge brought into the region by the space scientists and engineers remained enclosed in a privileged enclave, circulating back through global workshops, conferences, and publications that helped build careers, but circulating over and beyond the reach of indigenous local actors, who were largely excluded from its benefits. When knowledge moves across borders, it can have very uneven effects at the sites where it is deployed, trading on existing patterns of exploitation and exclusion for its accomplishments.

In chapter 7 Neil M. Maher, an environmental historian, describes NASA's use of the *Landsat* Earth observation satellite to build an international community of trained scientists and engineers who could interpret *Landsat*'s data and use it to intervene in their environment. NASA set up training courses and organized symposia that brought together local communities, who learned how to use satellite data to detect deforestation and desertification, to measure crop yields and identify crop diseases, and to recognize the signatures of valuable natural resources in satellite images. The internationalization of *Landsat*'s data temporarily secured additional financial support for the program from a skeptical Congress and promoted an image of a benevolent and generous United States abroad. Local scientists, engineers, and governments were empowered by knowing how to acquire and make use of *Landsat* data themselves, overcoming fears that it violated their national sovereignty by putting another tool for their exploitation in American hands (e.g., by enabling US corporations to invest in extracting natural resources detected on foreign soil or by manipulating commodity markets when *Landsat* detected low yields from fields abroad). Transnational networks of *Landsat* users distributed knowledge of the planet globally, making nature and its changes more legible to policy-makers and facilitating measures to intervene in its activities.

Part III brings together four chapters dealing with the global movement of individuals and explores the axes along which their hybrid transnational identities were constructed. In chapter 8 Adriana Minor focuses on the negotiations over nationality that are never far from the surface in transnational interactions. She describes the travels of the distinguished Mexican physicist Manuel Sandoval Vallarta, who was recruited as an assistant professor at MIT in 1926. He was actively involved in improving US-Mexican scientific relations in the interwar period. In 1942 he became the head of the Committee on Inter-American Scientific Publication, which was based at MIT and was supported by the Office of the Coordinator of Inter-American Affairs. (This office was established by President Roosevelt in 1941 with banker Nelson Rockefeller at the helm to promote a positive image of the United States in Latin America.) Vallarta's prestigious academic position in the United States enabled him to leverage his Mexican nationality to build a hemispheric community of scientists around English-language publications. Wartime conditions recalibrated his role. Vallarta, as a Mexican national, was excluded from government-sponsored war research, which was taking a heavy toll on his MIT colleagues. The chair of the Physics Department, John Clarke Slater, insisted that he contribute to the American war effort by taking on an increased teaching load in Cambridge. Vallarta refused: as he saw it, his wartime duties lay in his work for the Committee on Inter-American Scientific Publication. Slater was incensed, accusing him of divided loyalties and insisting that he choose: was his primary allegiance to his home country or to MIT? Vallarta's fragile hybrid identity was splintered, and he finally resigned in 1946. He reaffirmed his Mexican identity and developed an active role in multiple international forums as a science diplomat representing the interests of the Mexican government.

The social work required to maintain knowledge-in-movement is necessarily present in all the essays presented here and is an explicit topic in many of them. In chapter 9 Michael J. Barany ingeniously uses the three personae of Rockefeller Foundation officer Harry "Dusty" Miller as an entry point to the labor of making transnational science. Each name, the informal Dusty, the formal Miller, and the "bureaunym" HMM, represents a different but related network of relationships—with colleagues in the foundation, with government offices at home and abroad, and with the US and international scientific community. Barany's close analysis of desk work, of the archival paper trail left by Miller's efforts to identify suitable grantees and to prepare the way for their journeys, highlights the combination of bureaucratic discipline, diplomatic savvy, and friendly familiarity with leading scientists that Miller could draw on. The file on the brilliant young Uruguayan mathematician José Luis Massera illustrates Miller's capacity to secure a visa for a man who was an outstanding scientist but also an active communist. Barany deftly describes the material culture and day-to-day practice involved in the personal and bureaucratic work of a foundation officer constructing hemispheric transnational cooperation in the 1940s and 1950s in line with Roosevelt's Good Neighbor policy in Latin America.

In chapter 10 Olival Freire Jr. and Indianara Silva deal with the transnational movement of scientists, mostly physicists, between Brazil and the United States that began in World War II and continued through the Cold War and the Brazilian dictatorship. They track the intersection between scientific mobility and foreign policy, charting the tensions between the idealism of scientific internationalism and government restrictions on transnational movement in the name of national security. Their chapter begins in World War II with the Office of the Coordinator of Inter-American Affairs, an organization also discussed by Minor in chapter 8. Along with other major American foundations, it mobilized scientists and engineers (as well as film stars, radio, movies, and print) as cultural ambassadors for American democracy. Capitalizing later on these networks, in the early Cold War a physicist like David Bohm could settle in Brazil, which provided a safe haven for an eminent physicist fleeing McCarthyism. The military dictatorship, which began in 1964, changed that. While overtly critical of the regime, successive US presidents covertly supported it. American physicists welcomed Brazilian colleagues into the United States and even lobbied successfully for their release from prison—unless they were deemed to be "dangerous communists." This chapter's discussion of the role of US foundations in fostering transnational scientific travel complements Barany's chapter. It also reinforces Daniels's argument in chapter 1, illustrating the importance of visas and passports as instruments used by the national security state to control the movement of "knowledgeable bodies."

In chapter 11 Josep Simon uses the internationalization of an American-inspired physics curriculum with its associated teaching aids in the late 1950s and the 1960s as a platform to reflect on the specificity of a transnational approach to the writing of history. He traces the origins of the Physical Science Study Committee in the context of early Cold War America and describes its octopus-like extension into, above all, Europe and Latin America. The committee fused the skills of physicists, high school teachers, instrument makers, and filmmakers with those of professional testing services to produce a curriculum that was quickly translated into Spanish and Portuguese for use in Latin America. Simon traces the existing global network of physicists who served as vectors for the transmission of the materials into dozens of countries. Above all, he emphasizes the importance of hybrid identities as constitutive of some of the transnational actors who facilitated the curriculum's promotion worldwide. The success of the committee's physics courses outside the United States relied on people who combined multiple language skills with an already-existing transnational professional experience, physicists who had diverse national and cultural identities that sensitized them to other worldviews and enabled them to build bridges and establish dialogue with educational reformers abroad.

Part IV deals with the postwar nuclear regime. The adventures of an International Harvester truck equipped with a mobile radioisotope research exhibition are the centerpiece of Gisela Mateos and Edna Suárez-Díaz's chapter. They tell the story through the eyes of the Austrian truck driver, Josef Obermayer, who reported back regularly to his superiors at the International Atomic Energy Agency's technical assistance program in Vienna. Crossing borders meant maneuvering a huge vehicle designed for the American Great Plains along narrow, winding, muddy roads in six Latin American countries. It meant loading it on small cargo ships and carrying it on trains too narrow for its axle width. It meant dealing with strikes at customs posts, coping with dilapidated communications infrastructures, and adjusting itineraries in the aftermath of earthquakes. It meant securing the often-reluctant support of local officials and pressuring them to provide the financial resources needed to demonstrate the benefits of the peaceful atom in remote villages as well as in major capitals. It meant the clash of "modernity" with the customs and traditions of the "undeveloped" "third world," mediated by Herr Obermayer, whose determination to fulfill his mission inured him to technical breakdowns, personal discomfort, and endless frustration with the "inefficiency" of local administrative practices. As the International Atomic Energy Agency's truck meandered from one town or city to the next, it raised the profile of the peaceful atom in the countries it visited. It also enhanced the social capital of the local technicians, scientists, and dignitaries who greeted it and operated its instruments, linking them to a global network that included a national nuclear facility in Oak Ridge, Tennessee, and an international atomic energy organization in Vienna. Transnational knowledge networks are built, not from thin air, but from the material stuff that makes travel possible. Knowledge circulates, not "by itself" through a "frictionless global space," but thanks to the mundane, everyday labor of bureaucrats and truck drivers.

The horrifying prospect of nuclear Armageddon has not stopped rival powers from acquiring nuclear arsenals if they can. It has also stimulated transnational exchanges between nonstate actors, notably physicists themselves, who have served as technical experts advising governments on the security implications of proposed disarmament measures. In chapter 13 Zuoyue Wang describes the changing landscape of back-channel discussions between American and Chinese nuclear scientists since the 1980s, focusing on a face-to-face encounter between eminent Stanford physicist and arms control expert Wolfgang Panofsky and a high-level Chinese delegation in Beijing in 1988. In a delicate historical context like this, marked by a widespread sensitivity in China to perceived Western imperial arrogance, the mere fact that Panofsky traveled to China (and was willing to criticize some aspects of his government's policy) signaled his sincere wish to maintain a dialogue with his opposite numbers. The reciprocal exchange of information and ideas was restricted to broad issues of policy and to technical matters that could be shared in a quasi-public domain. It was facilitated by the presence of some Chinese scientists who had studied in the United States in the 1950s and by the long tradition of scientific internationalism in physics that naturalized such encounters and helped to overcome mutual suspicion. Wang suggests that these transnational encounters improved Sino-US relations in arms control—though that was thanks to the chemistry in the "contact zone." It created a climate of trust that could easily have disintegrated through government intervention or personal recrimination. These networks last only as long as government leaders see the point in maintaining them. They can easily fall prey to changing international relationships over which the actors have no control.

The collection of essays presented here attests, we hope, to the rich seams opened by taking a transnational approach to the production and movement of scientific and technological knowledge across borders. They not only exploit classical aspects of the transnational approach while concentrating on knowledge as the object of analysis but also create synergies that open new avenues of investigation. A brief afterword coauthored with Michael Barany addresses various features of this study that define its specificity and perhaps the specificity of any transnational approach that takes scientific and technological knowledge as its core object. We note the centrality of the United States deriving from its significance as a global scientific and technological power; the need to engage—somewhat ironically—with US exceptionalism and the resurgence of borders in a transnational history precisely because the United States has created a modern bureaucratic system that places a high value on knowledge as a national resource and is determined to regulate cross-border movement in the name of national security and economic competitiveness; the importance of the expansion, not to say imposition, of English as the lingua franca of scientific and technical communication, including its role in constructing transnational networks; the construction of hybrid selves that combine one's identity as a knowledgeable body with national and political allegiance; and, in a self-reflexive moment, we touch on the multiple identities of the authors who produced this volume. Above all, we draw attention to the intellectual and also personal and political stakes involved in taking a transnational approach, all the more so in the current political conjuncture, which is marked by xenophobic and racist chauvinism. It is sometime said that transnational history cannot speak to the construction of a sense of personal identity, and of belonging, that national histories address. We argue instead that it matters all the more at the moment precisely because national narratives have degenerated into nationalistic programs that harbor xenophobia and racism. This project is not simply an intellectual experiment devoid of political overtones. On the contrary, it derives from and, we believe, advances distinctively cosmopolitan political convictions rooted in the challenges of our times. The intellectual and broader political interest of a transnational approach is surely established: I hope that others will build on what we have done, adding more struts to this "concept under construction."

My thanks to all those who participated in the workshop from which this collection has emerged, especially Michael Barany. Thanks too to Dan Amsterdam and to Zuoyue Wang for their helpful suggestions on the introduction.

Notes

1. C. A. Bayly, Sven Beckert, Matthew Connelly, Isabel Hofmeyr, Wendy Kozol, and Patricia Seed, "*AHR* Conversation On: Transnational History," *American Historical Review* 111, no. 5 (2006): 1441–1464; Ann Curthoys and Marilyn Lake, eds., *Connected Worlds: History in Transnational Perspective* (Canberra: Australian National University Press, 2005); Akira Iriye, *Global and Transnational History: Past, Present and Future* (Basingstoke: Palgrave Macmillan, 2013); Emily S. Rosenberg, *Transnational Currents in a Shrinking World* (Cambridge, MA: Belknap Press of Harvard University Press, 2014); Pierre-Yves Saunier, *Transnational History* (Basingstoke: Palgrave Macmillan, 2013); Ian Tyrell, "American Exceptionalism in an Age of International History," *American Historical Review* 96, no. 4 (1991): 1031–1055; Ian Tyrell, "Reflections on the Transnational Turn in United States History: Theory and Practice," *Journal of Global History* 4, no. 3 (2009): 453–474. Among historians of science and technology, see Martin Kohlrausch and Helmut Trischler, *Building Europe on Expertise: Innovators, Organizers, Networkers* (New York: Palgrave Macmillan, 2014); John Krige, "Hybrid Knowledge: The Transnational Coproduction of the Gas Centrifuge for Uranium Enrichment in the 1960s," *British Journal for the History of Science* 45, no. 3 (2012): 337–357; Michael Neufeld, "The Nazi Aerospace Exodus: Towards a Global, Transnational History," *History and Technology* 28, no. 1 (2012): 49–67; Asif A. Siddiqi, "Competing Technologies, National(ist) Narratives, and Universal Claims: Towards a Global History of Space Exploration," *Technology and Culture* 51, no. 2 (2010): 425–443; Simone Turchetti, Néstor Herran, and Soraya Boudia, "Introduction: Have We Ever Been 'Transnational'? Towards a History of Science across and beyond Borders," *British Journal for the History of Science* 45, no. 3 (2012): 319–336; Erik van der Vleuten, "Towards a Transnational History of Technology: Meanings, Promises, Pitfalls," *Technology and Culture* 49, no. 4 (2008): 974–994; Erik van der Vleuten and Arne Kaijser, eds., *Networking Europe: Transnational Infrastructures and the Shaping of Europe, 1850–2000* (Sagamore Beach, MA: Watson Publishing International, 2006); Zuoyue Wang, "Transnational Science during the Cold War," *Isis* 101 (2010): 367–377; Zuoyue Wang, "The Cold War and the Reshaping of Transnational Science in China," in *Science and Technology in the Global Cold War*, ed. Naomi Oreskes and John Krige (Cambridge, MA: MIT Press, 2014), 343–370.

2. Rosenberg, *Transnational Currents in a Shrinking World*, 2.

3. Bernhard Struck, Kate Ferris, and Jacques Revel, "Introduction: Space and Scale in Transnational History," *International History Review* 33, no. 4 (2011): 573–584.

4. James Secord, "Knowledge in Transit," *Isis* 95, no. 4 (2004): 654–672, at 655.

5. Wiebe E. Bijker, Thomas P. Hughes, and Trevor Pinch, *The Social Construction of Technological Systems* (Cambridge, MA: MIT Press, 2012). See also the focus on knowledge in Martin Kohlrausch and Helmuth Trischler, *Building Europe on Expertise: Innovators, Organizers, Networkers* (New York: Palgrave Macmillan, 2014).

6. Charles Bright and Michael Geyer, "Regimes of World Order: Global Integration and the Production of Difference in Twentieth-Century World History," in *Interactions: Transregional Perspectives on World History*, ed. Jerry H. Bentley, Renate Bridenthal, and Anand A. Young (Honolulu: University of Hawai'i Press, 2005), 202–237, at 204. See also Frederick Cooper, "What Is the Concept of Globalization Good For? An African Historian's Perspective," *African Affairs* 100 (2001): 189–213.

7. Bayly et al., "*AHR* Conversation On: Transnational History."

8. Michael McGerr, "The Price of the 'New Transnational History,'" *American Historical Review* 96, no. 4 (1991): 1056–1067, at 1064.

9. To cite Nick Cullather, "Development? It's History," *Diplomatic History* 44, no. 2 (2000): 641–653, at 650. He is referring, respectively, to Michael Adas, *Machines as the Measure of Men: Science, Technology, and Ideologies of Western Domination* (Ithaca, NY: Cornell University Press, 1989); and James C. Scott, *Seeing like a State: How Certain Schemes to Improve the Human Condition Have Failed* (New Haven, CT: Yale University Press, 1998).

10. Rosenberg, *Transnational Currents in a Shrinking World*.

11. Travel is of course widely studied by historians of science and technology. See, e.g., David J. Arnold, *The Tropics and the Travelling Gaze: India, Landscape, and Science, 1800–1856* (Seattle: University of Washington Press, 2016); Marie-Noëlle Bourget, Christian Licoppe, and H. Otto Sibum, eds., *Instruments, Travel and Science: Itineraries of Precision from the Seventeenth to the Twentieth Century* (London: Routledge, 2002); Mary Louise Pratt, *Imperial Eyes: Travel Writing and Transculturation* (London: Routledge, 1992).

12. See, typically, Roy MacLeod, ed., *Nature and Empire: Science and the Colonial Enterprise*, Osiris, 2nd ser., vol. 15 (Chicago: University of Chicago Press, 2000).

13. Kapil Raj, *Relocating Modern Science: Circulation and the Construction of Knowledge in South Asia and Europe, 1650–1900* (New York: Palgrave Macmillan, 2007); Kapil Raj, "Beyond Postcolonialism . . . and Postpositivism: Circulation and the Global History of Science," *Isis* 104, no. 2 (2013): 337–347.

14. Raj, "Beyond Postcolonialism," 345.

15. Joseph Masco, "'Sensitive but Unclassified': Secrecy and the Counterterrorist State," *Public Culture* 22, no. 3 (2010): 433–463; Geraldine J. Kzeno, *"Sensitive but Unclassified" Information and Other Controls: Policy and Options for Scientific and Technical Information*, CRS Report for Congress, Dec. 2006, accessed Jan. 11, 2018, https://fas.org/sgp/crs/secrecy/RL33303.pdf.

16. Akira Iriye, "Internationalizing International History," in *Rethinking American History in a Global Age*, ed. Thomas Bender (Berkeley: University of California Press, 2002), 47–62, at 52.

17. Charles Maier, *Among Empires: American Ascendancy and Its Predecessors* (Cambridge, MA: Harvard University Press, 2006), 106–107.

18. Krige, "Hybrid Knowledge."

19. For a beautiful example, see Alison Sandman, "Controlling Knowledge: Navigation, Cartography, and Secrecy in the Early Modern Spanish Atlantic," in *Science and Empire in the Atlantic World*, ed. James Delbourgo and Nicholas Dew (New York: Routledge, 2008), 31–51.

20. Charles S. Maier, *Once within Borders: Territories of Power, Wealth, and Belonging since 1500* (Cambridge, MA: Belknap Press of Harvard University Press, 2016), responding to critics at H-Diplo Roundtable: *H-Diplo Roundtable Review* 19, no. 3 (Sept. 18, 2017), http://www.tiny.cc/Roundtable-XIX-3.

21. Frederick Cooper, *Colonialism in Question: Theory, Knowledge, History* (Berkeley: University of California Press, 2005), 90–91.

22. See https://www.trumanlibrary.org/whistlestop/50yr_archive/inaugural20jan1949.htm.

23. Cited in Fiachra Gibbons, "US Is an Empire in Denial," *Guardian*, June 2, 2003, reporting on a lecture by Niall Ferguson.

24. On modernization, see David C. Engerman, "American Knowledge and Global Power," *Diplomatic History* 31, no. 4 (2007): 599–614 and references therein. See also Daniel Immerwahr, *Thinking Small: The United States and the Lure of Community Development* (Cambridge, MA: Harvard University Press, 2016).

25. Successively, Stuart W. Leslie, "Atomic Structures: The Architecture of Nuclear Nationalism in India and Pakistan," *History and Technology* 31, no. 3 (2015): 220–242; Nick Cullather, "Miracles of Modernization: The Green Revolution and the Apotheosis of Technology," *Diplomatic History* 28, no. 2 (2004): 227–254, at 234, 227; Matthew Connelly, *Fatal Misconception: The Struggle to Control World Population* (Cambridge, MA: Belknap Press of Harvard University Press, 2008).

26. Ricardo D. Salvatore, *Disciplinary Conquest: U.S. Scholars in South America, 1900–1945* (Durham, NC: Duke University Press, 2016).

27. Pratt, *Imperial Eyes*, 4.

28. Robert K. Merton, "The Normative Structure of Science," in *The Sociol-*

ogy of Science: Theoretical and Empirical Investigations (Chicago: University of Chicago Press, 1973), 267–280.

29. Geert J. Somsen, "A History of Universalism: Conceptions of Internationality of Science from the Enlightenment to the Cold War," *Minerva* 46 (Sept. 2008): 361–379.

30. Michael J. Mulkay, "Norms and Ideology in Science," *Social Science Information* 15, nos. 4–5 (1976): 637–656; Ian I. Mitroff, "Norms and Counter-Norms in a Select Group of Apollo Moon Scientists: A Case Study of the Ambivalence of Scientists," *American Sociological Review* 39, no. 4 (1974): 579–595.

31. See John Krige, "National Security and Academia: Regulating the International Circulation of Knowledge," *Bulletin of the Atomic Scientists* 70, no. 2 (2014): 42–52; John Krige, "Regulating the Academic 'Marketplace of Ideas': Commercialization, Export Control, and Counterintelligence," *Engaging Science Technology and Society* 1 (2015): 1–24.

32. This "art" is on display in John Krige, *Sharing Knowledge, Shaping Europe: U.S. Technological Collaboration and Nonproliferation* (Cambridge, MA: MIT Press, 2016).

33. Nicholas Dew, "*Vers la Ligne*: Circulating Knowledge around the French Atlantic," in Delbourgo and Dew, *Science and Empire in the Atlantic World*, 53–72; Bruno Latour, "Circulating Reference: Sampling and the Soil in the Amazon Forest," in *Pandora's Hope: Essays on the Reality of Science Studies* (Cambridge, MA: Harvard University Press, 1999), 24–80; David Livingston, *Putting Science in Its Place: Geographies of Scientific Knowledge* (Chicago: University of Chicago Press, 2003); Joseph O'Connell, "Metrology: The Creation of Universality by the Circulation of Particulars," *Social Studies of Science* 23(1993): 129–173; Simon Schaffer, "Golden Means: Assay Instruments and the Geography of Precision in Guinea Trade," in Bourget, Licoppe, and Sibum, *Instruments, Travel and Science*, 20–50.

34. Introduction to Delbourgo and Dew, *Science and Empire in the Atlantic World*, 11.

35. Michael D. Gordin, *Scientific Babel: The Language of Science from the Fall of Latin to the Rise of English* (Chicago: University of Chicago Press, 2015), chap. 8.

36. Janet Martin-Nielsen, "'This War for Men's Minds': Birth of a Human Science in Cold War America," *History of the Human Sciences* 23, no. 5 (2010): 131–155.

37. Ibid., 131. This point is stressed by Steven Shapin in his review of Gordin's *Scientific Babel*: "Confusion of Tongues," *London Review of Books* 37, no. 23 (2015): 23–26.

38. Robert Anderson, *Nucleus and Nation: Scientists, International Networks, and Power in India* (Chicago: University of Chicago Press, 2010); Jahnavi Phalkey, *Atomic State: Big Science in Twentieth Century India* (Ranikhet:

Permanent Black, 2013). On Bhabha's vision, see John Krige, "Techno-utopian Dreams, Techno-political Realities: The Education of Desire for the Peaceful Atom," in *Utopia/Dystopia: Conditions of Historical Possibility*, ed. Michael D. Gordin, Helen Tilley, and Gyan Prakash (Princeton, NJ: Princeton University Press, 2010), 151–175.

39. Leslie, "Atomic Structures." Seaborg is quoted on 223.

40. http://www.nsf.gov/statistics/seind14/index.cfm/etc/tables.htm#chp2, table 2.13.

41. Neil G. Ruiz, "The Geography of Foreign Students in U.S. Higher Education: Origins and Destinations," Brookings Institute, http://www.brookings.edu/research/interactives/2014/geography-of-foreign-students#/M10420.

42. John C. Burnham, "Transnational History of Medicine after 1950: Framing and Interrogation from Psychiatric Journals," *Medical History* 55 (2011): 3–26, at 8, 9, 25.

43. Gabrielle M. Spiegel, "The Task of the Historian," American Historical Association presidential address, 2009, https://www.historians.org/about-aha-and-membership/aha-history-and-archives/presidential-addresses/gabrielle-m-spiegel.

44. I thank Jessica Wang for drawing my attention to the importance of inter-imperial scientific cooperation.

45. McGerr, "The Price of the 'New Transnational History.'"

PART I

The US Regulatory State

CHAPTER ONE

Restricting the Transnational Movement of "Knowledgeable Bodies"

The Interplay of US Visa Restrictions and Export Controls in the Cold War

Mario Daniels

The complex process of globalization has created new dependencies between multiple historical actors. In particular, these dependencies have challenged the notion of the national border as a meaningful barrier to the circulation of knowledge, people, and goods. This is not to say that national borders do not matter. On the contrary, a multitude of instruments (classification, visas and passports, and export controls, to name the most prominent) have been put in place by national states to protect their economic competitiveness, national security, the national identities of their peoples—and also the knowledge produced by national innovation systems.

Beginning in the 1940s, the US government imposed a variety of controls on the circulation of knowledge, mapping the perceived dangerousness of bits of information onto a multi-tier system of regulations over access and transmission. Classification is the best known and most widely studied of the instruments used to control the sharing of knowledge.[1] Classificatory regimes have (and still do) controlled the exchange of information using the criteria of nationality, political allegiance, and a "need to know." A differentiated and complex system embeds information control in the control over people, restricting circulation to those nationals who have security clearance and curtailing its communication to foreign individuals and entities within and without the US borders. The key criterion prohibiting any transmission of knowledge was whether it could reach the adversary beyond the national border, enabling him to harm the United States.

As for regulating the mobility of people, since World War I, and in the United States even more so since World War II, passports and visas have become the centerpiece of modern border policing. They, like classification, are a "documentary regime" that regulates the entry to and exit from nation-states.[2] In fact, a wide and constantly redefined array of complex visa criteria establishes a fine-grained system of decision-making about who should be let in and who should be refused entry to national territory.[3]

Finally, as regards the movement of "things"—running the gamut from commercial merchandise to rifles and high-tech weaponry, from machine tools to complex production systems, from mundane scientific instruments to high-speed computers—the United States has since the 1940s set up an incredibly complex export control regime, based on bureaucratic methods that define strategically relevant technology and trace their global movements using a sophisticated paper trail.[4] The "export license" is the regulatory equivalent of visas and passports and serves to open up a temporary gate in the border to let through the exported commodity. And it impacts not only trade. Export controls also cover the exchange of what has been legally defined as *unclassified* "technical data." Classification is thus one but by no means the only instrument controlling the circulation of knowledge. Export controls directed at the transnational movement of unclassified technical data restrict the movement of not only printed information but also intangible know-how, adding another layer to the US knowledge control efforts.

This article argues that the bureaucratic tools to keep tabs on the circulation of information, things, and people are not separate systems of control. They are in fact closely intertwined and complementary. To show the interplay of the regimes, I will analyze how visa regulations are directly and closely linked to the US export control regime.[5] At first sight they seem to regulate different things. In practice, by virtue of the broad sweep of export controls, the two sets of regulations converge where they both target scientific-technological knowledge. In combination with export controls, visa regulations are not simply directed at suspect aliens. More fundamentally they target "knowledgeable bodies," people who could acquire sensitive formal and tacit knowledge that could subvert US interests.

Basically there were—and still are—two bureaucratic procedures that determine whether or not a foreign national can legally enter the United States. They bring the regulatory strands of visa policy and export controls together. First, export control regulations provide knowledge-based criteria for the decision to issue a visa. Second, if a visa is issued, export controls are used to circumscribe and police the knowledge acquisition activities of foreign scientists as they travel *within* the United States.

This analysis of bureaucratic practices will to a certain extent correct as well as enrich a story line that still dominates the Cold War literature. It moves beyond the work of Galison, Dennis, Wellerstein, and others by focusing the argument on *unclassified* sensitive knowledge. It complements the work of Wang by focusing on the restrictions on *transnational* movement in the decade after the end of World War II. Of course, Cold War historians are familiar with the constraints on the travel of foreign communist or left-leaning researchers in the McCarthy period. And as is once again evident today, political and national allegiance alone can be reason enough to deny a visa to a potential visitor to the United States. But ideological vetting is only one, and not necessarily the most important, dimension of travel control. Thus, the historiographical focus on the story of ideological, especially anticommunist discrimination, relevant and important as it is, distracts one from the other, often much more mundane, day-to-day practices that inform decisions to grant visas to scientists, engineers, and researchers who want to enter the United States. Consequently, the scholarly literature has largely been oblivious to how the implementation of export controls in combination with visa controls obstructed the circulation of people bearing sensitive knowledge in their heads and hands, a lacuna this chapter will fill.

One should not be deluded by the massive circulation of knowledge in our interconnected world into thinking that export controls are peripheral to visa adjudications. For, as John Krige and I have shown elsewhere, the vast unregulated space that embodies "scientific internationalism" was coconstructed with the export control regime, a regime that deliberately excluded certain categories of knowledge, notably basic science, from government control.[6] It is the bureaucratic application of this criterion (and others) that gives scholars from nontargeted countries the impression that getting a visa is a mere formality.

Indeed, after World War II, US policy-makers turned the idea of scientific internationalism into a tool of US national security and foreign policy.[7] In the Cold War mind-set, national security did not stop at the American national border. Central to this concept was the building of an international order that created the "West" as a bulwark against the perceived threat of an expansionist Soviet Union.[8] Part of this project of constructing a Western alliance was the sharing of scientific-technological knowledge to strengthen economic prowess and military power through cooperative innovation and technology transfers.[9] Yet, as much as international cooperation and US national interest went hand in hand, there was a constant tension between them.[10] Since the United States claimed the role of international leadership and understood being ahead of everybody else as the very base of its power, it jealously monitored what knowledge it shared and what it denied—not only with enemies but even among allies. The communication of knowledge was allowed as long as it buttressed US interests, but it was stopped as soon as it was deemed to threaten the American lead.[11] Thus, at the same time as the United States fostered scientific internationalism and, by extension, the internationalization of science and technology, it carefully defined its boundaries. Export controls and visa regulations were central bureaucratic instruments to enforce these limits in practice.

In this chapter I will not only show how this remarkably creative interplay of visa and export control regulations worked in practice but also offer, more importantly, a bureaucratic and intellectual history of the visa and export control systems as regimes of knowledge regulation, situating their development in the gradually changing Cold War context from the 1940s to the 1960s. My central point is that the US national security state made—and still makes—the decision to allow a scientist or student to visit the United States not least on the basis of considerations about what the individual scientist *knows* but, more importantly, about what he or she could potentially *learn* in the United States and take back with her to her home country (see also Krige, chapter 2).

The McCarran Acts and the "Uranium Curtain": Visa Restrictions in the Early Cold War

Increasingly since the late 1940s, American scientists who wanted to travel abroad were subjected to special scrutiny regarding not only their political leanings but also, and even more important, their expertise and the knowledge they potentially could share beyond US borders. US physicists and chemists were in fact the only professional group singled out for a particularly close security screening by the Bureau of Security and Consular Affairs within the State Department's Passport Office.[12] Scientists applying for a passport were "screened as to their loyalty to the United States, keeping in mind what the individual knows."[13] If doubts existed about the political beliefs of a scientist and if his knowledge seemed to be too sensitive to be shared abroad, the State Department could deny him a passport or issue a limited passport allowing travel only to certain, clearly stated destinations. This happened time and again, not surprisingly mainly to scientists with leftist political leanings. Then, in the late 1950s, a series of Supreme Court decisions established step-by-step the concept of a US civil "right to travel." Henceforth, passport denials could not be used as an instrument of knowledge transfer control.[14]

US visa policy, by contrast, was not affected by these legal changes. Beginning in the late 1940s visas had become an increasingly powerful instrument of national security policy in general and of managing an anticommunist border regime in particular. The use of visas to fight political groups extends back to the early twentieth century. In 1903 Congress voted for legislation that denied admission to the United States of "anarchists, or persons who believed in or advocated the overthrow by force or violence" of the US government. This was the first federal legislation that directly linked immigration and national security. This link became even more pronounced during World War I, when the United States for the very first time introduced a general requirement of visas to enter the country. The Immigration Act of October 1918 and its amendments in the 1920s codified the targeting of anarchists, "saboteurs," and other "subversive aliens" for the next decades.[15]

After World War II, this legislation was increasingly used to prevent communists from entering the United States, and in the late 1940s it began to visibly affect US scientific exchanges with foreign countries.[16] These travel restrictions became even more pronounced with the enactment of the two McCarran acts (so called after their main sponsor, Senator Pat McCarran, D-NV)[17] passed during the Korean War in a climate of strident anticommunism, to replace the Immigration Act of 1918. The new legislation, officially called the Internal Security Act of 1950 (McCarran act) and the Immigration and Nationality Act of

1952 (McCarran-Walter act), denied visas especially to communists—a broadly defined category—whose presence in the United States was deemed to be "prejudicial to the public interest, or endanger the welfare, safety, or security of the United States." In addition to political subversion, the acts were directed against the threats of sabotage and espionage.[18]

These stipulations complicated the lives of the international scientific community. The Federation of American Scientists reported in 1952 that "at least 50 per cent of all the foreign scientists who want to enter the United States" faced visa denials or delays of their visa applications of four months up to one year. French scientists were even more seriously affected: 70–80 percent of them experienced difficulties, mainly because about 70 percent of all French scientists were members of the Association des travailleurs scientifiques, an organization that US national security bureaucrats deemed "subversive." The Federation of American Scientists had collected information on about sixty cases but estimated that altogether three times as many scientists had had visa problems.[19] Conferences were particularly affected by these difficulties, and from 1950 onward some scientific organizations began to plan to hold their meetings outside the United States to spare foreign colleagues the "embarrassment" of the visa procedure.[20]

Even though businessmen, journalists, artists, and—also of interest for our argument—technical assistance team members felt the effects of the stricter visa policy, it seems that no professional group was as much affected by visa restrictions as were scientists.[21] This was an expression of the deep suspicion that the US government and the public had developed toward scientists in the early Cold War. The perceptions of the role scientists played in postwar politics were deeply ambivalent. On the one hand, the scientific community had earned enormous prestige through its accomplishments during World War II, in particular the development of the atomic bomb. On the other hand, many scientists had leftist political leanings and advocated publicly ideas of "scientific internationalism." Many critiques saw in this the signs of dangerous political ambiguity at best, and of divided loyalty out of sync with the ideological Cold War mainstream at worst. Scientists developed powerful weapons to enhance US national security—but could they be trusted to keep national military secrets? The widely publicized atomic espionage cases involving scientists like Klaus Fuchs and Alan Nunn May threw these concerns into sharp relief, as did the mysterious defection of nuclear physicist Bruno Pontecorvo to the Soviet Union in 1950.[22]

Indeed, espionage became one of the dominant lenses through which the US national security community perceived the dangers of scientific exchanges. The term "espionage" became a highly emotionally charged political cipher for all forms of uncontrolled sharing of security-sensitive information, an intellectual paradigm that shaped the understanding of the power and perils of knowledge dissemination.[23]

In contrast to what we know about US passport policy, it is not clear to what extent the US visa practices of the late 1940s and early 1950s had established formalized mechanisms to address the dangers of unwanted knowledge transfer. But it is very likely that some measures were in place. For example, the guidelines entitled "Control of Visits to Industrial Facilities in the Interest of Internal Security" issued by the Office of Defense Mobilization informed their readership, "Through visa, immigration and naturalization, and related procedures, the Government endeavors to exclude from the United States foreign nationals whose background indicates that they might engage in espionage, sabotage, or otherwise present a threat to the national security."[24] Similarly, the FBI also referred to visa restrictions to all Iron Curtain country citizens as well as the denial of passports to US communists as countermeasures against Soviet espionage. It considered all Soviet travelers, explicitly including scientists and students, as agents sent by the Soviet intelligence services with specific intelligence collection missions, including scientific and industrial espionage.[25]

The scientific community clearly felt the effects of this counterintelligence approach. In the special issue of the widely read *Bulletin of the Atomic Scientists* on the repercussions of visa restrictions on the scientific community, Edward Shils described how in the enforcement of the McCarran acts "[s]cientists in certain fields of work like nuclear physics, electronics, and other fields are especially suspect. Some of the recent victims of the American visa policy . . . have remarked how alarmed consular officers have become when they learned that the applicant was a physicist." The visa problems encountered by the University of Chicago when it organized an international congress on nuclear physics in 1951 were for Shils "only one more illustration of this contradictory belief in the supreme importance of scientific knowledge and the terrible fear of scientists as unreliable, untrustworthy vessels of this crucial knowledge."[26] Shils called the visa and passport hurdles that the State Department put in the way of exchange a "Paper Curtain."[27] British and French scientists grasped the knowledge focus of the regulations much better. They referred to a "Uranium Curtain,"[28] hinting at a connection between visa policy and the information control regime of the Atomic Energy Act.[29]

As lucid and sharp as the analyses in the *Bulletin of the Atomic Scientists* were, they had one blind spot. Time and again they objected that the visa restrictions impaired the communication of unclassified information. Implicitly accepting the need for classification in the interest of national security, the authors deplored the tendency of federal regulations to blur the line between the realms of openness and classification. Since the vast majority of scientific interactions pertained to unclassified information, it seemed "obvious" to Shils that if visa restrictions seem "necessary, then all of the elaborate precautions taken to protect our secret research and installations must be functioning very poorly indeed. The whole system of security clearance, guards, classification etc.—must be utterly ineffective if foreign scientists could, if they wished, break through it,"[30] acquiring classified knowledge that should be denied them. For Shils and his colleagues visa restrictions had been put in place to plug leaks in the control system for classified information. What they did not see was that worries about the dangers of freely circulating *unclassified* information were at the very heart of the policies of visa and export control. That becomes startingly clear when we listen to *Bulletin* author Victor Weisskopf's description of the advantages of face-to-face interactions between members of the international scientific community: "International exchange of ideas and discussion is indispensable because details of scientific research are never written down in the actual publications. Frequently, only conversations can reveal a special technique or a special design which foreign scientists have used to make their experiments work. . . . There is a long list of discoveries which can be traced directly to international gatherings. Indeed, some of these discoveries had direct applications to the production of weapons such as radar or the atomic bomb."[31]

That was exactly the point that the national security community was trying to make when it warned of the unchecked communication of ideas and advocated travel controls. Personal contacts were dangerous because they could convey knowledge other sources could not—and they could lead to innovations that could shorten the lead time of American military technology. It is here that visa regulations and export controls met. As an influential 1950 State Department report phrased it in its critique of regulations affecting the flow of unclassified scientific information: "The interchange of persons and of information is a closely related means of accelerating scientific creative thinking."[32] Whether that was for good or for ill depended on the point of view one took.

The Origins of Technical Data Export Controls

Indeed, the exchange of people and the exchange of information were closely related. Accordingly, travel restrictions for scientists and engineers codeveloped with the export control regime over "technical data" in the framework of the burgeoning American national security state. Whereas the public paid most attention to the high-profile cases of atomic espionage and similar betrayals of *classified* military information, the national security bureaucracy was also deeply concerned about what the Soviets learned from *unclassified* information freely available within the American open society. Immediately after the end of World War II, the US security apparatus was alarmed to see that the Soviet Union was systematically collecting every unclassified book, scientific journal, printed patent, government report, etc. that it could get hold of and that could be helpful to its research and development efforts. As early as March 1946 a member of the State Department's U.S.S.R. Committee warned of the Soviets' "'total exploitation' of all American technical information. This is gathered here indiscriminately from all available sources and sifted and analyzed in the Soviet Union."[33] The US national security bureaucrats were keen adherents of the "mosaic theory" and emphasized that pieces of information, innocuous in isolation, could become technologically meaningful, even militarily dangerous, in aggregate.[34]

To answer the threat from Soviet information collection, the United States constituted interdepartmental working groups and special offices in the Departments of Commerce and Defense. During the late 1940s and throughout the 1950s they discussed ideas of how to control the spread of unclassified but sensitive "technical data" in the interest of national security without infringing to an unacceptable extent on international exchange and democratic freedoms. These boards, completely overlooked by the scholarly literature to date, constituted one of the central clearinghouses for the formulation of US information policy in general and especially of export controls.

Commonly, export controls are understood as an implement of trade policy, applied to regulate the exchange of goods—for example, as an instrument of economic sanctions against adversaries. Moreover, export controls targeting the transfer of technology are widely recognized as an integral part of the fight against the proliferation of weapons of mass destruction.

But there is another central objective. US export controls aim at the permanent, constant, everyday regulation of a very broad array of scientific-technological *knowledge.* After science and technology had successfully been mobilized for waging and winning World War II, it became the credo of the national security establishment that the relentless production and utilization of technology, especially cutting-edge technology, formed the bedrock of military power and, closely related, national economic prowess. US global hegemony, this story went, depended on technological superiority, on being technologically ahead of everybody else. Knowledge was indeed power, and national security was a function of technological lead time.

To maintain its lead, not only did the United States strive to sustain a culture of permanent innovation, but officials also very carefully watched over practices of transnational knowledge sharing. They understood not only that successful technology transfer was a learning process that enabled the recipient to replicate a technological product, but also that the acquired knowledge potentially boosted the innovation capabilities of the absorbing society, helping it to catch up with the United States. Hence, the US export control regime aimed at stopping or at least slowing down the learning process of enemies as well as allies if it seemed to endanger US scientific and technological preeminence and so US security.

Since knowledge was a main target of export controls, their scope was quickly stretched beyond the regulation of physical goods to include the transmission of "technical data." This term included all kinds of knowledge sources: manuals, blueprints, scientific papers, statistical data, images—as well as people in the know, their skills, ideas, and conversations.

Seen in this light export controls had much in common with the US classification system that reached maturity at the very same time during the 1940s and 1950s. But export controls began where secrecy ended. They regulated *unclassified*, theoretically openly accessible information that appeared to give the competitor an undue advantage if shared too freely. And in contrast to classification, export controls covered not only knowledge that was directly controlled by the US government but also all kinds of private intellectual productions. Thus, the export control regime carved out a potentially vast gray zone between openness and secrecy and blurred the lines between state and civil society.

After 1945 export controls became the crucial concept with which to think about the regulation of information. Printed papers became, next to machines and weapons, a regulated product. Traveling scientists and engineers came to be seen as exporters and the valuable knowledge in their heads as a national security asset in need of oversight if shared with non-US citizens. When the first of the working groups on information control, the Unclassified Technology Information Committee, met for the first time in January 1947, it discussed, among other topics, "scientific papers," "releases of trade and technical press," and the "dissemination of knowledge through exchange of scientists" as examples of "exports" of "technological information [that] aids and increases the military potential of the recipient country."[35] Along the same lines, the successor board, the Interdepartmental Committee on Industrial Security, a division of the State–Army–Navy–Air Force Coordinating Committee, discussed the export control of "unclassified technology" and the regulation of "technological publications" and "visits of scientists" in the same context as "basic problems in industrial security." In regard to traveling scientists, the committee's agenda asked: "Should any passport or visa rules be amended in order to restrict the movements of alien and American scientists during the present 'peace-time'? . . . What standards should be used in determining those scientists who can move freely and those who cannot?" The committee knew well that these issues had potentially serious repercussions. Self-critically it asked, "Would such restrictions in 'peace-time' infringe upon civil rights?" and "Would such restrictions interfere with American scientific or technological progress?"[36] At stake was the technological as well as the moral leadership of the US liberal democracy in the Cold War. Concerns of knowledge protection had to be balanced against the political ideals and principles of the U.S. Constitution. That was not an easy task.

Because of such qualms, the US government at first did not thoroughly implement control over the export of technical data even though the Export Control Act of 1949 gave a broad statutory basis for such regulation.[37] To avoid the appearance of using export controls as an undemocratic tool of censorship, the Truman administration opted for "a voluntary plan to control the export of non-secret technical data, the dissemination of which abroad might be harmful to national security." This program solicited the cooperation of the business and scientific communities, asking them to consult with the Department of Commerce before they shared any information that related to "advanced technology" with companies or colleagues abroad.[38] This appeal was widely ignored, but the outbreak of the Korean War offered the opportunity to introduce, on March 1, 1951, mandatory controls of data exports to the Soviet Union.[39]

These rules were not revoked at the end of the war. On the contrary, in 1955 they were even extended and included data exports to countries that were considered to be politically friendly to the United States. The definitions the regulations applied were incredibly broad. Controllable "technical data" was understood as "*any* professional, scientific or technical information, including *any* model, design, photograph, photographic negative, document or other article or material, containing a plan, specification, or descriptive or technical information of *any* kind which can be used or adapted for use in connection with *any* process, synthesis, or operation in the production, manufacture, or reconstruction of articles or materials."[40] This referred to unclassified, unpublished information. Classified data was covered by separate sets of rules, based on Executive Order 10501 of 1953 and the Atomic Energy Act of 1954.[41]

The definition of what constituted an "export" was also very far-reaching. For the government an export was "any release of technical data for use outside the United States (except Canada) . . . includ[ing] the actual shipment out of the United States as well as *furnishing of data in the United States* to persons with the knowledge or intention that the persons to whom it is furnished will take such data out of the United States."[42] Examples of exports included "blue prints, specifications, technical aid contracts, manufacturing agreements, patent license agreements, instructional or training material, training of foreign personnel, personal delivery by U.S. personnel sent abroad, etc."[43] This clearly had huge implication for travelers from and into the United States whether they were businessmen, scientists, or engineers. If these regulations were taken literally, a broad spectrum of business and scientific exchange would become potentially subject to government regulations.

In a protest note against the data control regulations, the Engineering College Research Council of the American Society for Engineering Education spelled out their possible, serious consequences for the everyday activities of American universities. The society argued that much of the teaching and research done in universities and colleges was based on unpublished material and thereby fell under the jurisdiction of the new controls. Indeed, almost every kind of scientific communication would require a government license. Every conversation with a colleague who would leave the country afterward was an export. Every foreign student, every foreign visitor at a conference, but also every American who went abroad to present his or her research would be an exporter. And every letter to a colleague and every contract with a foreign institution would come under suspicion of being an export, which had to be approved by the government. In other words, the export controls on technical data "would require if conscientiously adhered to, that the entire programs of colleges and universities for teaching and research in science and engineering should be conducted under conditions of 'trade secrecy.' . . . Greater damage to higher education in this country, and to the country's international relations, can hardly be imagined." The engineers therefore requested that the Department of Commerce exempt universities and colleges from the export controls on data.[44]

On the surface, the protest was successful, as the export regulations were amended in April 1955 and explicitly excluded "instruction in academic institutions and academic laboratories." Also, they stipulated an exemption for "[t]he dissemination of scientific information not directly and significantly related to design, production, and utilization in industrial processes. Information thus exempted, includes correspondence and attendance at or participation in meetings."[45] This was an attempt to distinguish harmless, even desirable exchanges of scientific knowledge from dangerous "leaks" which informed unfriendly nations about developments of strategic importance in the United States.[46]

But despite these changes, many of the basic problems the protest note addressed were not resolved. What exactly was, in practice, the difference between "educational" information and technical data? At what point was scientific information "significantly related to . . . industrial processes"?[47] The definitional fuzziness that would time and time again cause the U.S. academic a serious headache was unavoidable, given the conceptual complexities of "knowledge." Moreover, flexible definitions allowed for constant adaptation of the regulations to the changing relationship with the enemy as well as to technological change. Avoiding rigidity was thus crucial for making the tools for monitoring and policing international exchanges of scientific-technological information work effectively. This became especially important when the United States carefully opened up exchanges with the Soviet Union.

Exchanges with the Soviets: Policing Visiting Enemies

Administering scientific-technological exchanges with enemies posed a challenge that led to forging a stronger and increasingly formalized link between technical data export and travel control policies. As the United States vigilantly opened its borders, it also tightened the bureaucratic grip on the movement of people and information.

In the years following Stalin's death in 1953, the political, scientific, and cultural relations between the United States and the Soviet Union became closer at a surprising pace. Milestones of this thaw in Cold War tensions were the Four Power conference in Geneva in 1955, the Atoms for Peace conference in the same city a few weeks later, and the directive of National Security Council NSC 5607 of June 1956 that laid out the framework of East-West exchanges for the next two decades or so. The Statement of Policy that accompanied NSC 5607 made clear that exchanges were seen as an instrument to liberalize the Soviet Union through contact with Western values, thus weakening the stature of communism in the confrontation between two world orders. The first US-Soviet exchange agreement was signed on January 27, 1958, the culmination of negotiations that had begun only a few months earlier in fall 1957.[48]

Travel and export regulations needed to be adapted to allow for more Soviets to visit the United States and for a meaningful conversation about scientific and technological matters for the benefit of both partners. At the same time, visitors from the Soviet Union posed a security risk. Accordingly, NSC 5607 directed "the Secretary of State and the Attorney General to continue to cooperate in developing appropriate internal safeguards with respect to the admission of Soviet and satellite nationals to the United States."[49] The existing export and travel regulations were probably the most important of these safeguards. They were barriers to too much openness, which could put American knowledge at risk and lead to the strategic advantage of the visiting enemy. The challenge for the national security bureaucrats was to find the right balance between exchange interests and security concerns. The main actors in this process were the State Department and the Commerce Department. The two departments, step-by-step, paved the way for a closer coordination of travel and export controls simultaneously with the negotiations for the first exchange agreement with the Soviet Union.

In November 1957, members of Commerce and State met to discuss the problem of visits to US plants under the auspices of the upcoming East-West Contacts Program "for the purpose of removing any obstacles to the process of the program that might result from our present Department of Commerce technical data controls."[50] State had brought up the idea of waiving the data regulations for officially authorized exchange visitors. Loring E. Nacy, director of the Bureau of Foreign Commerce (which was in charge of Commerce's export control), was opposed. He warned of the dangers of a wholesale abandonment of data controls, fearing that data that the bureau had not licensed for access by the Soviet bloc under the established policy would be released during such plant visits.[51]

In talks in December 1957 State and Commerce came to an agreement. The export control regulations for technical data were to be amended by a passage addressing explicitly the visits to "U.S. plants, laboratories and facilities for the purpose of general inspections, sales discussions, and private or government sponsored exchange visits." They were exempt from regulation, provided that visitors were "not furnished with detailed explanations, engineering drawings or models of such a specific character as to constitute the basis for copying of the product, design, etc., on the part of the visitor." This meant that only general information and basic science could be shared without a license. Correlatively, communicating applied, sophisticated technical data was still subject to licensing. This formula seemed to be a viable compromise for the State Department, which was "reluctant to take full responsibility for the elimination of technical data controls" for visits. However, it put the onus of the implementation of data controls on industry. Host firms had to define in practice where the line between basic and applied, innocuous and dangerous information was situated.[52] Export controls of technological exchanges were thus in fact dependent on self-censorship by US industry and on trade secrecy.

One main goal of this strategy was to alert the private sector to its obligation to take technical data control regulations seriously, especially since the "general unawareness throughout U.S. industry" still continued to "plague" Commerce as late as 1960.[53] The same set of rules applied to academic and industrial conferences and symposia, trade fairs and exhibits. Even though there were some doubts as to whether significant technology transfers occurred at such events—and Commerce wanted to avoid giving the impression of censorship—the department drafted a standard letter to be sent to conference organizers to inform them that oral and written presentations were within the reach of US export controls.[54]

All these measures were directed at Soviet visitors who had already entered the country. In 1957 Loring Nacy also advocated a review of exchange proposals and visitors' itineraries in advance of the actual trip. The machinery was already available: the Standing Committee on Exchanges of the Intelligence Advisory Committee vetted pretravel applications to enter the United States. Commerce wanted to become a member in order to give export control regulations their due weight and to voice "the Department's views on possible technology loss."[55]

The Standing Committee on Exchanges was established in February 1956, a few months after the summit and the Atoms for Peace conference in Geneva and even before the promulgation of NSC 5607. The chairman, William Bundy, and the executive secretary, Guy Coriden, came from the CIA. Other members were from the State Department, the Army, Navy, Air Force, Joint Chiefs of Staff, and the Atomic Energy Commission.[56] Its main objective was to advise the Department of State "on all intelligence aspects" of exchange programs and to "assess the probable net advantages from an intelligence standpoint, considering both intelligence and technological gains or losses."[57] The committee therefore had both defensive and offensive functions. Not only did it try to limit what visitors from Iron Curtain countries could learn about US technology, but visits to those countries were also seen as missions to gather intelligence for the United States. For this purpose, travelers to the East were briefed and, after their return, debriefed about their trips. "[M]aximizing the intelligence yield from East-West delegation exchanges" was the avowed goal. Risks and advantages of the exchange program were weighed against each other.[58]

In July 1959 Commerce finally became part of the exchange committee after it had reminded CIA director Allen Dulles and the secretary of state, Christian Herter, of the necessity to monitor data export regulations in East-West exchanges. Commerce claimed not sole but "principal responsibility of securing the necessary cooperation of U.S. business in the implementation of any approved exchanges in the technical/industrial fields, including the itineraries of the visiting groups in the U.S."[59]

The exchange committee systematically screened exchange program proposals from the Soviet Union, Eastern Europe, and the United States, including US industry, for intelligence value and risk. Already between March and September 1956, it had screened thirty-five exchange proposals, six of which the committee advised the State Department not to implement.[60] Whereas in the beginning the main concern was industrial exchanges, science entered the picture with the exchange agreement of January 1958.[61]

An example from 1962, related in the CIA's journal *Studies in Intelligence*, shows what effects these screening mechanisms of the State Department and the intelligence community had in practice. In that year, the Soviets proposed several different exchanges in the field of computer science and technology. The director of the USSR Academy of Sciences Computing Center lobbied the Computing Center at New York University to accommodate two Soviet students for a two-month visit. A Soviet student asked to attend the Western Joint Computer Conference in Los Angeles. The University of Illinois was asked to welcome a student in computer technology. The Soviet embassy in the United States, circumventing the Department of State, contacted IBM to ask whether a Soviet educational exchange delegation could visit the IBM headquarters at Rochester. And an exchange visitor, sponsored by the American Council of Learned Societies, and with alleged ties to Soviet intelligence services, asked for permission for an itinerary that would let him learn about the use of computer technology in the field of economic planning. "Although the Department of State of necessity handled each of these proposals separately vis-à-vis the Soviets, inside the government they were treated as a concerted Soviet effort to get needed information on all aspects of U.S. research in automation and computer technology." After assessments by the intelligence community, certainly including the exchange committee, the State Department refused three of these requests, taking no action for the student who wanted to go to the University of Illinois and limiting the itinerary of the Council of Learned Societies guest to universities doing unclassified research only.[62] Similar exchange visits to IBM facilities in the following year were deemed acceptable only if IBM made sure that it protected its classified contractual work for the government and followed technical data export regulations.[63] The appeal to basic science provided the needed political instrument to make the judgment call.[64] The aim of these actions, one analyst wrote, was to "isolate our visitors from applied research and development and restrict their exploration to basic science" because "[a]lmost any scientific or industrial field can be related to war and weaponry."[65]

Conclusion: Why and How Visa and Export Controls Matter after 9/11

The visa and export control regime is not just of interest to Cold War historians. There is a strong line of continuity, leading from the 1940s right up to the present. The basic structure of the two-pronged regime as described in this chapter was put in place in the first two decades of the Cold War and was barely changed in the 1970s and 1980s. However, in the late 1990s, and then with much more momentum after September 11, 2001, the time-tested control, surveillance, and policing mechanisms of the Cold War saw a revival, affecting the global movement of scientists and students once again. Today, a very similar regime as the one analyzed above affects transnational scientific exchanges on a daily basis, as part of the bureaucratic machinery of the homeland security state.

Two strands of national security concerns—the fear of terrorism and the rise of China as an economic and military competitor—converged in the decade after the end of the Cold War. Their common denominator was fear of losing sensitive knowledge to enemies who could use it against the United States, in the form of either military weaponry, economic competition, or a terrorist attack (e.g., with biological agents like anthrax).[66]

Several programs reminiscent of the combined visa and export control regime of the Cold War have been set up since the late 1990s to keep closer tabs on international scholars and students coming to the United States. The State Department installed a special screening process, called Visas Mantis, for visa applicants from certain target countries with a scientific or engineering background. On the basis of a Technology Alert List (echoing a similar alert list of the 1970s), drawn from the export control regulations and enumerating technologies deemed to be especially sensitive to national security, the State Department in cooperation with the intelligence community assesses the risks of what a visitor could learn at US educational and research institutions. If the dangers appear too great, no visa is issued. In 2004 twenty thousand students and scholars were screened according to the Visas Mantis procedure, and in about 2 percent of the cases the visa was denied.[67]

At the same time, the Departments of Commerce and State began to enforce more rigorously the data export control regulations, thus restricting the activities of students and scholars after the issuance of a visa. The regulation of oral, visual, and written communication of technological-scientific data within the United States is today called "deemed export" and has become, despite vociferous criticism against its stifling effects on international scientific relations, a set of rules that universities and research labs have to reckon with when they have exchange relations with foreign countries.[68] The consequences for universities and their faculties if they do not rigorously police these regulations was dramatically demonstrated in 2008 when Professor John Reece Roth from the University of Tennessee in Kentucky was sentenced to four years in prison at the age of seventy-two for including graduate students from China and Iran in his unclassified research without paying attention to the International Traffic in Arms Regulations, the State Department regulations for weapons technology (see chapter 2 of this volume).[69]

Finally, in the post-9/11 world, every foreign student's and exchange scholar's movements and educational activities are documented in a central database, the Student and Exchange Visitor Information System, which universities have to feed with information.[70] For example, if a foreign student decides to change her or his major from history of science and technology to nuclear physics or biomedicine, the university has to report back to the Department of Homeland Security.[71] The system, however, is apparently not used for the purposes of Visas Mantis or the enforcement of "deemed export" regulations, because of a lack of interagency cooperation—a fact repeatedly criticized as creating "a loophole that foreign nationals could exploit in order to gain inappropriate access to controlled U.S. technology."[72]

The history of the system of border policing of the movement of scientists, engineers, students, and commercial travelers by an interplay of visa and export control regulations since the 1940s shows how and why national borders matter for the transnational history of science and technology. Obviously, the US postwar national security state, as well as its latest incarnation, the counterterrorist homeland security state, perceived traveling scientists crossing the US border as a crucial vector for the transfer of technical and scientific skills and ideas. From a national security perspective, the mobility of "embodied" knowledge beyond borders posed a serious threat to American scientific, technological—and so military and economic—preeminence. It had to be regulated and, if need be, stopped. The "border" invoked by the national security community was not identical to the geographical border of the country. Rather, the border was that site where knowledge, presumably having a US identity, was shared with a foreign entity, be it a state or a foreign person. Thus, in the logic of the visa and export control regime, the US border could be found within the United States, on the Stanford University campus, in the office of an American professor. At the same time, the physical border mattered when it was being crossed by a foreigner, coming to or leaving the United States with knowledge in his or her head and hands. The Iron Curtain demarcated the US border as well—it was an ideological border between competing systems as well as a geographic border cutting through Europe.

The combination of two bureaucratic border-policing regimes—travel documentation and export licensing—mobilized to target such a highly elusive thing as embodied scientific-technological knowledge was a remarkable act of administrative creativity, based on sophisticated reflections on the nature of scientific and technological knowledge and its inter- and transnational transfer. The combined visa and export control regime shows that borders mattered for the sharing of knowledge because knowledge mattered to US national security bureaucrats.

Notes

1. For introductions to the history of government secrecy as well as secrecy in science and technology, see Peter Galison, "Removing Knowledge," *Critical Inquiry* 31, no. 1 (2004): 229–243; Peter Galison, "Secrecy in Three Acts," *Social Research* 77 (2010): 941–974; Timothy L. Ericson, "Building Our Own 'Iron Curtain': The Emergence of Secrecy in American Government," *American Archivist* 68, no. 1 (2005): 18–52; Harold C. Relyea, "Government Secrecy: Policy Depths and Dimensions," *Government Information Quarterly* 20 (2003): 395–418; Michael Aaron Dennis, "Secrecy and Science Revisited: From Politics to Historical Practice and Back," in *Secrecy and Knowledge Production*, ed. Judith Reppy, Cornell University Peace Studies Program, Occasional Paper 23 (Ithaca, NY: Cornell University Peace Studies Program, 1999), 1–16; Sissela Bok, "Secrecy and Openness in Science: Ethical Considerations," *Science, Technology, and Human Values* 7, no. 38 (1982): 32–41; Jonathan Felbinger and Julia Reppy, "Classifying Knowledge, Creating Secrets: Government Policy for Dual-Use Technology," in *Government Secrecy*, ed. Susan Maret (Bingley, UK: Emerald Group, 2011), 277–299; Alex Wellerstein, "Knowledge and the Bomb: Nuclear Secrecy in the United States, 1939–2008" (PhD diss., Harvard University, 2010); Harold C. Relyea, "Information, Secrecy, and Atomic Energy," *NYU Review of Law and Social Change* 10, no. 2 (1980/81): 265–286.

2. Craig Robertson, "The Documentary Regime of Verification: The Emergence of the U.S. Passport and the Archival Problematization of Identity," *Cultural Studies* 23, no. 3 (2009): 329–354. See also the nice "Typology of Papers" presented by John Torpey, *The Invention of the Passport: Surveillance, Citizenship and the State* (Cambridge: Cambridge University Press, 2000), 158–167.

3. Eric Neumayer, "Unequal Access to Foreign Spaces: How States Use Visa Restrictions to Regulate Mobility in a Globalised World," *Global Migration Perspectives* 43 (2005), accessed Apr. 7, 2018, http://www.iom.int/jahia/webdav/site/my/jahiasite/shared/shared/mainsite/policy_and_research/gcim/gmp43.pdf.

4. Harold J. Berman and John R. Garson, "United States Export Controls—Past, Present, and Future," *Columbia Law Review* 67, no. 5 (1967): 791–890; Michael Mastanduno, "Trade as a Strategic Weapon: American and Alliance Export Control Policy in the Early Postwar Period," *International Organization* 42, no. 1 (1988): 121–150; Alan P. Dobson, "The Changing Goals of the U.S. Cold War Strategic Embargo," *Journal of Cold War Studies* 21, no. 1 (2010): 98–119; Gernot Stenger, "The Development of American Export Control Legislation after World War II," *Wisconsin International Law Journal* 6, no. 1 (1987): 1–42.

5. During the early Cold War the passport was equally important. This essay, however, focuses on visas. I explore the complex history of passport controls in a forthcoming paper: Mario Daniels, "Controlling Knowledge by Controlling People: Travel Restrictions of U.S. Scientists and National Security in the Early Cold War," *Diplomatic History*.

6. Mario Daniels and John Krige, "Beyond the Reach of Regulation? 'Basic' and 'Applied' Research in Early Cold War America," *Technology and Culture*, in press.

7. For discussions of "scientific internationalism," see Joseph Manzione, "'Amusing and Amazing and Practical and Military': The Legacy of Scientific Internationalism in American Foreign Policy, 1945–1963," *Diplomatic History* 24, no. 1 (2000): 21–55; Geert J. Somsen, "A History of Universalism: Conceptions of the Internationality of Science from the Enlightenment to the Cold War," *Minerva* 46 (2008): 361–379; Patrick David Slaney, "Eugene Rabinowitch, the *Bulletin of the Atomic Scientists* and the Nature of Scientific Internationalism in the Early Cold War," *Historical Studies in the Natural Sciences* 42, no. 2 (2012): 114–142; William I. Hitchcock, "The Marshall Plan and the Creation of the West," in *The Cambridge History of the Cold War*, ed. Melvyn P. Leffler and Odd Arne Westad, vol. 1 (Cambridge: Cambridge University Press, 2010), 154–174.

8. Melvyn P. Leffler, "The American Conception of National Security and the Beginnings of the Cold War, 1945–48," *American Historical Review* 89, no. 2 (1984): 346–381.

9. For the bigger picture, see John Krige, *American Hegemony and the Postwar Reconstruction of Science in Europe* (Cambridge, MA: MIT Press, 2006).

10. Such tensions between national interest and scientific internationalism

were, of course, not entirely new. See Paul Forman, "Scientific Internationalism and the Weimar Physicists: The Ideology and Its Manipulation in Germany after World War I," *Isis* 64, no. 2 (1973): 150–180.

11. John Krige, *Sharing Knowledge, Shaping Europe: U.S. Technological Collaboration and Nonproliferation* (Cambridge, MA: MIT Press, 2016).

12. *Report of the Commission on Government Security Pursuant to Public Law 304, 84th Congress, as Amended* (Washington, DC: Government Printing Office, 1957), 467.

13. RG 40, UD Entry 56, box 1, Office of Strategic Information to Joint Operating Committee, Mar. 10, 1955, National Archives and Records Administration (hereafter NARA), College Park, MD.

14. Daniels, "Controlling Knowledge by Controlling People."

15. *Report of the Commission on Government Security*, 527–528.

16. *Science and Foreign Relations: International Flow of Scientific and Technological Information*, Department of State Publication 3860 (May 1950), 78–79.

17. Michael J. Ybarra, *Washington Gone Crazy: Senator Pat McCarran and the Great American Communist Hunt* (Hanover, NH: Steerforth Press, 2004).

18. The Immigration and Nationality Act of 1952 amended and extended Section 212, pertaining to visa denials, of the Internal Security Act of 1950. Both versions are reprinted in *Bulletin of the Atomic Scientists* 8, no. 7 (1952): 257–258.

19. Victor F. Weisskopf, "Report on the Visa Situation," *Bulletin of the Atomic Scientists* 8, no. 7 (1952): 221–222, at 221.

20. Edward Shils, "Editorial: America's Paper Curtain," *Bulletin of the Atomic Scientists* 8, no. 7 (1952): 210–217, at 212. See also *Whom We Should Welcome: Report of the President's Commission on Immigration and Naturalization* (Washington, DC: Government Printing Office, 1953), 67.

21. *Whom We Should Welcome*, 66; Weisskopf, "Report on the Visa Situation," 222.

22. Jessica Wang, *American Science in an Age of Anxiety: Scientists, Anticommunism, and the Cold War* (Chapel Hill: University of North Carolina Press, 1999); David Kaiser, "The Atomic Secret in Red Hands? American Suspicions of Theoretical Physicists during the Cold War," *Representations* 90, no. 1 (2005): 28–60; Lawrence Badash, "From Security Blanket to Security Risk: Scientists in the Decade after Hiroshima," *History and Technology* 19, no. 3 (2003): 241–256; Gregg Herken, "'A Most Deadly Illusion': The Atomic Secret and American Nuclear Weapons Policy, 1945–1950," *Pacific Historical Review* 49, no. 1 (1980): 51–76; Simone Turchetti, *The Pontecorvo Affair: A Cold War Defection and Nuclear Physics* (Chicago: University of Chicago Press, 2012); Frank Close, *Half-Life: The Divided Life of Bruno Pontecorvo, Physicist or Spy* (New York: Basic Books, 2015); Robert Chadwell Williams, *Klaus Fuchs: Atom Spy* (Cambridge, MA: Harvard University Press, 1987).

23. For the concept and history of this "espionage paradigm," see Katherine S. Sibley, "Soviet Military-Industrial Espionage in the United States and the Emergence of an Espionage Paradigm in US-Soviet Relations, 1941–45," *American Communist History* 2, no. 1 (2003): 21–61.

24. RG 40, UD Entry 56, box 2, Director of Office of Defense Mobilization to Secretary of Commerce, Dec. 10, 1954, attachment "Control of Visits to Industrial Facilities in the Interest of Internal Security," NARA.

25. Federal Bureau of Investigation, *Soviet Intelligence Travel and Entry Techniques*, Apr. 1953, i–iii, v, 1–2, 22–23. Even though the FBI focuses on Soviet citizens, it uses Alan Nunn May, born in the United Kingdom, as one example. CREST files, CIA-RDP65-0076R000400080001-9, NARA.

26. Shils, "Editorial: America's Paper Curtain," 213. Shils also writes, along these lines, "The transformation of science into a subject of crucial importance to national defense has changed patronizing distrust into active and harassing suspicion—recently exaggerated and aggravated by the Fuchs, Nunn May, and Pontecorvo cases" (ibid.).

27. The term was also used to criticize the spread of government secrecy in the early Cold War. In this context the "paper curtain" was between the US government and the people. *Availability of Information from Federal Departments and Agencies: Twenty-Fifth Intermediate Report of House Committee on Government Operations* (Washington, DC: Government Printing Office, 1956), 3.

28. *Whom We Should Welcome*, 67.

29. Wellerstein, "Knowledge and the Bomb."

30. Shils, "Editorial: America's Paper Curtain," 212–213. For references to unclassified information, see also ibid., 211; John Toll, "Scientists Urge Lifting Travel Restrictions," *Bulletin of the Atomic Scientists* 14, no. 8 (1958): 326–328 (in regard to passport denials); Victor F. Weisskopf, "Visas for Foreign Scientists," *Bulletin of the Atomic Scientists* 10, no. 3 (1954): 68–69, 112, at 68.

31. Weisskopf, "Report on the Visa Situation," 222.

32. *Science and Foreign Relations*, 76. This report accordingly criticizes visa regulations, the Export Control Act of 1949, and other forms of control of unclassified information in the same short chapter (ibid., 76–85).

33. RG 40, Entry UD 76, box 3, minutes of meeting of the U.S.S.R. Committee, State Department, 2, Mar. 21, 1946, NARA.

34. This concept is being discussed as a controversial legal tool in the "War against Terrorism" after 9/11: David E. Pozen, "The Mosaic Theory, National Security, and the Freedom of Information Act," *Yale Law Journal* 115, no. 3 (2005): 628–679; Jameel Jaffer, "The Mosaic Theory," *Social Research* 77, no. 3 (2010): S. 873–882; Benjamin M. Ostrander, "The 'Mosaic Theory' and Fourth Amendment Law," *Notre Dame Law Review* 86, no. 4 (2011): 1733–1766; Orin S. Kerr, "The Mosaic Theory and the Fourth Amendment," *Michigan Law Review*

111 (2012): 311–354; Susan N. Herman, *Taking Liberties: The War on Terror and the Erosion of American Democracy* (Oxford: Oxford University Press, 2011), 128–129, 140–141, 202.

35. RG 40, Entry UD 76, box 2, Unclassified Technological Information Committee, minutes, 2, 10, Jan. 20, 1947, NARA. The State Department U.S.S.R. Committee mentioned above also advocated making "more use of visa controls" to "slow down the flow of U.S. technology to the U.S.S.R." RG 40, Entry UD 76, box 3, minutes of meeting of the U.S.S.R. Committee, State Department, 7, Mar. 21, 1946, NARA.

36. RG 40, Entry UD 59, box 8, Interdepartmental Committee on Industrial Security, "Basic Problems of Industrial Security," 1–3, Oct. 14, 1948, NARA.

37. The Export Control Act of 1949, 63 Stat. 7, chap. 11, sec. 3a, states that "the President may prohibit or curtail the exportation from the United States . . . of any articles, materials, or supplies, including technical data."

38. "Voluntary Controls Put over Export of Knowhow," *Washington Post*, Nov. 11, 1949, 2; "'Voluntary' Plan Bars Data Export," *New York Times*, Nov. 11, 1949, 11; "Plan to Control Export of Advanced Technical Data," in Department of Commerce, *Foreign Commerce Weekly* 37, no. 9 (Nov. 28, 1949): 43; Department of Commerce, *Export Control and Allocation Powers: Ninth Quarterly Report to the President, the Senate and House of Representatives* (Washington, DC: Government Printing Office, 1949), 3.

39. Department of Commerce, *Export Control and Allocation Powers: Fifteenth Quarterly Report* (Washington, DC: Government Printing Office, 1951), 1.

40. Export Regulations, part 385: Exportations of Technical Data, *Federal Register* 19, no. 253 (Dec. 31, 1954): 9384–9386, at 9384 (§385.1a) (emphasis added).

41. RG 40, UD 56, box 1, Executive Order 10501, Nov. 5, 1953, "Safeguarding Official Information in the Interest of the Defense of the United States," NARA; Atomic Energy Act of 1954, 68 Stat. 919.

42. Export Regulations, part 385: Exportations of Technical Data, 9384 (§385.1c) (emphasis added).

43. Ibid., 9385 (§385.4d vi).

44. RG 40, Entry UD 59, box 3, Engineering College Research Council of the American Society for Engineering Education, "Recommendation to the U.S. Department of Commerce to Exempt Colleges and Universities from Export Control of Technical Data," Feb. 23, 1955, NARA.

45. RG 40, Entry UD 59, box 3, Department of Commerce press release, Apr. 16, 1955, NARA.

46. Ibid.

47. For problems of definition, see Frank E. Samuel, "Technical Data Export Regulations," *Harvard International Law Club Journal* 6, no. 2 (1965): 125–165, esp. 135–136; J. N. Behrman, "U.S. Government Controls over Export of Tech-

nical Data," *Patent, Trademark, and Copyright Journal of Research and Education* 8 (1964): 303–315, esp. 304–305.

48. On the history of the East-West exchange programs, see Yale Richmond, *U.S.-Soviet Cultural Exchanges: Who Wins?* (Boulder, CO: Westview Press, 1978); Yale Richmond, *Cultural Exchange and the Cold War: Raising the Iron Curtain* (University Park: Pennsylvania State University Press, 2003); Glenn E. Schweitzer, *Scientists, Engineers, and Two-Track Diplomacy: Half a Century U.S.-Russian Interacademy Cooperation* (Washington, DC: National Academy Press, 2004); Robert F. Byrnes, *Soviet-American Academic Exchanges, 1958–1975* (Bloomington: Indiana University Press, 1976); Herbert Kupferberg, *The Raised Curtain: Report of the Twentieth Century Fund Task Force on Soviet-American Scholarly and Cultural Exchanges* (New York: Twentieth Century Fund, 1977). For the Atoms for Peace conference, see John Krige, "Atoms for Peace, Scientific Internationalism, and Scientific Intelligence," *Osiris* 21, no. 1 (2006): 161–181. For the general development of exchange relations as well as a reprint of NSC 5607 and the accompanying policy statement, see Richmond, *U.S.-Soviet Cultural Exchanges*, 1–9, 133–137.

49. NSC 5607, in Richmond, *U.S.-Soviet Cultural Exchanges*, 134.

50. RG 489, A1 Entry 1, box 7, John C. Borton (Department of Commerce) to Henry Kearns (Department of State), "Change in Export Control Regulations to Facilitate Visits to U.S. Plants under East-West Contacts Program," Dec. 20, 1957, NARA.

51. RG 489, A1 Entry 1, box 7, Loring E. Nacy (Commerce) to Henry Kearns (State), "Visits by Soviet Bloc Nationals to U.S. Plants under the East-West Contacts Program," Nov. 14, 1957, with attachment "Proposed Amendment to P.D. 1192," and draft letter, Secretary of Commerce to Secretary of State, Nov. 15, 1957, with attachment "Proposed Amendment to P.D. 1192" (a different version), NARA. The Commerce Department's need for the visa application information for meeting its export control responsibilities is spelled out in RG 489, A1 Entry 7, box 7, report on meeting held by representatives of State and Commerce on Oct. 14, 1957, NARA.

52. RG 489, A1 Entry 1, box 7, Borton to Kearns, "Change in Export Control Regulations to Facilitate Visits to U.S. Plants under East-West Contacts Program," Dec. 20, 1957, NARA.

53. RG 489, A1 Entry 1, box 7, Statement Meyer (Commerce), "Our Position re: Enforcement of the Export Control Program on Unclassified Technical Data as It Relates to the East/West Exchange Program and Industrial Conferences and Trade Shows," Nov. 21, 1960, NARA.

54. RG 489, A1 Entry 1, box 7, Frank Sheaffer (Commerce) to F. D. Hockersmith (Commerce), Jan. 11, 1961, and draft letter to the International Conference of the American Nuclear Society and the International Conference on Strong Magnetic Fields at MIT, May 15, 1961, NARA.

55. RG 489, A1 Entry 1, box 7, Nacy to Kearns, "Visits by Soviet Bloc Nationals to U.S. Plants under the East-West Contacts Program," Nov. 14, 1957, NARA. Quotation in draft letter, Secretary of Commerce to Secretary of State, Nov. 15, 1957.

56. [CIA] Informal History: Intelligence Involvement in the East-West Exchanges Program [ca. 1974], 3, accessed Jan. 15, 2017, https://www.cia.gov/library/readingroom/docs/DOC_0001495225.pdf; Standing Committee on Exchanges, July 16, 1956, accessed Jan. 15, 2017, https://www.cia.gov/library/readingroom/docs/CIA-RDP85S00362R000600030005–7.pdf.

57. Director of Central Intelligence Directive no. 2/6, Committee on Exchanges (Coordination and Exploitation of East-West Exchange Program), Apr. 3, 1963 (showing the redactions of the version of 1959), attachment to United States Intelligence Board memorandum, "Proposed Amendments to Director of Central Intelligence Directives," Apr. 3, 1963, accessed Jan. 15, 2017, https://www.cia.gov/library/readingroom/docs/CIA-RDP86B00269R000200060080–5.pdf.

58. IAC Standing Committee on Exchanges, First Semi-annual Report, Oct. 4, 1956 (quotation on p. 1), CREST files, CIA-RDP61-00459R000300050005-3, NARA. For the close connection between scientific exchanges and scientific intelligence, see also Krige, "Atoms for Peace." From a CIA perspective, see Guy E. Coriden, "The Intelligence Hand in East-West Exchange Visits," *Studies in Intelligence* 2, no. 3 (1958): 63–70; CREST files, CIA-RDP78-03921A000300210001-1, NARA.

59. Secretary of Commerce Lewis L. Strauss to Allen Dulles, June 11, 1959, and Strauss to Secretary of State, May 8, 1959, with attachment "Suggestions for Modifying Present Procedures for Implementation of the East-West Exchange Program" (quotation in the attachment), accessed Jan. 15, 2017, https://www.cia.gov/library/readingroom/docs/CIA-RDP80B01676R000800010018–6.pdf.

60. IAC Standing Committee on Exchanges, First Semi-annual Report, Oct. 4, 1956, appendix A.

61. IAC Standing Committee on Exchanges, Third Semi-annual Report, p. 4, Feb. 11, 1958, and annex A to this report: "Interim Evaluation of the Intelligence Aspects of the East-West Exchange Program," 5, Feb. 11, 1958, CREST files, CIA-RDP61-00549R000300050002-6, NARA.

62. James McGrath, "The Scientific and Cultural Exchange," *Studies in Intelligence* 7, no. 1 (1963): 25–30, at 27–28; CREST files, CIA-RDP78T03194A000200010001-2, NARA.

63. RG 489, A1 Entry 1, box 4, Ralph Jones, Department of State, Soviet and Eastern European Exchanges Staff, to IBM, Feb. 21, 1963, NARA. A similar letter mentions the involvement of the Department of Defense in making decisions about exchanges: RG 489, A1 Entry 1, box 4, Frank G. Siscoe, director, So-

viet and Eastern European Exchanges Staff, to the president of the System Development Corporation, Mar. 29, 1963, NARA.

64. Daniels and Krige, *Technology and Culture.*

65. McGrath, "Scientific and Cultural Exchange," 30.

66. For an introduction to the current debates on science, technology, and national security after 9/11, see National Research Council, *Science and Security in a Post 9/11 World: A Report Based on Regional Discussion between the Science and Security Communities* (Washington, DC: National Academy Press, 2007).

67. Government Accountability Office, *Border Security: Streamlined Visas Mantis Program Has Lowered Burden on Foreign Science Students and Scholars, but Further Refinements Needed*, 5–7, Feb. 2005, accessed Jan. 27, 2017, http://www.gao.gov/assets/250/245374.pdf.

68. Deemed Export Advisory Committee, *The Deemed Export Rule in the Era of Globalization: Report to the Secretary of Commerce* ("Augustine Report"), Dec. 20, 2007, accessed Jan. 27, 2017, https://fas.org/sgp/library/deemedexports.pdf; John Krige, "National Security and Academia: Regulating the International Circulation of Knowledge," *Bulletin of the Atomic Scientists* 70, no. 2 (2014): 42–52; Benjamin Carter Findley, "Revisions to the United States Deemed-Export Regulations: Implications for Universities, University Research, and Foreign Faculty, Staff, and Students," *Wisconsin Law Review*, 2006, 1223–1274.

69. Daniel Golden, "Why the Professor Went to Jail: Is John Reece Roth a Martyr to Academic Freedom or a Traitor?," *Bloomsburg Business Week*, Nov. 1, 2012, accessed Jan. 27, 2017, https://www.bloomberg.com/news/articles/2012–11–01/why-the-professor-went-to-prison.

70. Alison Jackson Tabor, "It's Not Just a Database: SEVIS, the Federal Monitoring of International Graduate Students Post 9/11" (PhD diss., University of Kentucky, 2008); Julie Farnam, *U.S. Immigration Laws under the Threat of Terrorism* (New York: Algora, 2005), 97–129.

71. See the statement of Asa Hutchinson, undersecretary for border and transportation security, Department of Homeland Security, in *The Conflict between Science and Security in Visa Policy: Status and Next Steps*, hearing before the House Committee on Science (Washington, DC: Government Printing Office, 2004), 59.

72. Inspector Generals of Departments of Commerce, Defense, Energy, Homeland Security, State, and the CIA, *Interagency Review of Foreign National Access to Export-Controlled Technology in the United States*, vol. 1, 22, Apr. 2004, accessed Jan. 27, 2017, http://www.dodig.mil/audit/reports/fy04/04–062.pdf. Similar: Government Accountability Office, *Export Controls: Agencies Should Assess Vulnerabilities and Improve Guidance for Protecting Export-Controlled Information at Universities*, 18–19, Dec. 2007, accessed Jan. 27, 2017, http://www.gao.gov/new.items/d0770.pdf.

CHAPTER TWO

Export Controls as Instruments to Regulate Knowledge Acquisition in a Globalizing Economy

John Krige

In August 2008 John Reece Roth, a physics professor at the University of Tennessee, appeared before court for violating export control regulations. The most serious charges against him were that he had used a Chinese (and an Iranian) graduate student on an Air Force contract to improve the performance of drones and that he had visited China several times to discuss his research. Roth, who had spent many years working for NASA before moving to academia, insisted on his innocence. He had had security clearance in the space agency and understood that the circulation of knowledge was sometimes controlled for national security purposes. But this was not classified work. What is more, his military project had not progressed much beyond the stage of fundamental research. He could not see how export controls—that regulated the circulation of goods, he thought—were relevant to what he was doing. The prosecutor dismissed Roth's pleas. Roth had been warned that he was in violation of the International Traffic in Arms Regulations, which also controlled the circulation of technical data to foreign nationals on American soil. His refusal to comply was indicative of academic hubris, of a "cavalier mindset" that balked at intrusive bureaucratic requirements. This was galling. As the prosecutor put it in summing up: "We don't want the perception, of course, that these are just unimportant bureaucratic regulations. That is very important. . . . This is the type of case where the sentence can have a very significant deterrent impact. This sentence will be noticed by those who are engaged in military contracts and anybody who might be inclined just to blow these type of arms export controls off."[1] The jury agreed, and Roth was jailed for four years beginning January 2012. He was in his midseventies, walked with difficulty, and was hard of hearing at the time of his incarceration.[2]

Export controls barely graze the working world of researchers in the humanities and the social sciences. They are, however, part of the fabric of everyday life of people engaged in cutting-edge research in science and engineering in academia and in corporations that develop and market technologically sophisticated goods and services. This chapter draws on the transnational approach developed in the introduction to focus on export controls as regulating the movement of scientific and technological knowledge in these social milieus under the shadow of the American national security state. It complements the discussion of their role in visa policy in the previous chapter, tracking the expanding reach of the regulatory state into the heart of the process of the production of new knowledge: sharing knowledge in face-to-face learning between "expert" and "apprentice."

Leaving aside their role in regulating international trade, export controls can, at little risk of exaggeration, be understood as an epistemologically rich bureaucratic corpus of legal injunctions that are sensitive to the multiple practices whereby knowledge can be transferred. The changing scale and scope of their implementation are markers of changing perceptions of the threats that the loss of knowledge poses to American scientific and technological leadership and, ipso facto, to America's national security and economic competitiveness in the global arena. This chapter highlights the growing concern about the dangers to American global leadership and security posed by the transfer of knowledge in situations in which American scientific and technological expertise was shared with foreign students, engineers, and project managers. In particular, it focuses on the "contact zone" where US citizens and foreign nationals met to share and to exchange knowledge "across borders" as defined by the nationality and political allegiance of the people concerned. The performance of knowledge sharing and acquisition at these sites was profoundly reconfigured by export controls that shaped what you could say, what you could show, and what you could do with another person as you huddled over an instrument in a laboratory, gave a paper at a conference, demonstrated the advantages of high-tech equipment to a visitor from abroad, or helped a foreign client improve the performance of a complex technological system.

The period covered here extends from the 1970s to the early 2000s, a period marked by major transformations in the geopolitical landscape. The public face of American foreign policy regarding major communist powers oscillated from mutual accommodation and détente in the 1970s to defining the Soviet Union as an "evil empire" in the early 1980s; from a massive expansion in trade of all kinds, including arms, with the People's Republic of China (PRC) in the late 1980s to sensational accusations in the 1990s that the PRC had mounted a widespread effort to obtain US military technologies by legal and illegal means. Each phase of trade liberalization was followed by a demand for tighter restrictions, including restrictions on the circulation of knowledge. More to the point, the restrictions on sharing knowledge became increasingly broad in scope, to include the transfer of tacit knowledge; increasingly invasive, reaching to the core of the process of knowledge acquisition; and increasingly nationalistic, appealing to loyalty to the United States as a key consideration for sharing sensitive knowledge with foreign nationals in corporate and academic research settings. The recalibration of American power in a new age of globalization has led to the quiet but steady expansion of the depth and breadth of the regulatory powers of the national security state—further fueled and legitimated by the terrorist attacks of 2001. The mobilization of export controls to manage knowledge acquisition is indicative of the tight coupling between US national security, economic strength, and access to new knowledge, not in the asymmetric conditions of knowledge/power that marked the early and middle Cold War, but in a global marketplace in which knowledge is a commodity sought after by multiple, competing state and nonstate actors.

Blowback against Détente: The Bucy Report

From its inception the national security state has targeted the international circulation of knowledge-bearing people, information, and things to countries of concern.[3] People with expert knowledge who could bolster the scientific and technological capacity of communist rivals were prime targets. The recognition that trained scientists and engineers were a national resource and that their know-how was a precious national asset that could not be freely shared influenced both the scientific and the corporate worlds. Controls put in place in wartime were reactivated as the fractured lines of superpower rivalry emerged. One of these was the Export Control Act of 1949 administered by the Department of Commerce to regulate items identified on a Commerce Control List (as it was called as of 1965). Another was the Mutual Security Act of 1954 (the precursor of the act that sent Roth to jail), administered by the Department of State to control the export of items on the US Munitions List.

The Export Control Act was initially conceived as an extension of World War II measures to control the export of strategic materials that were in short supply and to ensure the availability of particular goods needed for the postwar reconstruction of Europe.[4] By the time the act was passed in February 1949, national security had become an additional consideration, all the more so after the Korean War broke out in June 1950. The measure was extended repeatedly and without amendment every two or three years until 1962. Its near-embargo character was, however, incompatible with the trade liberalization called for by the policy of détente with the Soviet Union. In 1969 the Export Control Act was replaced by the less restrictive Export Administration Act.

Critics of the new openness to the communist bloc were quick to raise the alarm. The Department of Defense (DoD) was particularly concerned by the visits of high-level Soviet delegations and engineers, who negotiated generous technology-sharing deals with leading American aircraft industries like Boeing and Lockheed.[5] The DoD's Defense Science Board was instructed to take action. In 1974 it established a task force to review the export control system for advanced technologies with defense implications, notably jet engines, airframes, solid-state devices, and scientific instruments. It was chaired by Fred Bucy, then executive vice president of Texas Instruments, and it was populated with senior officials from the DoD and private industry.

The so-called Bucy Report, delivered in February 1976, called for a conceptual shift in thinking about export controls. It highlighted the value to the enemy of the acquisition of tacit knowledge learned when new, technologically sophisticated equipment was sold to them. The threat to American economic and military supremacy lay not only in the use of the equipment as such but also in the training and learning-by-doing that were necessary to build and maintain generic equipment and in the acquisition of nonformalizable know-how that would enable Soviet engineers to develop new generations of materiel on their own.

The Bucy Report identified "the highest and most effective level of technology transfer" as being "the export of an array of design and manufacturing information [or know-how] plus significant teaching assistance which provides technical capability to design, optimize, and produce a broad spectrum of products in a technical field."[6] Export controls needed to embrace know-how, the report insisted, along with what it called "keystone" equipment. Keystone equipment was that unique kind of equipment that optimized the performance of a production process and allowed it to be modified and improved further. The example used most frequently was integrated-circuit inspection or test equipment.

Bucy insisted on the immense benefits to a foreign country of shared "technology," a term that was conflated with know-how. As he put it, the transfer of design and manufacturing know-how "confers a new capability on a receiving nation to produce goods to meet both present and long-term needs."[7] It also carried immense risks for the donor nation. Technical know-how was "intangible . . . , carried in the mind as well as embodied in machinery and equipment."[8] Thus, "once released, technology can neither be taken back nor controlled. Its release is an irreversible decision."[9]

Bucy identified several "active" mechanisms for technology transfer, including the interpersonal or face-to-face transfer of know-how. It involved learning: as the Bucy Report put it, this most active form of technology transfer was typically "an iterative process: the receiver requests specific information, applies it, develops new findings, and then requests further information," particularly about the use of keystone equipment.[10]

By extending the focus of controls away from the product as such, Bucy hoped to streamline the export control process. He argued that although controls should concentrate on the military significance of an item, it was not enough to focus on military use. End use was not a good selection criterion for export controls both because seemingly benign end users (like universities) were often fully integrated into the Soviet military system and because seemingly benign devices might have valuable military applications. What mattered was the potential for a product to cross the boundary between the civilian and military sectors of an economy (i.e., its "dual-use" functionality). At bottom, Bucy's concern was that many new, revolutionary technological advances in the late 1970s were being produced in the civilian sector before being "spun-on" to the defense market. Export controls should target not only obviously military items, then, but the Soviet military-industrial complex as a whole; they should be treated as tools of economic warfare. This is why he made so much of know-how and why he emphasized its intangibility. The "same" item could cross the civil-military divide: if you knew how to design and manufacture a chip for a computer, you could more easily design and manufacture a hardened chip for a nuclear-tipped guided missile.

The DoD set out to construct a Militarily Critical Technologies List shaped by Bucy's thinking, shifting "emphasis from product control to control of technology (specific products, equipment, and arrays of know-how)."[11] However, instead of simplifying the export control mechanism, the Militarily Critical Technologies List turned out to be far more comprehensive than the Commerce Control List it was supposed to replace. Its scope expanded steadily both because of the ambiguity in the concept of military "significance" and because each branch of the executive concerned with the issue sought to cover every base to avoid being blamed later for overlooking some dual-use item that could be deployed in the field.[12] The Militarily Critical Technologies List's 620 section titles were published in the *Federal Register* in October 1981; there were "literally thousands" of elements specified under those titles, and they were still classified.[13]

This new emphasis on know-how as the target of export controls had revolutionary implications for universities. As was noted in chapter 1, by the late 1950s universities were exempt from controls on the circulation of technical data in the Export Control Act, at least insofar as they concentrated on producing basic, unclassified, published or publishable research. Know-how was a different matter. In 1982 one witness to a five-day-long hearing of a congressional subcommittee dealing with technology transfer to the Soviet bloc explained that an electrical engineering student could learn how to manufacture a microprocessor chip in one year at MIT or Stanford: "Starting with a blank notebook, during that year, the student would have used computer-aided design to design the microprocessor, he will have used computer-aided layout to lay out the processor on silicon, manufactured the chip either in the laboratory or in collaboration with a manufacturer, tested the circuit, packaged the circuit, mounted the microcomputer on a printed circuit board, and made the resulting computer work."[14] Acquiring know-how was an essential aspect of an engineering education in the United States: universities were being drawn inexorably into the orbit of export controls.

The Export Administration Act of 1969, renewed in 1974 and 1977, was comprehensively rewritten in 1979 to take account of Bucy's suggestions. The 1979 act, amended from time to time toward increased liberalization, remains the basis of the Commerce Department's export control system today. The Export Administration Regulations (EAR) that implement the act now define "technology" as *information* that is "necessary for the "development," "production," or "use"' of an item on the Commerce Control List, which numbers over three thousand "dual-use" commodities. A license is required to "export" specific information necessary for any of the following modes of use: "operation, installation (including on-site installation), maintenance (checking), repair, overhaul, and refurbishing . . . of any such item." This information is shared between people via "technical assistance" that "may take forms such as instruction, skills, training, working knowledge, consulting services" and "may involve transfer of 'technical data.'" For the EAR the "immutable mobiles" carrying this data "may take forms such as blueprints, plans, diagrams, models, formulae, tables, engineering designs and specifications, manuals and instructions written or recorded on other media or devices such as disk, tape, read only memories."[15] Other modes of communication by which knowledge can be exported are, to quote,

> (i) Visual inspection by foreign nationals of U.S.-origin equipment and facilities;
> (ii) Oral exchanges of information in the United States or abroad; or
> (iii) The application to situations abroad of personal knowledge or technical experience acquired in the United States.[16]

Knowledge and knowledge acquisition, quite apart from goods and services, are today one key target of export controls.

The gradual encroachment of export controls into the system of knowledge circulation was a response to the threat to US technological leadership posed by the policy of détente in the 1970s. America's unrivaled technological supremacy in the early Cold War was boosted by state-driven technological innovations that stimulated the domestic market; national security was protected by "embargoes" on the export of goods and services that America's partners in Europe and Asia mostly accepted. By the "middle" Cold War US allies could compete for communist markets and refused to allow American foreign policy to dictate their trade agreements. Détente was an ideological, political, and economic response to this situation. It liberalized export control to the Soviet bloc, opening a window of opportunity for American, Western European, and Japanese firms to trade with the Soviet military-industrial complex. The Soviet authorities, for their part, mobilized scientists, engineers, and design and production units to acquire foreign technology and know-how by every possible means, cutting the American technological lead in strategic sectors from as much as a decade to as little as a couple of years by the late 1970s—or so the opponents of trade liberalization in Washington claimed. In parallel, the political economy of innovation shifted, with private firms playing an increasing role in producing new goods that were equally sought after in both civilian and military markets. As the technological playing field became more level, and controls on "dual-use" goods more difficult to enforce in the name of national security, knowledge—specifically, design and manufacturing know-how of advanced technological products—was singled out as one of the most important factors securing a competitive edge over rivals. There was a new emphasis on the need to regulate knowledge acquisition and not only the exports of sensitive goods. In this charged climate, heightened by a reassessment of the Soviet threat, it was only a matter of time before research universities, as prime sites of learning, became entangled in the web of export control restrictions being spun by the national security state.

The Reagan Administration Targets Universities: The Fundamental Research Exclusion

The Soviet invasion of Afghanistan in December 1979 transformed political and economic relations between the two superpowers. Trucks rolling into Afghanistan that were produced at the Kama River plant with the help of American and Western technology emphasized to its critics the folly of trade liberalization as an instrument of political reform. Multiple sources, including a Soviet KGB defector who handed over thousands of secret documents to the French in 1981–1982 (the Farewell Dossier), described the massive, coordinated Soviet effort to acquire Western technology legally and illegally. As the assistant secretary of commerce Lawrence Brady put it in March 1981, "Operating out of embassies, consulates, and so-called 'business delegations' KGB operatives have blanketed the developed capitalist countries with a network that operates like a gigantic vacuum cleaner sucking up formulas, patents, blueprints and know-how with frightening precision." They had exploited the "soft underbelly" of America's open society, he said, including "the desire of academia to jealously preserve its prerogative as a community of scholars unencumbered by government controls." CIA director William Casey insisted in March 1982 that US-Soviet scientific exchange was "a big hole. We send scholars or young people to the Soviet Union to study Pushkin poetry: they send a 45-year old man out of their KGB or defense establishment to exactly the schools and professors who are working on sensitive technologies."[17] The Soviets, said Lara Baker, of the Los Alamos National Laboratory, had taken advantage of "the best technology transfer organization in the world . . . the U.S. university system."[18]

This assault on universities was an attempt to push back on the modus vivendi established between the research community and the national security system in the 1950s. At that time the universities had managed to carve out a space for the unimpeded exchange of knowledge across borders by appealing to the distinction between "basic" and "applied" science.[19] This served as a politically satisfying response to congressional and indeed presidential concerns that an exaggerated emphasis on the need for secrecy was chipping away at the First Amendment, was damaging international collaboration, and was anathema in an open society anyway. In consultation with American Nobel Prize winners who were invited to testify on The Hill, the distinction between applied and basic science was mapped onto knowledge that should be subject to control and that which could circulate freely within and across borders, securing access by the American research community to the global pool of knowledge without undermining national security. This was now deemed too lax. The gap between basic and applied research had narrowed, and the country was dealing with a far more menacing Soviet Union, which had taken huge strides thanks, in part, to the acquisition of an enormous amount of knowledge made in America.

Dale Corson (president emeritus, Cornell University) was asked to chair a panel to investigate the "legitimate concerns" that scientific openness might harm national security by providing adversaries with "militarily relevant technologies." The National Academies published their findings in 1982 in a lengthy report entitled *Scientific Communication and National Security*.[20] The panel focused its energies on new pressures being put on universities by the federal government. These had arisen, the panel suggested, in part because much "basic" research of military significance was dual-use, was undertaken without any concern for traditional controls like classification, and was close to application; as the Corson Report put it, there was a "perception that American universities were shifting towards research that is closer to technological frontiers." These "newly emerging technologies, particularly those that evolve directly from scientific research," were prime targets of Soviet bloc collection efforts, according to the US intelligence agencies.[21]

The Corson panel honed in on the risk of leakage via scientific communication in the research community. The report distinguished four very different types of technical information and their mode of transfer that might be of interest to foreign governments. Advances in scientific theory and insights into the progress being made in specific scientific fields were two of them. These were shared in written and oral communications in the community. Then there was the knowledge embodied in scientific and technical equipment, whose transfer involved "the physical transportation of objects." Finally, there was experimentation and procedural know-how, detailed information that was mostly obtained "through direct observation and experience with scientific and technical techniques." This mode of transfer was "a leading concern of the U.S. intelligence community." The Corson Report explained that the "transfer mechanism for such detailed information involves neither documents nor equipment, but more typically is the 'apprenticing' experience that takes place, among other means, through long-term scientific exchanges that involve actual participation in ongoing research."[22] Small wonder then that permissions for specific foreign scientific visits had been abruptly denied by the State Department. And that universities had been asked to help monitor and enforce restrictions on the movement of foreign scientists and of foreign students on campus.[23]

The report included in an appendix several fiery exchanges between senior university faculty and the government over the new efforts to extend export controls into academia. One, signed collectively by the presidents of Caltech, Cornell, MIT, Stanford, and the University of California, was sent to the secretaries of commerce, state, and defense. It objected strongly to the new "construction" of export controls that appeared to "contemplate government restrictions" on international scientific exchange.[24] In particular, nationality had become a key criterion for access to knowledge on campus, in violation of the fundamental values of the US university system. As the letter from the five presidents explained, it seemed that in the broad scientific and technical areas defined by the existing export control regulations, a situation might arise in which "faculty could not conduct classroom lectures when foreign students were present, engage in the exchange of information with foreign visitors, present papers or participate in discussions at symposia and conferences where foreign nationals were present, employ foreign nationals to work in their laboratories, or publish research findings in the open literature." The Corson Report, in short, was a response to a very specific controversy, to wit, the attempts by "the government" to extend controls beyond "the classification of documents and an export licensing system for physical products." The government now wanted to target also the transfer of "technology" or information by "scientific communications and [by] foreign scientific visitors." The regulatory state was extending its reach into the core of transnational scientific exchange, of face-to-face interaction, a zone in which sensitive but unclassified knowledge was communicated to foreign nationals who, it was feared, would use their new insights to the detriment of US national security.

The members of Corson's panel were fully apprised of the vast range of militarily significant technology acquired by the Soviet bloc, and they were careful to highlight the multiple legal and illegal channels through which such transfers took place. But their conclusion was unambiguous as regards the threat that universities posed to national security: "discussions of the Panel with representatives of all U.S. intelligence agencies failed to reveal specific evidence of damage to U.S. national security caused by information obtained from U.S. academic sources."[25] It followed that "the vast majority of university research, basic or applied, should be subject to no limitations on access or communications."[26] In those few special cases when secrets needed to be kept, classification was to be used. There was, the panel admitted, a "gray area" between openness and secrecy, an area in which unclassified basic research merged imperceptibly and rapidly into application. Even here openness was to be preferred unless, for example, the end product had "identifiable, direct military applications," was dual-use, and involved process- or production-oriented techniques.[27] Risks of leakage in these gray areas were to be dealt with by clauses in the research contract with the university that prohibited foreign nationals from participating in them and that allowed for prepublication review of research findings by a federal agency contracts officer. The Corson panel was opposed to invoking export controls to deal with gray areas in federally funded university research.

The underlying philosophy that guided the Corson panel's thinking —in fact, the philosophy that guided the thinking of the research community whenever controls over knowledge circulation loomed on the horizon—was that security lay in accomplishment and that "an essential ingredient of technological accomplishment is open and free scientific communication."[28] Unrestricted basic research fueled scientific and economic advance, as well as military progress. The risk was there, but it was a risk worth taking. American industry was able to exploit new technology more rapidly than could its adversaries. The goal, after all, was to stay ahead, to open up the widest possible technological gap between friend and foe alike. That goal was best achieved by tapping into the global pool of knowledge, not by restricting access to it by building a wall of secrecy, the relentless quest of the intelligence agencies.

The Corson Report was injected into an ongoing political debate that pitted academia against powerful forces in the executive. Some, like Admiral Bobby Inman, the deputy director of the CIA, were willing to accept the Corson panel's findings in testimony before Congress. Nevertheless, deep down they felt that preemptive action against future threats was imperative. Inman warned the research community at the 1982 annual meeting of the American Association for the Advancement of Science that "unless scientists controlled the 'hemorrhage' of sensitive research information, there would be a 'tidal wave' of repressive legislation aimed at controlling the publication and release of such information."[29] There was a "clear trend" by the Soviet bloc to target universities and research institutes more aggressively, and something had to be done now to stop it.

No significant steps forward were taken for a couple of years. The policy issues were so complex, and the diverse stakeholders so reluctant to yield, that a working compromise was evasive. The deadlock was broken by the DoD-University Forum, a consultative group cochaired by Richard DeLauer, undersecretary of defense for research and engineering, and Donald Kennedy, the president of Stanford University and one of the five signatories of the letter to the secretaries of commerce, state, and defense mentioned previously. DeLauer let it be known that he agreed broadly with the Corson panel and saw no need to restrict university research unless it was classifiable.[30] An interagency working group that was trying to define a coherent export control policy at the time was called upon to draft a policy on those lines. Their proposal was accepted by George Keyworth, the president's science adviser, and by senior DoD officials. It formed the basis of National Security Decision Directive 189 (NSDD189), signed by President Reagan on September 27, 1985.

NSDD189 stipulated that "to the maximum extent possible, the products of fundamental research remain unrestricted."[31] When national security required it, classification was to be used to control dissemination. Fundamental research actually went beyond the range of work covered by the term "basic research." It was a term of art that was invented for the purpose. It related specifically to federally funded activity in science, technology, and engineering at colleges, universities, and laboratories and was defined as "Basic and applied research in science and engineering, the results of which ordinarily are published and shared broadly in the research community, as distinguished from proprietary research and from industrial development, design, production and product utilization, the results of which are ordinarily restricted for proprietary or national security reasons." This Fundamental Research Exclusion (FRE), as it is called, was welcomed by academia as throwing a "loop" around academic research "within which loop there should be no restrictions on dissemination or participation."[32] In essence, it imposed no controls on the production of knowledge, as regards neither the equipment used nor the person using it, focusing only on the sensitivity of the output.

The Corson Report focused on scientific communication. It was not directly concerned with the needs of American industry. These were considered subsequently and most notably by the National Academies panel mandated by the Omnibus Trade and Competitiveness Act of 1988. The panel was formally established in October 1989, just a month before the fall of the Berlin Wall, whose implications obviously weighed heavily on their deliberations.[33] Their main concerns were two: to strengthen controls on the proliferation of weapons of mass destruction and to enable US firms to contribute proactively to the conversion of the Soviet defense industry into a market-oriented civilian enterprise. To that end they wanted a sharper line to be drawn between direct military and dual-use technologies. End use would be a key criterion for implementing controls on the export of goods, allowing companies to trade freely in commercial items. Indeed, the panel went so far as to suggest that American national security policy should turn its back on the "denial regime" that had defined its relationship with the communist bloc for more than forty years and "[m]ove progressively toward the removal of export controls on dual-use items to the Soviet Union and the East European countries for commercial end-uses that can be verified."[34] By the early 1990s, then, the United States was proactively promoting the global deregulation of trade, unambiguously military items and weapons of mass destruction excepted. Even China benefited from this loosening of restrictions in the 1980s—until it began to replace the Soviet Union in the American imagination as a major threat to national security and economic preeminence.

Clinton's Chinagate: Monitoring Satellite Launches "from the Cradle to the Grave"

The opening to the PRC initiated by President Nixon's ping-pong diplomacy in the early 1970s was followed by a massive expansion in trade, including technology and arms, between the PRC and the United States during President Reagan's second term.[35] These closer ties were not unchallenged, of course, and successive presidents had to navigate through a minefield of conflicting pressures, between those encouraging engagement with Beijing to secure access to markets and those calling for reprisals against the PRC in the name of violations of human rights (notably, the Tiananmen Square massacre in 1989), of disregard for nonproliferation agreements, and of threats to US national security.

The last issue came to a head during Clinton's second term when members of the Republican Party had a majority in both houses of Congress. Several scandals, which came to be known collectively as Chinagate, bedeviled the Clinton administration.[36] Two directly concerned export controls. One involved the granting of export licenses that allowed American satellite manufacturers to use Chinese launchers. The other concerned the theft from US nuclear laboratories of top-secret technology for miniaturizing nuclear weapons. The impact of this latter charge was amplified by an inflammatory report by a congressional committee chaired by Congressman Christopher Cox (R-CA). Its task was to investigate charges that the PRC had acquired sensitive technology and information that enhanced their nuclear armed intercontinental ballistic missiles and the manufacture of weapons of mass destruction. The report claimed that a Taiwanese-born Chinese American, Wen Ho Lee, had stolen the "crown jewels" of US nuclear security from the Los Alamos National Weapons Laboratory.[37]

Chinagate entangled the loss of militarily sensitive knowledge to

China in a bruising political confrontation intended to discredit the president as a legitimate leader of the nation. There were accusations that the Chinese government and foreign businesses had illegally funded the Democratic Party to influence the presidential election. Sex scandals, of which Clinton's affair with White House intern Monica Lewinsky is the most notorious, further damaged his public image. The ensuing politicization of export controls, along with accusations that the executive branch was lax on compliance, called forth new legislation in Title XV of the Strom Thurmond National Defense Authorization Act for Fiscal Year 1999. This title had particularly important implications for firms that launched satellites in China and led to tight controls on the sharing of knowledge between Western consultants and Chinese engineers.

In 1995 and 1996 two US manufacturers of civil communications satellites, Hughes Space and Communications International Inc. and Space Systems/Loral, assisted Chinese engineers in figuring out why their Long March rocket had failed, destroying its American-built payloads.[38] The firms justified their sharing of technical knowledge with the PRC by arguing that such assistance fell under the export license they had received from the Department of Commerce in February 1994. The Justice Department was not convinced and began a criminal investigation into whether the two firms had violated export control laws. At least three classified studies concluded that they had compromised national security by helping the Chinese improve the accuracy and reliability of their future ballistic missiles, including the guidance systems. Congress and the executive branch also noted that after a string of failed launches between 1992 and 1995, China had had an unbroken sequence of twenty-seven successful launches for commercial and government/military clients. It seemed too much of a coincidence. In January 2002 Space Systems/Loral agreed to pay $20 million to settle the charges of illegal technological transfer brought against it. Fifteen months later Boeing, which had acquired Hughes in 2000, agreed to pay $32 million in fines for its multiple export violations. The firms were punished for allowing scientists and engineers to share knowledge with their Chinese counterparts to identify the technological flaws that had led to the launch failures.

The legal framework for the penalties was provided by the Arms Export Control Act of 1976. The act stipulates that the export of items or technologies that can be used to enhance military capabilities, broadly understood, requires a license from the Directorate of Defense Trade

Controls in the State Department. The directorate implements the International Traffic in Arms Regulations by reference to a Munitions List that specifies what is subject to control, from rockets to high-speed computers to advanced (i.e., five-axis) machine tools—as well as satellites and space-qualified components. More important for my argument here, it also insists that a license is needed to share "documentation, designs or presentations concerning them [i.e., items on the list], as well as any services pertaining to them, including operations, maintenance, and repair."[39] Neither Hughes nor Loral had acquired such a license when asked for help after the launch failures.

Hughes and Loral were penalized for helping Chinese engineers gain a better understanding of the performance failures of the Long March launcher.[40] In December 1992 Hughes engineers estimated that aerodynamic buffeting of the rocket's fairing accounted for the failure to orbit its *Optus B2* satellite built for Australia, and they passed on their findings to the PRC. The local engineers hoped to solve the problem simply by adding rivets. This proved inadequate to deal with high-altitude crosswinds when, in January 1995, Hughes's *Apstar 2* launch also failed. Corrective measures along the lines first proposed by Hughes were then implemented—corrective measures that a congressional committee said had helped improve the "reliability" of PRC rockets used for both civil and military launches.

A year later there was a similar experience with a Loral *Intelsat 708* satellite. The new Long March LM-3B rocket carrying the comsat tilted off its launch tower at takeoff and smashed into a hillside village close to Xichang, killing or injuring more than one hundred people. The PRC engineers suspected that the launch had been aborted because of a fault in the inner part of the inertial measurement unit of the rocket guidance system. Telemetry data did not fully confirm this, however. The insurance company that had agreed to cover the imminent launch of another *Apstar* satellite (typically for about $50 million) demanded that an independent review committee look into the matter. The committee comprised representatives from the PRC, Hughes, Loral, Daimler Benz Aerospace, and retired experts who had worked for British Aerospace, General Dynamics, and Intelsat.[41] It placed great weight on the telemetry data and established that the follow-up frame, not the inner part of the inertial measurement unit, was responsible for the accident.

The independent review committee came to its conclusions after two major "face-to-face" meetings with their Chinese counterparts, the first in Palo Alto, the second in Beijing. The latter was particularly controversial. Four Chinese engineers were present at the first three-day meeting. Twenty-two Chinese engineers and officers attended the second, which lasted from April 30 to May 1, 1996. They were employed by several Chinese organizations that were involved in both civilian and military space activities, including the China Aerospace Corporation. This corporation and its subordinate companies, research academies, and factories developed and produced strategic and tactical ballistic missiles, space launch vehicles, surface-to-air missiles, cruise missiles, as well as military (e.g., reconnaissance and communications satellites) and civilian satellites. The meetings in Beijing were held in hotel rooms that were probably bugged by the Chinese intelligence services. The independent review committee also held unmonitored technical interviews with over a hundred PRC engineers and technical personnel and produced over two hundred pages of data and analyses.[42] The risk of sensitive information being shared with Chinese engineers who worked with both civilian and military space technologies was obvious.

Loral faxed a preliminary report of its findings to China in May 1996. The State Department learned that the firm had disclosed information that would probably improve the guidance system on Chinese missiles, without first having it reviewed for sensitive content and without an export license. Loral agreed not to share any more knowledge requiring government approval with the PRC. It denied that any of its employees dealing with China had acted illegally or damaged US national security. The firm also denied that the company chairman, Bernard Schwartz, who had contributed $1.5 million to the Democratic National Committee, was involved in any way.

The congressional assault on these lapses in knowledge transfer, undertaken before the fines were imposed, was part of a more general concern that President Clinton was putting business relationships with China ahead of US national security. The alleged violation of export controls and subsequent scares over the loss of top-secret information from the national weapons labs not only embarrassed the Clinton administration but also highlighted the growing fear of China as an emerging technological power.

It also had major legislative implications. Until the early 1990s communications satellites had been treated as defense articles, their export controlled by the International Traffic in Arms Regulations. Beginning in 1992 President George H. W. Bush and, later, President Clinton progressively moved them under the less stringent jurisdiction of the Department of Commerce, which licensed the export of dual-use items.[43] Clinton was particularly keen to reestablish engagement and dialogue with the PRC, which had been severely strained by the repression in Tiananmen Square in June 1989. In 1996 he authorized the Department of Commerce to have jurisdiction over the export of all commercial satellites, even those with potential military value.[44] The Strom Thurmond National Defense Authorization Act reversed the injunction. It included a short section devoted to controlling the export of satellite technology.[45] All "satellites and related items" were classified as military items. Comsats were placed under Section 38 of the Arms Export Control Act, administered by the State Department. In Section 1513 Strom Thurmond also removed the president's authority to change the jurisdictional status of satellites (and related items) even if they were for purely civilian applications: only Congress could do that. In so doing the act made satellites the only dual-use items that were required by law to be controlled as defense articles, even when they were being used to do space science.

American firms in the satellite business (as well as agencies like NASA) lobbied persistently to have this situation changed.[46] Henry Sokolski, a senior official in the first Bush administration who dealt with nonproliferation matters, identified the core of the problem. He explained that "intangible technology" was critical to the timely, reliable, and accurate placement of satellites into orbit *and* to the firing of missiles carrying nuclear warheads. The knowledge common to both included coupling load analysis, guidance data packages, upper-stage solid rocket propellant certification, upper-stage control design certification, lower-stage design certification, and general quality assurance.[47] The International Traffic in Arms Regulations strictly limited the sharing of these knowledges with foreign nationals. Indeed, as far as China (and Russia) are concerned, new procedures put in place in the 2000s are supposed to ensure that no "technology transfer" is possible when American-made satellites are launched on their rockets.

In response to the demands of the Strom Thurmond act, the DoD's Defense Threat Reduction Agency established a new Space Launch Monitoring division to keep an eye (at the companies' expense) on the goods and services sent abroad so as to prevent them from being exposed to foreign nationals. The surveillance teams included aerospace, satellite, and launch engineers, as well as a security expert.[48] They "monitor meetings and phone calls," and their personnel even "travel abroad to monitor the actual process of integration and launch" of American satellites from Chinese soil.[49] James Bodner, from the DoD, reassured a Senate subcommittee that the DoD now had a dedicated team of monitors at technical meetings and launch sites "to make sure that inappropriate tech transfers do not happen. . . . That is all they do. They monitor these things from the cradle to the grave." For China and Russia this stretched back to the design phase, for it was "in the design of the system in the first place [that] some of the most important tech data may be lost."[50]

There was really no alternative. Times had changed, as a senior official from the Department of Commerce reminded the Senate subcommittee. In the 1950s and 1960s the United States led the world in almost everything that mattered from a military standpoint. Export control was "easy" because there were no foreign competitors. All the United States had to do was "to make decisions about what we wanted the other guys to get, but we did not have to worry too much about what would happen if we said 'no.'"[51] Now these technologies were "ubiquitous." And now technological denial did not enhance national security: on the contrary, it simply reduced market share as others moved in to take the United States' place, further weakening the national defense industrial base. Maintaining markets required increased surveillance to stop, as best one could, the loss of intangible knowledge in a highly lucrative but militarily sensitive dual-use technological sector.

Controlling Knowledge Acquisition in Universities: The Renewed Emphasis on Deemed Exports

Among its many provisions the 1999 Strom Thurmond act also demanded increased transparency in the government's application of export controls. In particular, agencies that were responsible for their implementation were required to report annually to Congress for the next eight years.

In 2004 the inspectors general of the Departments of Commerce, State, Defense, Energy, and Homeland Security, along with the CIA, chose to assess the effectiveness of the deemed export policy specifically as an instrument to restrict the access of foreign nationals to controlled technology in the United States. Each agency provided its own report, from which highlights appeared in a joint report that was also submitted to Congress.[52]

I need to back up a moment to explain just what a "deemed export" is. The original version of the Export Control/Administration Act did not define the "export" of technical data to mean simply the transfer of such data and know-how from an American expert to someone else while in a foreign country. You did not have to leave the United States to "export" technical data. What mattered was where the data might travel to. Thus, in 1965 the export of technical data was defined in Export Regulation 385.1(b) as "any release of unclassified technical data for use outside the United States [including] the furnishing of data *in the United States* [my emphasis] to persons with the knowledge or intention that the persons to whom it is furnished will take such data out of the United States."[53] At the time, this clause was essentially intended to prevent American firms from sharing sensitive information with clients who had been sent to the United States to learn more about a product or process they had acquired. In 1994 the scope of the regulation was expanded by restricting the sharing of knowledge with a foreign national whether or not said individual was known to be taking such information out of the country. As stated by Clause §734.2(b)(2), an export was subject to the EAR if it took the form of "[a]ny release of technology or source code subject to the EAR to a foreign national. Such release is *deemed* [my emphasis] to be an export to the home country or countries of the foreign national." The "home country" was established with reference to a foreign national's most recently established legal permanent residency or most recently established citizenship.

In their joint report the inspectors general were disturbed by a lack of awareness in some universities and research centers of the requirements for the release of export-controlled technology.[54] The FRE was still in place. It had been essentially reaffirmed by Secretary of State Condoleezza Rice in 2001, immediately after the attacks on the World Trade Center, and it remained the key criterion appealed to by the research community to push back against government regulation of academic research. The inspectors general insisted that it had significant loopholes.[55] They were particularly distressed by the broad interpretation being given to the FRE as regards the participation in training and research of foreign nationals in the United States (for whom they used the abbreviation FNUS).

The broad interpretation of the FRE in academia (and by some officials in the Department of Commerce) held that normally a license was not needed for FNUS to use export-controlled technology. It did not matter who did the research or what research equipment they used to get the results: if those results were (to be) published in the open literature, no export license was required. In other words, in their interpretation, the FRE had no purchase on the practice of research as such, even if that research used export-controlled technology. The need for a license kicked in only if the results of that research were not to be published openly. This was the loophole the inspectors general wanted to close. For them, it was evident that FNUS should not be allowed to use export-controlled equipment without a license, even if they were doing fundamental research that was published without restraint.

The differing interpretations of the regulations arose from the meaning that was given to the definition of "use" in the EAR. Remember that the EAR defined using equipment as involving six different actions, namely, "operation, installation (including on-site installation), maintenance (checking), repair, overhaul, and refurbishing." The controversy arose because university officials interpreted the "and" at the end of this clause as meaning that the equipment was subject to export controls only if a foreign national performed *all six* actions with it. This of course hardly ever happened in practice. The inspectors general, by contrast, interpreted the phrase as a list, implying that a deemed export license was needed if FNUS used controlled equipment in *any one* of the six ways identified in the regulations. To clarify the matter, the Bureau of Industry and Security published a notice in the *Federal Register* in March 2005 soliciting comments on changing the definition of "use" in the EAR to align it with their interpretation: they suggested that the word "and" at the end of the use criteria be replaced by the couplet "and/or." This meant that any one of the six actions would qualify as "using" the equipment.

The universities fought back against the proposed definition of "use" through the Council on Governmental Relations. This was an association of 160 research universities and related entities (in 2005) that worked with the federal government to assess the impact of its policies on advanced research. The council pointed out that the Bureau of Industry and Security had given no reason why national security was not adequately protected by existing controls, namely, visa screening of foreign nationals and classification. They were emphatic that the FRE was to be understood as casting a protective shield around the process of knowledge production in all its stages if the research was fundamental—that is, if the findings were disseminated openly. And they pointed to the immense cost and impracticality of getting licenses for as many as six thousand people who would have access to thousands of items of dual-use equipment on university campuses. One university had an inventory of 50,000 research instruments costing over $5,000 each. Another, with multiple campuses, had some 140,000 pieces of equipment that would have to be assessed. The Bureau of Industry and Security backed down: to fall within the ambit of the EAR for using controlled equipment a researcher had to perform all six operations.

Then there was the question of the nationality or political allegiance of the user. This was one of the main issues that exercised a new advisory panel established by the Department of Commerce in 2006. The Deemed Export Advisory Committee (DEAC) was chaired by Norman Augustine, a former CEO of Lockheed Martin and a former undersecretary of the army in the Gerald Ford administration. The committee was tasked with finding ways to "ensure that the nation's Deemed Export policy continues to best protect United States national security while striving to promote the ability of United States industry and academic research to continue at the leading edge of technological innovation."[56]

The deemed export rule in place at the time did not apply to certain categories of FNUS. In particular Clause §734.2(b)(2) of the EAR stated, "This deemed export rule does not apply to persons lawfully admitted for permanent residence in the United States and does not apply to persons who are protected individuals [persons granted political asylum] under the Immigration and Naturalization Act (8 U.S.C. 1324b(a)(3))." The inspectors general objected to this provision. They did not think that green-card holders should be granted uncontrolled access to sensitive technology on the assumption that they had made a commitment to the United States and were unlikely to do it harm. As they pointed out, permanent residents were not obliged to become citizens (a step that would indicate a greater level of allegiance to the United States) and could travel freely back and forth to their home countries. US entities should therefore be required to get a license for employees or visitors who were foreign nationals to use such equipment "if they were born in a country where the technology transfer in question is EAR-controlled, *regardless of their most recent citizenship or permanent resident status* [my emphasis]."[57] The example of Qian Xuesen (Tsien Hsueshen) was probably at the back of their minds: it was mentioned in the highly inflammatory Cox Report. Qian Xuesen was one of about two thousand scientists who returned to China from abroad between 1949 and 1956. Born in Nationalist China, Qian Xuesen left for America in the 1930s, studied aeronautics at MIT between the wars, and stayed in the United States to rise to prominence at Caltech's Jet Propulsion Laboratory. In June 1950 he lost his security clearance and was put under virtual house arrest on charges of having communist sympathies. He was deported to the PRC five years later in exchange for American prisoners captured in the Korean War. There Qian Xuesen put his knowledge at the service of the Mao regime. He drew on Norbert Wiener's cybernetics to generalize the control theory of engineering and implanted operational research and scientific management in the PRC. He developed techniques for strategic planning, including the use of the PERT (Program Evaluation and Review Technique), developed for the Polaris missile program. Qian Xuesen is also credited with having played a major role in developing land- and sea-based missiles, as well as atomic weapons, nuclear-powered submarines, and satellites. In short, a great deal of what Qian Xuesen learned in the United States provided much of the bedrock on which Chinese technological strength in military and civil domains was built after the Soviets withdrew their technological support in the late 1950s.[58]

The DEAC grappled with the implications of targeting foreign nationals so broadly when thousands were graduating annually from American research universities. They cited National Science Foundation data showing that the "graduation rate of engineers who are United States citizens has actually declined by 20 percent over the most recent two decades, and two-thirds of the Ph.D.'s in engineering granted by United States universities are now being awarded to non-citizens" who were not born in the United States.[59] Many of these were doing research in dual-use domains that would enhance both America's economic strength and its national security. In the past many had stayed on to plow their knowledge back into the United States (people from outside the United States, predominantly Indians, had founded 52 percent of Silicon Valley companies and 39 percent of California start-ups between 1995 and 2005).[60] This was an argument for liberally interpreting deemed export regulations. Against that, it had to be admitted that now many were returning home, notably to China. Indeed, by 2005 the PRC had almost one million trained science and engineering researchers of its own.[61] This posed a major dilemma for security professionals. As the DEAC Report put it, "in today's post–Cold War globalizing, Internet-connected world, knowledge is a commodity that is exceptionally difficult to control if for no other reason than that it can be stored in the human brain, and humans are becoming increasingly mobile."[62] There was "a threat of industrial and defense espionage while such researchers are in the United States and the potential for knowledge portability when they repatriate."[63] On the one hand, deemed export regulations that were too permissive risked granting foreign nationals intent on harming the United States access to valuable science and engineering knowledge while in the country. But on the other, rules that were too restrictive might exclude the US research system from access to significant developments in the global pool of knowledge.

The DEAC made a number of interconnected proposals to deal with the risks posed by opening the doors of American research labs too widely. The committee was emphatic that a more determined effort should be made to identify and protect "elements of technical knowledge and military advantage that could have the greatest consequences in the national/homeland security sphere." It was far better to protect *select* highly sensitive military areas "rather than diffusing our efforts in the impracticable attempt to build high walls around large bodies of knowledge."[64] National security would be ensured once this protective shield around sensitive technology was in place. The FRE could then be maintained without trying to drive a wedge between the product of research and the process of acquiring it, and the debate over the interpretation of what it meant to use equipment could be rendered moot.

That being said, the DEAC agreed with the inspectors general that permanent residence was a weak criterion with which to assess the risk of harm being done to the United States by someone born abroad. In fact, they noted that most criminal violations of export control regulations that they knew of were committed by American citizens and US permanent residents who were not subject to deemed export rules. They were not persuaded that country of birth was all that important either. What mattered was not where you were born or whether you had a green card. What mattered was loyalty to the United States. Indeed, the lack of such loyalty was the first reason suggested for denying a license for a foreign national to have access to sensitive knowledge. Answering the question "Is the individual's loyalty tied to a country of concern?"[65] necessitated a "more comprehensive assessment of probable loyalties" than that currently in place. The "determination of the national affiliation of potential licensees [should] include consideration of country of birth, prior countries of residence, and current citizenship, as well as the character of a person's prior and present activities."[66]

There is a marked shift in thinking here, a shift from nationality to loyalty, from easily documented objective characteristics of an individual's life history to vaguely defined "prior and present activities" that are deemed indicative of political allegiance. True, export controls were always instruments of foreign policy, which was itself driven by ideological competition with the communist bloc during the Cold War. At that time, put simply, a country's degree of allegiance to communism was the filter used to decide whether a license was needed to export goods or to share technical data with a foreign national. In a global knowledge economy "loyalty to the United States" served the same ideological purpose. The problem is that establishing individual "loyalty" is far more arbitrary and open to bias than is citizenship or country of permanent residence (for which official documentary evidence is normally available) when it comes to deciding whether to grant an export license, or whether one is needed at all. Jessica Wang describes the loyalty-security system in the United States that took root in the late 1940s as assuming "the existence of a certain kind of private and authentic self accessible through indirect means, in which one's reading material, organizational affiliations, political acquaintances, tendency toward dissent, rejection of authority, and expressed political beliefs provided clues about reliability and allegiance." In this setting, she goes on, "stereotypical impressions of clean-cut, patriotic Americanness provided a cultural resource with which to navigate the loyalty-security system and its evaluation of personhood."[67] "Loyalty" is a far more invasive criterion than citizenship or permanent residency, is suitably flexible to be widely applicable, and is suitably charged ideologically to satisfy the proponents of "fortress America."[68]

I do not know whether or how the DEAC's proposals are being implemented.[69]The point to stress is that the unimpeded production and circulation of knowledge in the American research system now, as always, are subject to constantly negotiated regulatory constraints that depend on the nationality/loyalty of the individual concerned and the character of the unclassified knowledge and know-how in question. The notion that at least "fundamental" knowledge circulates "freely" in a global world is a fabrication, as John Reece Roth learned to his chagrin. His punishment was intended to make an example of him, an example whose deterrent effect was amplified by his being aged and infirm when he went to prison. The strategy seems to have worked. In the 1980s, as we saw, there was an outcry in some universities when the State Department took steps to control the itineraries of students and visitors from China and Russia to American campuses.[70] Those restrictions are now taken for granted as instruments to ensure national security (against bioterrorism, among other threats) and economic competitiveness with China. The relative freedom of circulation in academia that was possible even during the late Cold War is being chipped away: international scientific exchange is a negotiated and fragile social achievement.

Concluding Remarks

Export controls are notoriously arcane, labyrinthine, and ambiguous. They are also mostly ignored by historians and sociologists of science and technology, who probably take them to be instruments that regulate the global flow of trade. They are—but they are also important instruments that regulate the transnational flow of technology, understood as scientific and technical information, knowledge, and know-how. They are therefore of central importance in this project, which defines the circulation of knowledge as a key object for the transnational history of science and technology (see my introduction to this volume).

This chapter challenges the myth that in a global world knowledge flows by itself across borders. It has emphasized that, on the contrary, it is a social accomplishment that involves charting a path through the regulations on knowledge circulation imposed by a national security state determined to protect American scientific and technological leadership, and so its national security, without isolating itself from the global research community. It was one thing to devise policies to regulate trade. It was another to regulate the circulation of intangible knowledge passed on in a university laboratory, produced by a high-tech firm who employed a foreign national, or drawn on in an accident investigation with a foreign partner, all of which could involve a deemed export in an "apprentice-like" relationship between American experts and foreign nationals.

In fiscal year 2003 the Department of Commerce received 12,446 applications for export licenses, of which 846 (about 7 percent) were for deemed exports of knowledge, the majority of which were awarded.[71] It may be concluded from this that we are dealing here with a marginal phenomenon. But numbers are deceptive. The fines levied on the satellite manufacturers and the imprisonment of an aged and ailing university professor tell another story—as does the recent rapid expansion in the number of highly professional export officers and training programs on university campuses all over the country.[72] The national security state lays great store in controlling the flow of sensitive knowledge and know-how to foreign nationals. Penalties meted out to Hughes, to Loral, and to John Reece Roth have seemingly had the required deterrent effect. High-tech companies and researchers at major engineering schools are well aware of export controls and deal with them every day. Hopefully historians and sociologists of science and technology will also begin to treat them as important objects for further research on the transnational circulation of knowledge.

Notes

1. *US Government vs. John Reece Roth*, transcript of proceedings in the US District Court for the Eastern District of Tennessee, Northern Division, at Knoxville, TN, before the Honorable Thomas A. Varlan on May 13, p. 117, July 1, 2009.

2. Daniel Goldberg, "Why the Professor Went to Prison: Is John Reece Roth a Martyr to Academic Freedom or a Traitor?," *Bloomberg News*, Nov. 1, 2012, accessed Mar. 24, 2017, https://www.bloomberg.com/news/articles/2012-11-01/why-the-professor-went-to-prison.

3. Mario Daniels and John Krige, *Knowledge Regulation and National Security in Cold War America* (Chicago: University of Chicago Press, forthcoming).

4. This sections draws extensively on Harold J. Berman and John R. Garson, "United States Export Controls—Past, Present and Future," *Columbia Law Review* 67, no. 5 (1967): 790–890.

5. Michael Mastanduno, *Economic Containment: CoCom and the Politics of East-West Trade* (Ithaca, NY: Cornell University Press, 1992), chap. 6, is relied on extensively here. See also Fred Bucy, "Technology Transfer and East-West Trade: A Reappraisal," *International Security* 5, no. 3 (1980–81): 132–151; Defense Science Board Task Force on Export of U.S. Technology, *An Analysis of Export Control of U.S. Technology—a DoD Perspective* (Washington, DC: Office of the Director of Defense Research and Engineering, 1976), hereafter the Bucy Report.

6. Bucy Report, Finding I, 1.

7. J. Fred Bucy, "On Strategic Technology Transfer to the Soviet Union," *International Security* 1, no. 1 (Spring 1977): 25–43, at 28.

8. Mastanduno, *Economic Containment*, 189.

9. Bucy, "On Strategic Technology Transfer to the Soviet Union," 28.

10. Bucy Report, Finding II, 4.

11. "Export Control and the Universities," *Jurimetrics Journal*, Fall 1982, 40–49, at 40–41.

12. Mastanduno, *Economic Containment*, 213–216.

13. "Export Control and the Universities," 41.

14. Lara H. Baker Jr., "Transfer of High Technology to the Soviet Union and Soviet Bloc Nations," in *Hearings Before the Permanent Subcommittee on Investigations of the Committee on Governmental Affairs, United States Senate*, 97th Cong., 2nd Sess. (May 4–6, 11–12, 1982), 56.

15. "Technology," pt. 772, Definition of Terms, Export Administration Regulations, p. 41, https://www.bis.doc.gov/index.php/regulations/export-administration-regulations-ear.

16. "Scope of the Export Control Regulations," EAR §734.2(b).

17. Cited by Baker, "Transfer of High Technology to the Soviet Union and Soviet Bloc Nations," 55.

18. Ibid.

19. Mario Daniels and John Krige, "Beyond Regulation? Basic vs. Applied Science in Early Cold War America," *Technology and Culture*, in press.

20. *Scientific Communication and National Security, a Report Prepared by the Panel on Scientific Communication and National Security, Committee on Science, Engineering and Public Policy, National Academy of Sciences, National Academy of Engineering, Institute of Medicine* (Washington, DC: National Academy Press, 1982), hereafter the Corson Report.

21. Ibid., 11, 17. The commercialization of research helped along by the Bayh-Dole act and other such measures in the 1980s was already having effects.

22. Ibid., 15, for this and all quotations in this paragraph.

23. For details, see Harold C. Relyea, *Silencing Science: National Security Controls and Scientific Communication* (Norwood, NJ: Ablex, 1994), chap. 4.

24. Corson Report, appendix G.

25. Ibid., 19.

26. Ibid., 48.

27. Ibid., 49.

28. Ibid., 47.

29. Cited in Edward Gerjuoy, "Controls on Scientific Information Exports," *Yale Law and Policy Review* 3, no. 2 (Spring 1985): 447–478, at 460.

30. David A. Wilson, "Federal Control of Information in Academic Science," *Jurimetrics Journal*, Spring 1987, 283–296.

31. *National Policy on the Transfer of Scientific, Technical and Engineering Information*, National Security Decision Directive 189 (Sept. 21, 1985), accessed Mar. 28, 2017, https://fas.org/irp/offdocs/nsdd/nsdd-189.htm.

32. Wilson, "Federal Control of Information in Academic Science," 295.

33. National Academy Complex, *Finding Common Ground: U.S. Export Controls in a Changed Global Environment* (Washington, DC: National Academy Press, 1991).

34. Ibid., 2, 182.

35. Hugo Meijer, *Trading with the Enemy: The Making of the US Export Control Policy toward the People's Republic of China* (Oxford: Oxford University Press, 2016)

36. Robert D. Lamb, *Satellites, Security, and Scandal: Understanding the Politics of Export Control*, Center for International and Security Studies at Maryland (Digital Repository of the University of Maryland, Jan. 2005), 44ff.

37. *Report of the Select Committee on U.S. National Security and Military/Commercial Concerns with the People's Republic of China*, House Report 105-851 (the Cox Report), accessed Jan. 8, 2018, https://www.gpo.gov/fdsys/pkg/GPO-CRPT-105hrpt851/pdf/GPO-CRPT-105hrpt851.pdf; Dan Stober and Ian Hoffman, *A Convenient Spy: Wen Ho Lee and the Politics of Nuclear Espionage* (New York: Simon and Schuster, 2001).

38. The account that follows is based on the Cox Report, esp. chaps. 5–8. See also Lewis R. Franklin, "A Critique of the Cox Report Allegations of PRC Acquisition of Sensitive U.S. Missile and Space Technology," in *The Cox Committee Report: An Assessment*, ed. Michael M. May, with Alastair Iain Johnston, W. K. H. Panofsky, Marco Di Capua, and Lewis R. Franklin (Stanford, CA: Stanford University Center for International Security and Cooperation, Dec. 1999), 81–99, at sec. 3.2.1–3, http://iis-db.stanford.edu/pubs/10331/cox.pdf. See also Shirley A. Kan, *China: Possible Missile Technology Transfers from U.S. Satellite Export Policy—Actions and Chronology*, Congressional Research Service Report 98-485, updated Oct. 6, 2013, https://file.wikileaks.org/file/crs/98–485.pdf; Meijer, *Trading with the Enemy*, chap. 9.

39. David Damast, "Export Control Reform and the Space Industry," *Georgetown Journal of International Law* 42 (2010): 211–232, at 213.

40. Lamb, *Satellites, Security, and Scandal*, 30.

41. *Report of the Select Committee on U.S. National Security and Military/Commercial Concerns with the People's Republic of China.* In 1993 Daimler Benz Aerospace established a joint venture in China to coproduce satellites; see Jing-Dong Yuan, "United States Technology Transfer Policy towards China: Post–Cold War Objectives and Strategies," *International Journal* 51, no. 2 (1996): 314–338.

42. Kan, *China*, 6.

43. Joan Johnson-Freese, "Alice in Licenseland: US Satellite Export Controls since 1990," *Space Policy* 16 (2000): 195–204.

44. Lamb, *Satellites, Security, and Scandal*, 41.

45. Pub. L. 105-261, Title XV, "Matters Relating to Arms Control, Export Controls, and Counterproliferation," Subtitle B, "Satellite Export Controls."

46. On NASA, see John Krige, Angelina Long Callahan, and Ashok Maharaj, *NASA in the World, Fifty Years of International Collaboration in Space* (New York: Palgrave Macmillan, 2013), chap. 14.

47. Kan, *China*, 16.

48. Bianka J. Adams and Joseph P. Harahan, *Responding to War, Terrorism, and WMD Proliferation: History of DTRA, 1998–2008*, DTRA History Series (Fort Belvoir, VA: Defense Threat Reduction Agency, Department of Defense, 2008), 40.

49. Damast, "Export Control Reform and the Space Industry," 216.

50. *Oversight of Satellite Export Controls: Hearing Before the Subcommittee on International Economic Policy, Export and Trade Promotion, of the Committee on Foreign Relations, United States Senate*, 106th Cong., 2nd Sess. (June 7, 2005), 30–31 (statement of Mr. Bodner).

51. Ibid., 31–32 (statement of William A. Reinsch).

52. Office of Inspector General, Department of Commerce, Bureau of Industry and Security, *Deemed Export Controls May Not Stop the Transfer of Sensitive Technology to Foreign Nationals in the U.S.*, Final Inspection Report No. IPE-16176, Mar. 2004, accessed Mar. 28, 2017, https://www.oig.doc.gov/oigpublications/ipe-16176.pdf.

53. Berman and Garson, "United States Export Controls," 823–824.

54. Offices of Inspector General of the Departments of Commerce, of Defense, of Energy, of Homeland Security, of State, and the Central Intelligence Agency, *Foreign National Access to Export-Controlled Technology in the United States*, vol. 1, Report No. D-2004-062, Apr. 2004, accessed Mar. 26, 2017, http://www.dodig.mil/audit/reports/fy04/04-062.pdf.

55. See also John Krige, "National Security and Academia: Regulating the International Circulation of Knowledge," *Bulletin of the Atomic Scientists* 70, no. 2 (2014): 42–52; John Krige, "Regulating the Academic 'Marketplace of Ideas': Commercialization, Export Controls and Counterintelligence," *Engaging Science, Technology and Society* 1 (2015): 1–24.

56. Deemed Export Advisory Committee, *The Deemed Export Rule in the Era of Globalization: Report to the Secretary of Commerce*, i, Dec. 20, 2007, accessed Mar. 27, 2017, https://fas.org/sgp/library/deemedexports.pdf, hereafter the DEAC Report.

57. Ibid., 41.

58. The example is used in Bruce J. Casino, "'Deemed Export' Laws Restrict

Sharing Information with Foreign Nationals," *Journal of Research Administration* 36, no. 1 (2005): 16–20. On the case itself, see Iris Chang, *The Thread of the Silkworm* (New York: Basic Books, 1995). For richer analyses, see Jessica Wang, "A State of Rumor: Low Knowledge, Nuclear Fear and the Scientist as Security Risk," *Journal of Policy History* 28, no. 3 (2016): 401–446; Zuoyue Wang, "Transnational Science during the Cold War," *Isis* 101 (2010): 367–377; Zuoyue Wang, "The Cold War and the Reshaping of Transnational Science in China," in *Science and Technology in the Global Cold War*, ed. Naomi Oreskes and John Krige (Cambridge, MA: MIT Press, 2014), 343–370. See also John Krige, *Representing the Life of an Outstanding Chinese Aeronautical Engineer: A Transnational Perspective*, Technology's Stories, Mar. 2018, accessed Apr. 2, 2018, http://www.technologystories.org/chinese-engineer/.

59. DEAC Report, 12.

60. Ibid., 64.

61. Ibid., 69.

62. Ibid., 56–57.

63. Ibid., 69.

64. Ibid., 3.

65. "Seven Step Deemed Export Decision Process," in ibid., 89.

66. Ibid., 86. A call for public comments on this proposal was made in *Federal Register* 73, no. 97 (May 19, 2008): 29785–29787.

67. Wang, "A State of Rumor," 420–421.

68. *Beyond Fortress America: National Security Controls on Science and Technology in a Globalized World* (Washington, DC: National Academies Press, 2009).

69. The deadline for public comments (see n. 66) was extended to September 2008, after which the issue disappears without a trace in the *Federal Register*.

70. Samuel A. W. Evans and Walter D. Valdivia, "Export Controls and the Tensions between Academic Freedom and National Security," *Minerva* 50 (2012): 169–190.

71. Office of Inspector General, Department of Commerce, Bureau of Industry and Security, *Deemed Export Controls*, 4.

72. The Association of University Export Officers now has more than 170 members from over 110 different institutions of higher learning; http://aueco.org/membership/, accessed Mar. 27, 2017.

PART II
Colonial and Postcolonial Contexts

CHAPTER THREE

California Cloning in French Algeria

Rooting Pieds Noirs *and Uprooting* Fellah*s in the Orange Groves of the Mitidja*

Tiago Saraiva

In his *The Wretched of the Earth*, published in 1961, Frantz Fanon famously described the materialization of colonialism in space: "The colonist's sector is a sated, sluggish sector, its belly is permanently full of good things. The colonist's sector is a white folk's sector, a sector of foreigners. The colonized's sector, or at least the 'native' quarters, the shanty town, the Medina, the reservation, is a disreputable place inhabited by disreputable people. You are born anywhere, anyhow. You die anywhere, from anything. . . . It's a sector of niggers, a sector of towelheads."[1] This Manichaean colonial space had a global nature. Fanon wove a continuous thread connecting Muslim populations in the Maghreb and black workers in South Africa, Vietnamese peasants under French imperialism and African Americans under Jim Crow. And his violent urge for destroying the colonist world, for "nothing less than demolishing the colonist's sector, burying it deep within the earth or banishing it from the territory," did indeed reverberate worldwide.[2] While Jean Paul-Sartre brandished *The Wretched of the Earth* in Parisian cafés to accuse his fellow citizens of being no less than Nazis in their support of French colonialism in Algeria, Steve Biko, in the student residence of the University of Natal, circulated the book among his friends and comrades in the South African Students Association, the future leaders of the African National Congress.[3] According to the foundational myth of the Black Panthers, Bobby Seale and Huey Newton read Fanon's work in 1966 in a house in Oakland and, after being arrested for "blocking the sidewalk," launched the Black Nationalist Party.[4] *The Wretched of the Earth* traveled globally, pointing at a south detached from traditional geographical references. The south was to be found in Parisian *cités* or in the townships of Johannesburg, in West Philadelphia or in Martinique's Fort de France, Fanon's birthplace. In this global "disreputable place," this "Global South," as postcolonial scholars like to say, local dynamics of inequality are no more than a symptom of the more fundamental category of colonialism.[5]

Here, I want to explore the value of placing Fanon's dual colonial space in its actual historical location. The point is not to praise detailed local history at the expense of superficial global history but to demonstrate how engagement with materiality unveils alternative connected histories. Building on Fanon's powerful plea for bringing together apparently disparate colonial experiences, I hint at the value of transnational history grounded in concrete movements of people, ideologies, practices, and material artifacts. The chapter engages the orange orchards that surrounded the psychiatric hospital of Blida-Joinville in the Mitidja plain, southwest of Algiers, where Fanon worked from 1953 until 1956. It details the social and political dimensions of citrus production in Algeria under French rule, illuminating the role of cloning practices that originated in California in the making of the colonial relations denounced by Fanon.

Traditional horticultural practices of grafting and budding were at the origin of modern notions of cloning put forward by scientists working in the citrus belt around Los Angeles in the first decades of the twentieth century.[6] While historians of science and science studies scholars have been aware of such origins, their traditional overlooking of agriculture has led them to discuss cloning exclusively in relation to biomedical sciences and human reproduction.[7] This chapter points instead at the significance of the historical study of the agricultural settings from which cloning emerged and in which it became a practice with wide-ranging social consequences.

In what follows, I approach the travels of oranges from California into Algeria as movements of thick technoscientific things that bind science, technology, and politics together in a continuum.[8] This complicates the traditional narratives of plant transfers by environmental historians, who tend to take for granted the nature of what travels from one place to the other.[9] More recently, historians of capitalism have produced deservedly influential transnational narratives focusing on commodity chains, but they have disregarded, for example, the historical implications of the different varieties of cotton cultivated in different regions of the world.[10] In this text I detail what travels and what gets transformed when technoscientific things move: knowledge, social institutions, labor practices, political relations, as well as democratic ideals. By doing this, I aim to probe the value of history of science and technology in replacing generic notions such as the Global South with concrete transnational historical dynamics tying together different spatial realities.

Focusing on citrus instead of wine, the more common commodity in narratives of French colonization of Algeria, has two main advantages. First, it makes clear the importance of engaging the American example and its model of production for understanding colonial practices in the twentieth century, complicating stories based on the tension between European metropole and African colony. Crucially for the arguments of this volume, scientists were key actors in establishing this American connection. Second, while French-owned vineyards became the easy target of contemporary critics of the French presence in Algeria, calling attention to how a few large landowners controlled vast portions of the territory, the citrus story had the small white settler—the famous *pied noir*—at its center. In fact these *pieds noirs* became prime targets of the anticolonial struggle not simply because they dispossessed and exploited local farming communities but because they were deeply entrenched in the country and were far more numerous and difficult to uproot than the owners of large wine-growing properties. In other words, paying attention to citrus in Algeria allows us to better understand the dynamics of the violence against colonialism that found in Frantz Fanon one of its most eloquent advocates.

The Orange Groves of Blida and French Imperialism

Fanon arrived at the Blida hospital in 1953. The picturesque house with flowers around the doorway that welcomed Fanon and his wife belonged undoubtedly to the "white folk's sector."[11] The hospital, with its pavilions distributed between well-kept gardens and tree-lined avenues, seemed to be just another element of the colonial town denounced in the pages of *The Wretched of the Earth*. Famously, Fanon's therapeutic techniques were aimed at subverting such colonial dualism.[12] He distinctively established a direct connection between colonialism and mental disturbance, asserting how any discussion of individual neurosis led to an analysis of the social situation.[13] The patients who populated the psychiatric wards were the literal embodiment of colonialism, and their pathologies were the expression of the dehumanizing effects of colonial regimes. The methods Fanon had learned while studying in France—of reintegrating patients into society by reproducing inside the walls of the medical institution social life as it occurred in the outside world—didn't make much sense in a colonial situation: Why keep sending cured patients back into an environment that was responsible for producing mental disorder in the first place? What was the purpose of having *fellah*s (peasants) perform horticultural work in the modest vegetable garden of the hospital only to send them when cured to European-owned farms on which as wageworkers they were systematically transformed into a race of "disreputable people"?[14] The only possible way of producing a durable cure seemed to consist in fighting colonialism altogether. In 1956 Fanon resigned from his post, leaving Blida and the orange orchards surrounding it, and became a full-time member of the Front de liberation nationale (FLN), the guerrilla movement that would end French rule in the territory in 1962.

Blida is located on the southern border of the Mitidja plain, the exemplary space of the epics of the colonization of Algeria by the *pied noir*, the French settler.[15] The *mission civilisatrice* was manifested in the landscape through the transformation of a swampy, insalubrious region into an endless garden of vineyards and orange groves.[16] At the base of the Tell Atlas, the city of Blida was known as the "door to the South," placed at the crossroads between the mountainous, arid south and the Mitidja plain to the north. In 1953, the year Fanon arrived in Blida, some fifteen thousand Europeans were concentrated in the neighborhoods that grew up around the square of the European market in the northwestern part of the city. At the place d'Armes, "French Algerians could enjoy drinking sweet anisette in the cafes and have their shoes polished by little Arab boys."[17] Arabs and Berbers tripled the number of Europeans and occupied the districts in the old city center—the medina—to the south and to the east that survived the opening of new boulevards by the French, as well as more peripheral areas such as the Bou Arfa neighborhood on the south bank of the Oued El-Kebir river.[18]

Blida had served as a military outpost for the violent conquest by the

French of the southern and western provinces of Algeria in the 1830s. In the second half of the nineteenth century, the city would attract increasing numbers of Europeans because of its position as the main commercial and administrative urban center of the Mitidja plain. As for the native population, their growth was fueled by migrations from the southern mountainous areas. Inhabitants of the Atlas Mountains had typically combined cereal cultivation and horticulture (figs, olives) with sheepherding. The latter was the basis for trading with plains dwellers, who got their milk, meat, and wool by selling grain to shepherds. This was essential to complement the meager cereal production of the mountainous areas. Such exchanges between mountains and plains were interrupted by the seizure of lands in the Mitidja plain by French settlers through the typical colonial mix of violent military occupation and native indebtedness mechanisms.[19] Not only were plains dwellers pushed out of their lands into less fertile mountainous areas, but mountain shepherds also lost their primary source of cereal. The Atlas area would soon become synonymous with overpopulation and tragic famines, fueling colonial propaganda that contrasted native backwardness with the industriousness of European settlers.[20] The options left to Berbers and Arabs were either to become wageworkers in white-settler-owned farms or to join the increasing number of migrants responsible for the fast urbanization of Algeria. The influx of people from the mountainous south explains the fast expansion of peripheral neighborhoods of Blida such as Bou Arfa, the space of the uprooted people, of those "born anywhere, and anyhow."

In the years Fanon lived in Blida, citruses were second only to wine as an agricultural export from the colony.[21] During the final period of French Algeria, orange orchards had reached a maximum area of 45,000 hectares, and the tendency was to expand even more. By 1960 citruses were responsible for no less than 20 percent of the value of Algerian agricultural output. This success story of citruses and French colonialism in Algeria originated in no other place than Blida. The city had been praised since the sixteenth century as the main center for orange cultivation in the region because of its irrigated orchards cultivated by Moors from Andalusia fleeing the Spanish occupation of Granada in 1492.[22] Ibn Khaldun (1332–1406), a distinguished member of Andalusia's scholarly elite, had made oranges the symbol of his cyclical theory of history.[23] Oranges represented both urban sophistication and obsession with hedonist pleasures, a favorite among educated elites more concerned with beauty than with utility, and they announced the overthrow of decadent kingdoms by bellicose and ruder tribes. These tribes, in turn, would also fall under the spell of the orange tree, cultivate new orchards, and be replaced by barbarous warriors. Ibn Khaldun's historical sociology of the orange points to the important fact that the varieties of citrus disseminated by Arab expansion into the Maghreb were not primarily cultivated as food. It was the tree in its entirety that was praised and valued. Its green leaves, white flowers, and inebriating fragrance were all part of the reason the orange tree was equated with civilization. The essence extracted from its leaves was in fact a commodity as important as the actual fruit, the bitter orange.

Against Ibn Khaldun's predictions, Berber tribes of the mountainous south did not seize the lush gardens of Blida. Violent French military forces from the north did it instead. The imperialist invasion of Algeria in 1830 would transform the territory into the main proving ground for French enthusiasts of acclimatization theory, who attempted to adapt animals and plants to environments that were very different from the ones in which they had originated.[24] Relying on Lamarck's evolutionary hypothesis that progeny gradually adapts to new environmental conditions, no fewer than twenty-one "trial gardens" (*jardins d'essai*) were established following the military conquest.[25] Botanists were able to convince colonial officials of the importance of their undertaking by luring them with the promise of the immense riches to be expected from cultivating tropical commodities in great demand by metropolitan markets in the newly conquered lands. Acclimatization excited the imperial imagination of those who thought it would be possible to replicate in North Africa the profits of the plantation system of the tropical Caribbean islands.

Acclimatization of Citrus in Algeria and International Exchange Networks

Auguste Hardy was the director of the Algiers *jardin d'essai* from 1842 to 1867.[26] Not surprisingly he was one of the most active members of the Imperial Zoological Society of Acclimatization, the main French institution responsible for the advancement of acclimatization undertakings.[27] Instituting regular contacts with botanists around the globe, he introduced into his *jardin* in the outskirts of Algiers, among many other plants, eucalyptus, mandarins, and vanilla. The unsuccessful, painstaking efforts of adapting the latter from Martinique to Algeria offer a good example of the difficulties faced by Lamarckian projects of adapting plants to different geographies regardless of the conditions of cultivation in their place of origin, under the assumption of unlimited physiological flexibility of organisms. The failure of the direct transfer of one of the cash crops from the birthplace of Frantz Fanon—Martinique—into Algeria also suggests that the travels of colonialism are not an automatic historical phenomenon and demand more elaborate processes.

The case of mandarins, which Hardy introduced into Algeria in 1859 from other parts of the Mediterranean—namely, from Malta and Sicily—indicates a deeper knowledge of plant geography.[28] In the second half of the nineteenth century, botanists started to guide their acclimatization efforts by a more informed knowledge of local environmental conditions, and they put aside the project of introducing tropical plants into Algeria. In contrast to initial plans, the French colonization of Algeria didn't transform it into a new tropical area. Instead, for vast parts of its territory—namely, for the Mitidja plain—it meant the expansion of two typical Mediterranean landscapes: vineyards and orange groves.

After successfully planting mandarins in the Algiers *jardin d'essai*, Hardy promoted their cultivation in Blida, the first citrus area of the colony. French settlers were now the owners of the irrigated orchards surrounding the city.[29] Not only had the military occupation driven out most of the previous local proprietors, but military engineers had also enlarged the water infrastructure, building a dam to retain water flowing from the mountains and increasing the number of distribution canals and the irrigated area. Hardy's mandarins promised to reconstitute orchards badly damaged by the war with a variety whose fruit was highly prized in European markets, especially Paris.[30] Mandarin orchards cultivated for fruit production replaced sour orange orchards praised for their intense fragrance. It is important to note that despite all the colonial propaganda that blamed local populations for the decline of the territory since Roman times, the new colonial undertakings were built on top of a previous local experience.[31]

A description in 1887 of the orange orchards of Blida by Charles Joly, vice president of the French National Society of Horticulture, exulted over the four hundred hectares north and east of the city covered by citrus trees.[32] The groves, all French owned, rarely exceeded five hectares in size. Arabs and Berbers showed up in the tableau exclusively as gangs of pickers hired by late October and as groups of women sorting the fruit into the different grades. They were characterized only by their cost as wage laborers (2 francs/day for men, 1.5 francs/day for women), as if they had had no other role in the history of Blida orange groves. The author details the importance of the thirty kilometers of irrigation canals for the entire enterprise, emphasizing how the hydraulic infrastructure was key for the expansion of citrus production in the rest of the Mitidja, north of Blida.[33] Californian orange production attracted Joly's close attention when envisioning the models French settlers should follow in colonizing these new areas of the Mitidja. But pictures of new citrus orchards around Los Angeles had to share space with images of orange trees in Valencia (Spain) and detailed descriptions of Italian undertakings. Joly praised the fast expansion of citrus in California but didn't forget to mention that Spain was the first exporter of oranges to European markets, including to France itself, and that French settlers should thus be well acquainted with the orchards of the Valencia region.[34]

If we pay closer attention to the actual process of bringing mandarins to Blida, it is clear that we are not dealing with a simple replacement of a traditional variety cultivated by natives with a more productive one brought in by European botanists. Mandarin buds had to be grafted on top of sour orange rootstocks in order to resist gummosis, which had affected citrus orchards in many different parts of the world and had wiped out citrus production in the Azores islands, previously one of the main suppliers of the British market.[35] Gummosis traveled from the Azores into Spain and Italy and found its way into Algeria in the 1860s. The most probable explanation for the displacements of gummosis across the Mediterranean, from Italian citrus production areas to the Algerian Mitidja, is the very same acclimatization efforts by Hardy. The fungus that caused gummosis traveled along with the buds supplied by his international botanical exchange network.

While local populations had traditionally grown their trees from seed, the colonizers' mandarins had to be grafted on gummosis-resistant rootstocks. The asexual reproduction procedure of grafting a bud of mandarin on a sour orange rootstock led to more homogeneous orchards and smaller trees. Orchards with trees grown from seeds, the result of sexual reproduction, were more diverse. Not only were there larger variations in the fruit yield of each individual tree, but the quality of the fruit also varied. In the upcoming decades French scientists would tap into this diversity produced by local Arab and Berber horticulturalists to identify interesting new varieties and strains of citrus.[36]

Louis Trabut, the head of the botanical service of the government of Algeria since 1892, would extend the exchange networks of the garden started by Hardy and promote surveys of local soils and climates.[37] Military and civilian officers of the French colonial administration were part of his network of correspondents, which also included scientists from Australia, Japan, and the United States. Citruses, according to Trabut, were especially suited to exchanges since buds traveled so easily by mail. One had only to wax the buds' edges, cover them individually in paraffin paper, wrap the bundle in a damp newspaper, cover it with an impermeable tissue, and send it by post in metal tubes.[38] When Trabut welcomed Walter T. Swingle to Algeria in 1898, the USDA agricultural botanist used a similar procedure to send the offshoots of date palm trees from the Algerian oasis to the United States.[39] These would launch the Californian date industry in the Coachella valley. In the opposite direction, Trabut was able to introduce into Algeria the first navel oranges from California, the basis of the citrus boom in the United States.[40] But while Californian growers insisted on the advantages of developing their business around no more than two varieties (Washington navels and Valencias), Trabut pursued a program of constantly increasing the number of varieties available to French settlers in Algeria.

At the turn of the century, Trabut was promoting the dissemination of clementines, a hybrid of sour orange and mandarin first produced by Father Clement at the Misserghin orphanage in Oran Province.[41] The advantage of clementines over mandarins was their earliness, maturing from November until late December, when mandarins were still too green, and thus coinciding with the peak in the consumption of citrus in Europe during Christmas celebrations. Drawing on his extended network, Trabut also experimented in his Algiers garden with satsumas, Canton mandarins, Bombay mandarins, Cape mandarins, King of Siam mandarins, and tangelos.[42] He aimed at supplying French settlers with different citrus varieties to enable them to extend the presence of citruses from Algeria in European markets from October to March.

Trabut had clearly identified California as an important example for the development of citrus production in Algeria. He was particularly insistent about the value of Californian citrus varieties, especially the seedless Washington navel.[43] But his plant geography approach didn't differentiate California from other interesting citrus regions of the world. In his experimental plots in the outskirts of Algiers he grafted on sour orange rootstocks buds sent in from fellow botanists not only in California but also in the Cape, Florida, Brazil, Spain, Italy, and Australia. As in the description of Blida by Joly, California, for Trabut, was just another area of a global Mediterranean.[44] It was a Mediterranean climate region, which supplied horticultural resources that Trabut was able to access because of his membership in a global community of botanists.

Expanding Citrus Production in the Mitidja: Expropriation and Cooperatives

While the first settler orchards had been grafted on top of the Arab ones in Blida, the major expansion area for citrus would be further north in the Mitidja plain. Before the French occupation, the Ottoman rulers had divided the plain into *haouch*, large agricultural properties owned by the ruling elite in Algiers and exploited locally as latifundia by local tribes. Early French colonialists described the Mitidja under the Ottomans as an insalubrious, unreclaimed area awaiting the French civilizing mission. They tended to ignore that the region was not only already populated but exploited as well by local people for husbandry. In 1846 a law sequestered "Turkish property," ended the *haouch*, and transferred ninety-five thousand hectares to the colonial state, which distributed them among European settlers.[45]

The new town of Boufarik, located not coincidently at the geographic center of the Mitidja, between Algiers on the coast and Blida at the base of the Atlas, would overtake Blida in the first decades of the twentieth century as the main center for citrus production in Algeria. By 1913 the orange groves covered some four thousand hectares, and by 1928 the area had already doubled.[46] The city materialized the superposition of military conquest and rural settlement. Its perpendicular streets reveal the role of French military engineers in its urban planning, and they would be the model followed in dotting the plain with thirty-eight new settlement villages, the last one erected in 1897. By 1911 there were forty-four thousand Europeans and fifty-eight thousand Arabs and Berbers in the Mitidja. The latter had lost access to 80 percent of the land. *Fellah*, which derives from the Arabic word for "plowman," had traditionally referred to a smallholder, tenant farmer, or cultivator of common land: as described by an engaged anticolonial French scholar, the *fellah* "has dignity, he is connected through the land to his ancestors and he is not a slave, he is the master of his land."[47] Significantly enough for French settlers, *fellah* became a synonym for "agricultural wageworker." According to the geographer Marc Côte, no other land dispossession in the entire Maghreb was comparable to the one that took place in the Mitidja.[48]

Although white settler tales are always obsessed with the courage and independence of frontier men, the colonial state was the protagonist of the colonization of the Mitidja. The French colonial state expropriated the local population and produced a cheap wage labor; it built a railway line connected Blida with Algiers, with a local station in Boufarik; it put in place an expensive drainage and irrigation infrastructure designed by state engineers; it provided a safe market for agricultural exports. From 1920 until 1930 agricultural products represented no less than 86 percent of Algerian exports, sustaining some twenty-six thousand European landholders. Wine, from the province of Oran in the west, was by far the most important item, accounting for some 33–39 percent of that value. But while vineyards underwent an acute process of property concentration in the first three decades of the twentieth century, producing a distinct class of large landowners, citrus groves, especially those around Boufarik, were small-owner endeavors leading to a very different sociability.[49] More than two-thirds of Algerian orange groves occupied less than five hectares.[50] The proportion was even higher when considering just the Boufarik area.[51]

In 1922 a cooperative of citrus growers was founded in Boufarik.[52] Building on the French tradition of rural syndicates, its aim was to guarantee good prices in the market for its products by pooling together the fruit of its members while providing better credit conditions. Vineyards and orange groves represented important investments and increased dependence on credit: while mortgage debts represented some 9 percent of the total land value of French agriculture, that value was no less than 22 percent (27 percent by a different estimate) in Algeria.[53] The oscillation in export prices and the high values of debt had led to the concentration of vineyards in the hands of a few landowners. The Boufarik citrus co-op promised to counter such tendencies and make the market work as well for small settlers, the *pieds noirs*. Among its promoters was Amédée Froger, manager of the Domaine Saint-Charles and Boufarik's radical-socialist mayor from 1925 until 1956, an embodiment of the *mission civilisatrice* of the Third Republic and its promises of solving social unrest in France through the settlement of Algeria.[54] As president of the Federation of Algerian Mayors, he was one of the most outspoken representatives of small-settler interests.

The citrus orchards surrounding Boufarik whose owners participated in the cooperative were cultivated with the varieties that Trabut had been disseminating since the beginning of the twentieth century, such as mandarins, navels, and satsumas. But clementines were by far the most popular fruit, publicized as proof of the ingenuity of French colonizers and their unique horticultural sensibility.[55] Growers were attracted by the high retail price paid for clementines in Paris, overlooking the main limitation of the variety: its high variation in yield.[56] Citrus growers had formed a co-op to compensate for the ups and downs of the market, but they were growing a tree that offered high returns in one year and put them on the verge of bankruptcy the next. Only large owners could withstand such variations, and so citrus production seemed to lead to the same land concentration pattern of vineyards. The problem became only more acute during the Great Depression and the consequent abrupt decline of consumption of Algerian fruit in Europe.[57]

It was in this context that M. J. Brichet, Trabut's successor in the horticultural services of the Algerian colonial government, increasingly publicized California as the solution for sustaining the vibrant community of Boufarik growers.[58] Starting in December 1929, he would write a series of articles in the *Afrique du Nord agricole*, the main journal of Moroccan and Algerian export agricultural interests, warning against the profusion of citrus varieties in Algerian orchards and their uncertain value in the market.[59] The small Algerian clementine and its inconstant behavior were contrasted with the large Californian navel and its constancy in yield and quality.

Citrus Cloning and the Reproduction of Southern California in the Mitidja

Brichet's point was not limited to promoting Californian citrus varieties in the Mitidja. California was more than just a source of plant resources as it had been for Trabut. Brichet's comprehensive discussion of Archibald D. Shamel's scientific work on the variability of yields in citrus trees in Southern Californian orchards led him to a more sophisticated discussion of what had to be transferred in addition to the fruit. Washington navels and Valencias, the two citrus varieties that enriched Southern Californian growers, were seedless varieties, always reproduced through budding and grafting on a rootstock. Despite this asexual form of reproduction, the number of undesirable trees in Californian orchards was increasing, producing, according to Shamel, "irregular, light crops of inferior quality."[60] By 1919 Shamel's survey of Southern California orchards had identified no fewer than thirteen strains of the Washington navel, twelve strains of the Valencia orange, six strains of the Marsh grapefruit, eight strains of the Eureka lemon, and five strains of the Lisbon lemon varieties. The consequences for the wealth of the growers were obvious, as stressed in Shamel's assessment of the Washington navel orchards: "About 25% of the total number of trees studied in the original orchards in which these investigations have been conducted were found to be of undesirable strains having consistently low yields, or bearing fruits of poor quality, or both, such as those of the Australian, Unproductive, Corrugated, Pear-Shape, Sheep-nose, Flattened, Dry, and other inferior strains."[61] It was not enough to grow navels in order to guarantee the commercial success of an orchard. According to Brichet, Boufarik settlers would also have to import Californian practices of guaranteeing that only the proper strains of navels were reproduced.

In 1932 Brichet embarked on an agricultural mission to the United States to see what Algeria should learn from California.[62] In his report he exhaustively described the functioning of the Californian Fruit Growers Exchange, the citrus co-op responsible for harvesting, packing, shipping, and marketing the fruit of some fifteen thousand associates. This co-op was proof that Californian citrus growers, the large majority of whom owned small groves of between five and ten acres, when properly organized into associations, could actually prosper in a market economy and face the power of railway monopolies, the octopus embodying the evils of American capitalism. Although too many times Sunkist, the brand by which the co-op is known in America, is described as an example of the many problems associated with Californian agribusiness, a revisionist historiography has demonstrated its importance as a reformist social organization of the Progressive Era.[63]

Brichet's main point was that the good working of a co-op pooling the fruit of thousands of growers depended on the uniform production of its members. Growers delivering inferior strains of fruits to the co-op packinghouse would undermine the cooperative effort and had to be excluded. Thus, starting in 1917, Shamel had initiated in the Californian co-op a department of bud selection, which secured bud wood from selected trees and distributed it to propagators. Following Shamel's instructions, individual growers compiled performance records of each of their trees to identify the best trees in each orchard, from which buds were to be taken. The co-op became responsible not only for securing and distributing the buds but also for the "payment for the buds to the owners of the parent trees, the assembling, tabulating, and studying of extensive individual-tree data, the selection of the superior parent trees, collecting information as to the behavior of the buds and the trees grown from them, and the survey of new orchard areas for the location of additional parent trees."[64] From 1917 to 1935 the Californian co-op distributed no fewer than 1,402,950 selected buds from superior strains of the Washington navel orange and 2,338,004 of the Valencia orange.[65]

According to Brichet, events in Algeria went in the opposite direction. The buds sold by nurserymen to be grafted by citrus cultivators were reproduced without any control and so didn't originate from selected trees. "One makes a bud out of any branch," declared a desolate Brichet.[66] From his mission to California he concluded that the Boufarik co-op had to enlarge its range of action, and instead of limiting itself to keeping prices high for growers by controlling how much fruit was released into the market, it should also be responsible for the quality and quantity of fruits in the orchards to guarantee a stable source of income. In other words, the co-op also had to look after the trees in the orchard in order to look after its members. Directly inspired by his Californian experience, from 1937 onward Brichet was able to impose, through his horticultural services, the obligatory control of the buds sold by nurseries.[67] Every bud now had to include a tag identifying its ancestry, demonstrating that it had been propagated from selected trees. From then on buds could travel legally only when they had attached to them the identification of the trees they originated from.[68] The moment was particularly appropriate since a new boom of orange orchard cultivation in Algeria was propelled by the beginning of the Spanish Civil War, which meant the radical reduction of exports from the main supplier of European markets. From occupying some ten thousand hectares in 1937, the citrus area would increase to twenty-two thousand hectares in 1942.

The importation of the Californian method of controlling the reproduction of citrus trees demanded more than just a few trips to the United States. Fundamental to this process was the local scientific colonial infrastructure, represented in this case by the Boufarik experimental station that had been established in 1927 by Trabut.[69] In line with his vision of the role of agricultural experimentation discussed above, for Trabut the first task of the station was to acclimatize the maximum possible number of new citrus varieties in Algeria. Brichet pointed instead at the importance of using the Boufarik station to guarantee that only the best strains of a variety should be propagated.[70] While the French colonial project of sustaining small white settlers was made plausible by importing American standardization practices, it is also apparent that the increased presence of practices of American provenance in the colony was possible only by mobilizing resources already put in place by the colonial state. This points at a more general tendency of European colonial states to remake their practices by looking to the United States and of a globalization of America through old European colonial channels.

After cultivating the first rootstocks of the station orchard in 1929, the grafting of the different varieties was made in 1931. In the following years, with the horticultural services under Brichet's control, the main aim of the station was to identify the most productive clones in each variety through tree performance records.[71] The modern use of the term "clone" had been introduced by Herbert J. Webber, director of the University of California Citrus Experiment Station in Riverside, the center of citrus production in Southern California, to describe groups of plants that are propagated by the use of any form of vegetative parts such as bulbs, tubers, cuttings, grafts, buds, etc. and that are simply parts of the same individual seedling.[72] Although these were all very traditional practices used by farmers and gardeners all over the world for many centuries, Webber coined the new term to call attention to the importance of asexual forms of reproduction in modern forms of agriculture. No one disputed that the use of grafting and buds derived from ancestral practices, but the novelty of cloning relied on carefully documenting ancestry and progeny.[73] Following the procedure first developed by Shamel in California, each new citrus tree grown from a bud grafted on a rootstock (i.e., each new clone) in the Boufarik experiment station had a performance record identifying its productivity and quality of fruit. From the analysis of such performance records, one could then choose the clones from which buds could be cut and then propagated by nurserymen. From 1937 until 1950 the horticultural service in Algeria identified no fewer than 12,000 elite clones, from which 4,500,000 buds were produced.[74]

This period also witnessed the intense growth of the citrus area, which reached forty-five thousand hectares before independence in 1962. Such growth was based on these Californian cloning practices.

Cloning became central to the success of the Boufarik-Blida citrus co-op. Only by (re)producing equal amounts of fruit of the same quality every year from clones certified by the experiment station did the co-op survive and expand. The yield-variable clementines cultivated in the first decades of the twentieth century, although sold at high prices in Parisian markets, could sustain only large orchard owners able to compensate losses in one year with gains from another year. The constancy of the Californian navels promised instead to increase the profits of citrus cultivation for the five thousand or so owners of small orchards. Or, as eloquently summarized by René Dumont, in another report of a mission to the United States by a French agronomist, "only the large fruit is democratic":[75] while the small clementines were perceived as favoring only large landowners, the large Californian navels enabled small farmers to acquire, along with the owners of large farms, the profits to be had by marketing fruit to European consumers.

In the detailed discussion of how to reproduce Southern California cloning practices in the Algerian Mitidja, there was no mention of who actually performed the work of grafting selected buds on the rootstock or of who picked the oranges. En passant, French scientists, in their comparisons of the Californian and Algerian situations, spoke of the great advantages of having access to a much cheaper workforce than American growers did. Just as Californian growers totally erased the role of Chinese, Japanese, and Mexican growers in the making of the citrus orchards that made the fortune of the Los Angeles area in the beginning of the twentieth century, French *pieds noirs* were totally oblivious to the role of local *fellah*s. They ignored not only, as we saw above, that in Algeria citrus cultivation was initiated in Blida by Arabs from Andalusia. They also disregarded the fact that the very modern orchards cultivated following the Californian model relied on the horticultural skills of the Algerian peasantry. The budding and grafting were done by Berbers from the Atlas Mountains who had cultivated figs and olives as cash crops for many generations. The workers hired for such tasks, as well as the ones in charge of pruning, had a more permanent status than the pickers hired only during the harvest season. The latter were the large majority (in wine country the proportion was one permanent worker to five temporary), for whom unemployment was a constant presence and contracts lacked formality. They could easily be dispensed with on the basis of weather conditions or for not handling oranges carefully enough. Speed was important when picking fruit, but so was prevention of any bruising of the fruit skin caused by long nails or clumsiness.[76] Diligence and ability were as important as docility when pleasing the foremen.[77]

Conclusion: Uprooting Colonial Citrus Growers

The *pied noir* democracy of some 42,000 Europeans inhabiting the Mitidja plain depended on the direct exploitation and previous expropriation of 190,000 Arabs and Berbers, a major recruiting ground for the FLN and its guerrilla war for independence. In the monument erected in Boufarik in 1930 to celebrate the *pied noir* "Miracle of the Mitidja" of having converted a swampy area into a horticultural paradise, there were no references to the *fellah*. A gigantic sculpture nine meters high and forty-five meters long put French settlers next to military personnel, underlining the close association of military operations and agriculture reclamation. The metropolitan government spent at least forty million francs to commemorate the one-hundredth anniversary of French presence, about the same amount allocated to a social and educational plan for local populations, which nevertheless was opposed by French Algerian representatives.[78]

In 1954 the excluded from the "Miracle of the Mitidja" made themselves heard. On November 1 the FLN launched a coordinated series of attacks starting the war for independence, which would end only in 1962. One of the first bombs was exploded in the citrus co-op of Boufarik. If the co-op performed democracy and egalitarianism for its members, promising their reproduction on the basis of Californian cloning practices, it certainly produced something very different for the local Arab and Berber populations.

The explosion was followed by many other actions of uprooting orange trees in the following years. On December 28, 1956, Amédée Froger, the mayor of Boufarik and promoter of cooperatives among *pied noir* citrus growers, was murdered in Algiers. Some ten thousand people followed his funeral cortege through the streets of the capital city. A military escort opened the way for a forest of veteran (ancient combatants) flags, asserting the identification of Froger with the dearest French nationalist symbols. When passing in front of the American embassy the crowd yelled “Algérie française! Algérie française!” After that, chants of “La Marseillaise,” “Les Africains,” and “Ce n’est qu’un revoir,” all the anthems of hard-line *pieds noirs*, were punctuated by calls of “Death to the assassins.” Meanwhile, at the cemetery a hidden bomb exploded in a tomb next to Froger’s. The delayed start of the ceremony prevented major carnage. Right-wing extremists excited the horde: iron bars, guns, and knives suddenly multiplied. On the wharves, Muslim passersby were thrown into the water ten meters below; “suspects” were gunned down by the police in cold blood; and from one hundred to three hundred (nobody really knows the exact figure) lynchings were perpetrated throughout the night.[79]

In trying to make sense of the extreme violence of the decolonization process in Algeria, the sociologist Pierre Bourdieu characterized colonialism as a general phenomenon of uprooting.[80] Only recently has scholarship in social theory started to acknowledge the importance of Bourdieu’s years in Algeria during the war of independence to understand his interpretation of late modernity.[81] Suffice to note how Bourdieu contrasted the habitus of his Kabyle peasants—the object of his ethnographic work—who participated in a gift economy, with the rational, cold calculations of the capitalist world imposed by the French colonial order. For Bourdieu the violent resettlement that forced around two million Algerian peasants—roughly a quarter of the population of the country—into model villages, a policy imposed by the French military as antiguerrilla tactics, was, despite its apparently unique brutality, in a direct continuum with the former land expropriations that had forced the colonized to abandon their habitus and integrate into the modern world of cold labor relations that sustained the French-controlled agriculture of the Mitidja plain. Colonialism, notwithstanding the specific power relations it entailed through racism, was thus seen as a main harbinger of modernity replacing rooted premodern peasants with uprooted farm wageworkers or urban laborers.[82]

Bourdieu mentions citruses only en passant in his Algerian writings. Wine is his port of entry to the pied noir’s world.[83] This is certainly justified, since, as mentioned, vineyards were always more important than orange orchards in the settler economy, although the latter grew much faster than the first in the final period of the colony. And wine came in handy for Bourdieu’s arguments. In his *Sociology of Algeria* he described how large, six-hundred-hectare properties employing between three hundred and four hundred workers and producing some fifty thousand hectoliters per year undermined all the pied noir rhetoric built around the settler attachment to the land.[84] These same numbers were also used by Jean Paul Sartre in his introduction to Fanon's *Wretched of the Earth* to denounce how the land that used to feed the natives was now converted into large viticulture domains producing wine.[85] It was this line of reasoning that put Bourdieu and Sartre on the side of decolonization in the harsh intellectual French debate on the Algeria crisis and that left Albert Camus isolated as one of the few left-leaning intellectuals struggling for the coexistence of the two communities in the same land.[86]

Had Bourdieu, Sartre, or Fanon paid attention to the growing presence of citrus production in settler life they might have been more sympathetic to Camus's arguments. In contrast to the large wine estates, citrus orchards rarely exceeded ten hectares in size but sustained a much larger white community than viticulture. More important for the argument of this chapter, the California package didn't equate modernization with uprooting, mechanization, and large properties. Modernization, California style, was directed instead at rooting settlers in the land. Californian cloning practices were brought into Algeria explicitly to maintain a large population of white settlers on small holdings producing standardized fruit and organized in cooperatives. It is important to recognize these concrete ties connecting the orange orchards of the Mitidja with those of Southern California. On the basis of only generic notions of colonialism, it is hard to understand the extreme forms of violence unleashed by white settlers in Algeria. In other words, overarching modernization theories in which the blind force of capitalism uproots local populations are not totally satisfactory. The Californian-Algerian citrus connection suggests forms of modernization and American presence in the world with unexpected rooting effects.

Notes

1. Frantz Fanon, *The Wretched of the Earth*, trans. Richard Philcox (1961; repr., New York: Grove Press, 2004), 4–5.

2. Ibid., 6.

3. Homi K. Bhabha, "Framing Fanon," foreword to Fanon, *Wretched of the Earth*.

4. Ibid.

5. Fanon's own position on the importance of the local is more nuanced. Stefan Kipfer, "Fanon and Space: Colonization, Urbanization, and Liberation from the Colonial to the Global City," *Environment and Planning D: Society and Space* 25, no. 4 (2007): 701–726. For an introduction on "Global South" scholarship, see the special issue "The Global South and World Dis/order," ed. Caroline Levander and Walter Mignolo, *Global South* 5, no. 1 (2011).

6. Tiago Saraiva, *Cloning Democracy: Californian Oranges and the Making of the Global South* (forthcoming).

7. On cloning, see Jane Maienschein, *Whose View of Life? Embryos, Cloning and Stem Cells* (Cambridge, MA: Harvard University Press, 2003); Hannah Landecker, *Culturing Life: How Cells Became Technologies* (Cambridge, MA: Harvard University Press, 2007). Sarah Franklin has demonstrated the payoff of placing Dolly the sheep in her original agricultural context in *Dolly Mixtures: The Remaking of Genealogy* (Chapel Hill, NC: Duke University Press, 2007). For a discussion on the neglect of agriculture by historians of science, see Jonathan Harwood, "Introduction to the Special Issue on Biology and Agriculture," *Journal of the History of Biology* 39 (2006): 237–239.

8. On the notion of technoscientific things, see Tiago Saraiva, *Fascist Pigs: Technoscientific Organisms and the History of Fascism* (Cambridge, MA: MIT Press, 2016), 235–242; Ken Alder, "Introduction to Focus Section on Thick Things," *Isis* 98, no. 1 (2007): 80–83; John Tresch, "Technological World-Pictures: Cosmic Things and Cosmograms," *Isis* 98, no. 1 (2007): 84–99; Bruno Latour, "From Realpolitik to Dingpolitik," in *Making Things Public: Atmospheres of Democracy*, ed. B. Latour and P. Weibel (Cambridge, MA: MIT Press, 2005), 4–31; Lorraine Daston, ed., *Things That Talk: Object Lessons from Art and Science* (New York: Zonebooks, 2004).

9. For a general critique of such narratives, see William Beinart and Karen Middleton, "Plant Transfers in Historical Perspective: A Review Article," *Environment and History* 10, no. 1 (2004): 3–29. The classical example of the traditional narrative I am alluding to is Alfred W. Crosby, *The Columbian Exchange: Biological and Cultural Consequences of 1492* (Westport, CT: Praeger, 2003).

10. In an otherwise very informed book, Sven Beckert makes one single reference to the different genes of different cotton varieties, without grasping the historical importance of such differences. See Sven Beckert, *Empire of Cotton: A Global History* (New York: Alfred A. Knopf, 2015). This criticism is also valid for Edward Baptist, *The Half That Has Never Been Told: Slavery and the Making of American Capitalism* (New York: Basic Books, 2014). In a heated debate Olmstead and Rhodes have denounced this historical neglect of seeds at the expense of the more dramatic whip, as perpetrated by Baptist. See Marc Parry, "Shackles and Dollars," *Chronicle of Higher Education*, Dec. 8, 2016. For a full discussion of the significance of the evolutionary history of cotton, see Edmund

Russell, *Evolutionary History: Uniting History and Biology to Understand Life on Earth* (Cambridge: Cambridge University Press, 2011), 103–130.

11. For biographic details on Fanon, see David Macey, *Frantz Fanon: A Biography* (2000; repr., New York: Verso, 2012).

12. Ibid., 110–151; Hussein A. Bulhan, *Frantz Fanon and the Psychology of Oppression* (New York: Springer, 1985); Françoise Verges, "Chains of Madness, Chains of Colonialism: Fanon and Freedom," in *The Fact of Blackness: Frantz Fanon and Visual Representation*, ed. Alan Read (Seattle: Bay Press, 1996), 47–75.

13. Frantz Fanon, *Black Skin, White Masks*, trans. Charles Lam Markmann (New York: MacGibbon and Kee, 1967).

14. Nigel Gibson, "Thoughts about Doing Fanonism in the 1990s," *College Literature* 26, no. 2 (1999): 96–117.

15. Marc Côte, "L'exploitation de la Mitidja, vitrine de l'entreprise coloniale?," in *Histoire de l'Algérie à la période coloniale*, ed. Abderrahmane Bouchène et al. (Paris: La Découverte, 2014), 269–274; Georges Mutin, *La Mitidja décolonisation et espace géographique* (Paris: CNRS, 1977).

16. For general views on the French *mission civilisatrice*, see Alice L. Conklin, *A Mission to Civilize: The Republican Idea of Empire in France and West Africa, 1895–1930* (Stanford, CA: Stanford University Press, 1997); Diana K. Davis, *Resurrecting the Granary of Rome: Environmental History and French Colonial Expansion in North Africa* (Athens: Ohio University Press, 2007).

17. Macey, *Frantz Fanon*, 210.

18. Xavier de Planhol, "La formation de la population musulmane à Blida," *Revue de géographie de Lyon* 36, no. 3 (1961): 219–229.

19. For this dynamic and for the formation of a new sedentary group among Algerian "Arabophones," see Pierre Bourdieu, *Sociologie de l'Algérie* (1958; repr., Paris: PUF, 2018), 67–91.

20. Germaine Tillon, *L'Algérie en 1957* (Paris: Les Editions de Minuit, 1958).

21. Georges Mutin, "L'Algérie et ses agrumes," *Revue de géographie de Lyon* 44, no. 1 (1969): 5–36.

22. Louis de Baudicour, *La colonisation de l'Algérie* (Paris: Jacques Lecoffre, 1860); L. Trabut and R. Marés, *L'Algérie agricole en 1906* (Algiers: Imprimerie Algérienne, 1906).

23. Stephen Frederic Dale, *The Orange Trees of Marrakesh: Ibn Khaldun and the Science of Man* (Cambridge, MA: Harvard University Press, 2015).

24. Michael A. Osborne, *Nature, the Exotic, and the Science of French Colonialism* (Bloomington: Indiana University Press, 1994); Warwick Anderson, "Climates of Opinion: Acclimatization in Nineteenth-Century France and England," *Victorian Studies* 35, no. 2 (1992): 135–157; Christophe Bonneuil, "Mettre en ordre et discipliner les tropiques: Les sciences du vegetal dans l'empire français, 1870–1940" (PhD diss., Paris VII, 1997).

25. Michael A. Osborne, "The System of Colonial Gardens and the Exploitation of French Algeria, 1830–1852," in *Proceedings of the Meeting of the French Colonial Historical Society*, ed. E. P. Fitzgerald (Lanham, MD: University Press of America, 1985).

26. Louis Trabut and René Maire, "La station botanique de Maison-Carré en Algérie," *Revue de botanique appliquée et d'agriculture coloniale* 2, no. 7 (1922): 86–92; Paul Carra and Maurice Guait, *Le jardin d'essai du Hamma* (Algiers: Gouvernement Général de l'Algérie, Direction de l'Agriculture, 1952).

27. Michael Osborne, "The Société Zoologique d'Acclimatation and the New French Empire: Science and Political Economy," in *Sciences and Empires*, ed. P. Petitjean, C. Jami, and A.-M. Moulin (Dordrecht: Kluwer, 1992), 299–306.

28. Bonneuil, "Mettre en ordre," 183–186.

29. Charles Joly, *Les orangeries et les irrigations de Blidah* (Paris: Georges Chamerot, 1887); Baudicour, *La colonisation.*

30. Gouvernement général de l'Algérie, *Les fruits et primeurs d'Algérie* (Algiers: Imprimerie Algérienne, 1922).

31. On visions of the French colonial mission as remaking the granary of Rome, see Davis, *Resurrecting the Granary of Rome.*

32. Joly, *Les orangeries*, 6–8.

33. Ibid., 8–12.

34. Four years later Charles Joly would exhaustively discuss fruit production in California in *Note sur la production fruitière en Californie* (Paris: Gaston Née, 1891).

35. Louis Trabut, *L'arboriculture fruitière dans l'Afrique du Nord* (Algiers: Imprimerie Algérienne, 1921), 139–141, 182–184. On the Azores islands as citrus producers, see Fátima Sequeira Dias, "A importância da economia da laranja no arquipélago dos Açores durante o século XIX," *Arquipélago-Revista da Universidade dos Açores* 17 (1995): 189–240.

36. Trabut, *Arboriculture fruitière*, 34, 138–139.

37. René Maire, "Louis Trabut: Notice nécrologique," *Revue de botanique appliquée et d'agriculture coloniale* 98 (1929): 613–620.

38. Trabut, *Arboriculture fruitière*, 146.

39. Walter T. Swingle, *The Date Palm and Its Utilization in the Southwestern States* (Washington, DC: US Department of Agriculture, 1904).

40. Louis Trabut, "Les oranges précoces du groupe Navel en Algérie," *Bulletin des séances de la Société royale et centrale d'agriculture*, 1910, 867–895.

41. Louis Trabut, "Les hybrides de *Citrus nobilis*: La clémentine," *Revue de botanique appliquée et d'agriculture coloniale* 60 (1926): 484–489.

42. Ibid., 488.

43. Trabut, "Les oranges précoces du groupe Navel"; Louis Trabut, "Mutation par bourgeons chez les *Citrus*: La carpoxenie et la cladoxenie," *Revue de botanique appliquée et d'agriculture coloniale* 22 (1923): 369–377.

44. Jean-Baptiste Arrault, "A propos du concept de *méditerranée*: Expérience géographique du monde et mondialisation," *Cybergeo: European Journal of Geography*, Epistémologie, Histoire de la géographie, Didactique, document 332 (Jan. 3, 2006), accessed Feb. 14, 2017, doi:10.4000/cybergeo.13093.

45. Marc Côte, *L'Algérie, espace et société* (Paris: Amand Colin, 1996); Hildebert Isnard, *La réorganisation de la propriété rurale dans la Mitidja (1851–1867): Ses conséquences sur la vie des indigènes*, Mélanges d'historie algérienne (Algiers: A. Joyeux, 1947), 15–126.

46. Mutin, "L'Algérie et ses agrumes."

47. Michel Launay, *Paysans algériens: La terre, la vigne et les hommes* (Paris: Seuil, 1963), 203.

48. Julien Franc, *La colonisation de la Mitidja* (Paris: Champion, 1949); Mutin, *La Mitidja*.

49. Hildebert Isnard, *La vigne en Algérie* (Paris: Gap, 1954).

50. Henri Rebour, *Les agrumes en Afrique du Nord* (Algiers: Union des Syndicats des Producteurs d'Agrumes, 1950), 34; Mutin, "L'Algérie et ses agrumes," 14.

51. Larger groves were cultivated in new irrigated areas in Oran and the Chelif valley. See Mutin, "L'Algérie et ses agrumes."

52. "Coopérative agricole de la Mitidja," *L'écho d'Alger*, Oct. 20, 1922; "Une industrie agricole algérienne," *L'Afrique du Nord illustrée*, Dec. 20, 1924.

53. Corinne Desmulie, "L'agriculture coloniale en Algérie, 1930–1962: Objet et sources," *Colloque jeunes chercheurs en histoire économique*, Paris X, 2008.

54. Olivier Chartier, *Les ombres de Boufarik* (Paris: Flammarion, 2010). Chartier, a journalist, is the grandson of Froger. In this work he tries to come to terms with the problematic legacy of Froger: a small settler, radical-socialist mayor, who came to embody the reactionary racist positions among the *pieds noirs*.

55. Rebour, *Agrumes*, 37.

56. Ibid., 46.

57. For an informed discussion on the difficult context of the Great Depression for Algerian export agriculture, see Desmulie, "L'agriculture colonial."

58. Two authors have already explored this connection between California and French imperial undertakings in North Africa through citrus: Antoine Bernard de Raymond, "Une Algérie californienne? L'économie politique de la standardisation dans l'agriculture coloniale (1930–1962)," *Politix* 95 (2010): 23–46; Will D. Swearingen, *Moroccan Mirages: Agrarian Dreams and Deceptions, 1912–1986* (Princeton, NJ: Princeton University Press, 1987).

59. J. Brichet, "Un petit aperçu sur la production et le commerce mondiaux d'agrumes," *Afrique du Nord agricole*, Dec. 7, 1929; "Planteurs d'orangers, attention! Il y a Navel . . . et Navel," *Afrique du Nord agricole*, Dec. 6, 1930; "Tant que nous ne séléctionnerons pas nos orangers navels l'avenier de leur culture restera chimérique," *Afrique du Nord agricole*, Nov. 3, 1931.

60. A. D. Shamel, *Cooperative Improvement of Citrus Varieties* (Washington, DC: US Department of Agriculture, 1919), 250.

61. Ibid.

62. M. J. Brichet, *Mission algérienne agricole et commerciale aux États Unis (Mai–Juin 1932)* (Algiers: V. Heintz, 1932).

63. Douglas Cazaux Sackman, *Orange Empire: California and the Fruits of Eden* (Berkeley: University of California Press, 2005); David Vaught, "Factories in the Field Revisited," *Pacific Historical Review* 66, no. 2 (1997): 149–184; Tiago Saraiva, "Oranges as Model Organisms for Historians," *Agricultural History* 88, no. 3 (2014): 410–416.

64. Shamel, *Cooperative Improvement*, 265–275.

65. Hamilton P. Traub and T. Ralph Robinson, *Improvement of Subtropical Fruit Crops* (Washington, DC: US Department of Agriculture, 1937), 784–785.

66. Brichet, *Mission algérienne*, 206.

67. Rebour, *Agrumes*, 168.

68. Ibid., 191.

69. L. Blondel, *La station expérimentale de Boufarik* (Algiers: Imprimerie Moderne, 1951).

70. Brichet, *Mission algérienne*, 208.

71. Blondel, *Station expérimentale de Boufarik*, 11.

72. H. J. Webber, "New Horticultural and Agricultural Terms," *Science* 18 (1903): 501–503.

73. For a full discussion of Webber and cloning, see Saraiva, *Cloning Democracy*.

74. Rebour, *Agrumes*, 168.

75. Quoted in Raymond, "Une Algérie californienne?," 24.

76. Rebour, *Agrumes*, 363–378.

77. For the same set of attributes among vineyard workers, see Launay, *Paysans algériens*.

78. Jean-Pierre Peyroulou, "1919–1944: L'essor de l'Algérie algérienne," in Bouchène et al., *Histoire de l'Algérie à la période coloniale*, 319–346, at 330.

79. This is based on Chartier, *Les ombres de Boufarik*, 30–33.

80. Pierre Bourdieu and Abdelmalek Sayad, "Paysans déracinés: Bouleversements morphologiques et changements culturels en Algérie," *Études rurales* 12 (1964): 56–94; Bourdieu, *Sociologie de l'Algérie*; Paul A. Silverstein, "On Rooting and Uprooting: Kabyle Habitus, Domesticity, and Structural Nostalgia," *Ethnography* 5, no. 4 (2004): 553–578.

81. See Loïc Wacqant, "Following Pierre Bourdieu into the Field," *Ethnography* 5, no. 4 (2004): 387–414; Craig Calhoun, foreword to *Picturing Algeria*, by Pierre Bourdieu (New York: Columbia University Press, 2012); Azzedine Haddour, "Bread and Wine: Bourdieu's Photography of Colonial Algeria," *Sociological Review* 57, no. 3 (2009): 385–405.

82. Consciously and ironically, Bourdieu was appropriating a major theme of the French extreme Right, namely, of Charles Maurras and his Action française, who denounced the uprooting and demoralizing effects of capitalism in the French countryside, to make a critique of colonialism as violent uprooting modernity.

83. Bourdieu, *Sociologie de l'Algérie*, 120–129.

84. Ibid.

85. Jean-Paul Sartre, preface to Fanon, *Wretched of the Earth*, xliii–lxii.

86. James D. Le Sueur, *Uncivil War: Intellectuals and Identity Politics during the Decolonization of Algeria* (Philadelphia: University of Pennsylvania Press, 2001); Paul Clay Sorum, *Intellectuals and Decolonization in France* (Chapel Hill, NC: Duke University Press, 1977).

CHAPTER FOUR

Modalities of Modernization

American Technic in Colonial and Postcolonial India

Prakash Kumar

On the eve of World War I, Sirdar Jogendra Singh, an Indian landlord, purchased a 25 BHP (brake horsepower) tractor of an English make and hauled it to his sprawling farm in Kheri. The step was extraordinary for the times. It seems even more audacious considering that Singh was after all an Indian landlord, not one of the typical, prosperous planter-investors of European origin, who were known more commonly to import expensive machines to their plantations in the colony.[1] As Singh's tractor made its way to Aira Estate in Kheri District in the United Provinces of Agra and Oudh, the colonial Agriculture Department in the province prodded him along. Indeed, its highest officials made their way to Kheri, hoping to record details of the operation in order to gauge the utility of tractors for colonial agriculture. The district contained some of the more productive lands in the province, and the Agriculture Department had invested resources in building a network of irrigation channels and had invited settlers to the district. The eagerness of Jogendra Singh was not out of sync with the motivation of the colonial state in expanding arable, taxable land.[2]

Singh was a prosperous farmer who had inherited the sprawling twelve-thousand-acre estate. He rented out most of this land to tenants to farm on rent. But there was still a large part of the estate that had remained unsettled and that he wanted to bring under cultivation.

But this portion was annually inundated by floodwater from the local river, which deposited fertile alluvium, and given years of lack of effort at plowing them regularly, it was now covered with dense weeds of *kans* and *bainsura*. Given the sheer size of the landmass and the nature of the stubborn weeds, manual clearing would be cumbersome and would have taken months if not years. The use of the tractor seemed to make sense.[3]

Interestingly Singh, the benevolent landlord, justified his plans for turning Aira Estate into an arable tract as an act of generosity toward potential tenants who would come to live on the estate. As a prosperous landlord, he wrote, it was his "duty" to work for the amelioration of the extremely poor tenants in the area. "It is such a shame that the labouring classes in the villages in Oudh and other parts of the United Provinces should not have even enough food from year to year. . . . Going into a village in the United Provinces one is at once struck by the contrast between the physique of those who have had enough food and that of others who have only managed to exist somehow. The Brahman, the Rajput, and the Cowherd are well-built and strong, while the poor village drudge is only a skeleton."[4] Singh wanted to settle his estate with these "poor village drudges," who would gain employment through his astute use of tractors. In the process, they would be converted into rent-paying peasantry, the subjects of revenue measures put in place by the colonial state within the framework of law and liberalism.[5]

It is through texts such as Jogendra Singh's that the poorest of the agricultural lot, the so-called "village drudge," make an appearance and invite us to analyze the salience of foreign machines and of contingent modernization to them. These laboring classes would have significantly overlapped with the subordinates of colonial society, or the "subalterns," whom the South Asianists have identified as falling outside the ambit of the teleology of nation, class, and civilization. The task of recovering the subaltern's "particular forms of subjectivity, experience, and agency" has defined the project of subaltern scholars since the 1980s. Inflected by postcolonial criticism, South Asianists have expressed their discomfort with universals such as "modernization."[6] This chapter harnesses that discomfort to interrogate certain assumptions in the studies of transnational movement of technological knowledge. The fine work done by South Asianists highlights the complex relationship of modernization—as an ageless, timeless category, hinged to an assumed, ceaselessly "moving" technology—with diverse political and social identities. This relevance is analyzed in this chapter in modernizing impulses at three specific moments: the colonial moment, the Presbyterian moment, and the land-grant moment. In the process, the chapter brings to the fore the social limits of several universals that came into play at specific historical instances: the limit of the colonial state's technological imagination for bringing about productive change; the American Presbyterians' invoking of "human" values to turn "indigents" of India into productive small farmers; and the drive by American agronomists to spread intensification models for raising agricultural yield. The relentless desire of the modernizers itself provides clues to the blind spots. At other times, these actors voice their concern in fleeting instances. The transnational movement of machines and knowledges derived its authority by making these subjectivities invisible. This chapter questions the certitude of that process. Not questioning the pitfalls of such transnational processes would make us complicit in the same forgetfulness that the modernizers engaged in. Therefore, factoring in concerns expressed in South Asian historiography, this chapter initiates a dialogue between South Asia scholars and historians whose object is specifically the transnational movement of knowledge.

American Technic and Transnational History in South Asia

The historiography of colonial South Asia typically focuses on the locality of the "colonial-national" complex. And yet, South Asianists have not neglected the translocal. Indeed, one can argue that the metropolis or imperial, lying at the remote end of colonial locality, which is a constant referent in colonial South Asian historiography, makes the study of "colonial" inherently transnational. The need to connect South Asian history to the history of faraway places and to write histories on a wider scale was noted early on by the South Asianists.[7] David Ludden summed up the implications of the global turn very well for scholars of South Asia dealing with the questions of scale. Reemphasizing the necessity of finecombed analyses at the local level of the type area studies scholars provide, he noted the need to consider the "global and interactive" as well as the "local and multiple."[8] Ludden was only giving expression to the mainstreaming of a trend in South Asia that seeks to implicate not only the imperial but also the transnational in analyzing local, colonial trends.[9]

The focus on the America–South Asia connection and the deliberate conflation of colonial modernity with American modernization on the Indian subcontinent serve the goal of decentering from the area studies focus of South Asian historiography on the colonial and national axes.[10] The experience of agricultural modernization in India in the twentieth century was evidently complicated by the presence of American technology and expertise, which had a liminal presence in the colonial period but became much more robust in the postcolonial era. Through a focus on three "moments," as outlined above, this chapter highlights the entanglement of such efforts with colonial relations and with the postcolonial development regime. The occasional use of "technic" in my analysis aims to disturb the narrative of modernization in order to reach out to visions and worldviews that were being displaced by modernization, as against an abiding use of "technology" that traps us into analyzing "premodern" in the language of the modern. The use of the term "technology" makes us see the important processes of engagement, appropriation, and resistance. The deployment of "technic" enriches the analysis by gesturing toward the making of modern subjects and the enlisting of their consent. It underlines the growing attraction that modernization had for colonial and nationalist elites. Specifically, I draw attention to two things: first, the externality of American technic to the general colonial scheme of "improvement" and, therefore, to the extra labor that had to be expended to facilitate its entry into the colony; and second, the subtle maneuvering by Nehru's administration in the 1950s to nurture American technic in India in the face of charges of "American imperialism" from the political Left, both from the communist parties and from the socialist wing within the Indian National Congress. If anything, this extra effort points to the growing strength of attraction to modernization.

A few words are needed to clarify the theoretical borrowings. The preference for the term "technic" to refer to artifacts allows us to unlock the explanatory potential of the Foucauldian notion of biopower over populations. The framing of artifact in terms of technic (in contrast to technology) gives access to a distinct type of claim making by the colonial state.[11] To Gyan Prakash goes the credit for first deploying "technic" in a systematic and comprehensive way and for popularizing its usage in South Asian studies. In his rendering, technology is something concrete out there, whereas technic, in contrast, is something flowing through human consciousness and experience. Working with technic means ignoring the barriers between natural, human (including state), and technical forces.[12] This formulation is akin to Partha Chatterjee's use of the notion of "governmentalization of the state" that offers insights into the operation of "micropower" in a late Foucauldian tradition of the study of power. Chatterjee has used this notion to explore how "the regime secures legitimacy not by the participation of citizens in matters of state but by claiming to provide for the well-being of the population."[13] This conceptual apparatus dovetails with and enriches the argument being made in this volume in several ways. It emphasizes the movement of technology and knowledge across borders, it highlights the charm exerted by those knowledges as embedded in specific sets of social relations, and it draws attention to multiple sites of resistance by the "local" to transformative projects being driven by a range of state and nonstate actors both in India and in the United States.

Colonial Moves toward Tractors and Threshers

Let's return to our interlocutor, Jogendra Singh; pulling the Garrett tractor to Kheri was not going to be an easy task, and Singh clearly anticipated the difficulties. Kheri was not even accessible by a *kacha* road (dirt road). The nearest post office was nine miles away. Singh knew that carrying one hundred maunds of grains to the district headquarters usually cost Rs. 16, an expensive proposition. Under such circumstances getting a huge machine all the way to Kheri was a herculean task. But Jogendra Singh was determined. Taking the advice of the government's agricultural engineer, he selected the appropriate tractor model that would best answer his needs. He purchased it from local agents, Messrs. Burn and Company in Calcutta, and had it put on a train. The machine was offloaded at Lakhimpur and thereafter started the arduous twenty-one-mile journey, which it completed in a month and a half. Maneuvering through soft soil and *nala*s (streams), even dismantled once to be put on a boat to cross a deep stream and then reassembled on the other side by the company's mechanic, the machine reached Aira one evening, "followed by crowds of [curious] villagers."[14]

The task of acclimatizing a machine to a stubborn piece of land was formidable but nothing that the farmer-entrepreneur and his colonial backers were not ready to handle (fig. 4.1). The biggest hurdle came with the wheels getting stuck in soft soil around the water channels that the colonial government's Irrigation Department had built. Singh had all

FIGURE 4.1. The steam tractor and plows used in Kheri District.
Source: B. C. Burt, "Steam Ploughing Experiments in the Aira Estate, Kheri, United Provinces," *Agricultural Journal of India* 9 (1914): 1–6, pl. 1, facing p. 2.

that land recontoured in such a way that twenty-five acres of separate rectangular plots would have the water channel running on only one side of the land, so that the machine would always have a dry road to move along on the other side.

It is interesting that the "village drudge" who made a brief appearance in the expression of sympathy by Jogendra Singh never returns in the narrative of the benefits of tractor plowing at Kheri in colonial discourses. Singh admittedly attempted the tractor project in the name of "those who seemed so out of place in this God's world." But the subaltern seemingly ran into his moment of truth and is subsequently effaced from the calculations of Jogendra Singh. The colonial representative, B. C. Burt, who oversaw the plowing operation at Kheri, reached the conclusions required for his purposes. He authoritatively declared the utility of tractors for clearing weed-infested lands, saying, on the basis of the calculations made on the ground at Kheri, that the cost of "breaking up virgin land [with tractors] is reasonable" and "much less" than that of hand-digging. His pertinent advice going forward was that a 40 BHP tractor would serve the needs of clearing weeds better than the one used; in other words, a more powerful tractor would be even more effective. The officials did not make any definitive calculations that kept the "village drudge" in mind.[15]

To get to the larger point here—the Agricultural Departments of the colonial state made clear attempts to popularize the use of larger machines in land preparation, reaping, and threshing in the years preceding World War I and after. Districts like Kheri were suitable for deployment of tractors, as they had a surplus of cultivable wasteland. Kheri was, as one report said, "in excess of the cultivating capacity of the existing village population by ordinary methods." The agricultural officials tried hard to prove the viability of such operations, which they insisted would cover the working expenses of reclamation. The rent from the area in the coming years would all be part of profit on the initial investment in tractors. But any expectation that large landlords would readily take to such reclamation ran into the internal roadblocks of an attenuated capitalist modernity that colonialists themselves had introduced. The officials admitted that even if the rent in future years was ensured, "there was little hope" that the initial investment would be made by many, and thus, they reasoned that the state, not individuals, must continue to make efforts toward reclamation.[16]

The state officials were more optimistic about the adoption of reapers and threshers in productive tracts with assured irrigation, especially those of the Canal Colonies in the Punjab and the Central Circle of the United Provinces. Animal-driven reapers were found useful in the Canal Colonies and were energetically promoted quite early on in the pre–World War I era.[17] Officials constantly complained about the small size of holdings and how such land structure was holding back the adoption of useful machines. But a reaper that was small and easy to move from one field to another and could be pulled by bullocks without difficulty won favor. Here the officials also underlined the advantage of contracting out the machine to others to recover the cost of purchase and operation. The labor released from reaping could be engaged in gathering and binding the crop cut by the machine.

Colonial officials especially favored introducing threshing by steam engines in areas of wheat cultivation, the primary *rabi* (winter) crop in northern provinces.[18] The produce of several individual holders could be transported to a central place for threshing, thus bypassing the "problem" of small holdings. In addition, instead of using cattle to do the threshing, by pounding with their feet, the use of steam threshers would free up the cattle for the task of preparing land for the *kharif* (monsoon season) crop. Trials conducted at the state's Cawnpore Farm showed threshing of Pusa 8 wheat using steam engines to be "a practical proposition" that left little doubt that the innovation, if adopted widely, could break the vicious cycle of dependence on cattle for threshing and consequent nonavailability for the next crop.[19]

There was a larger context for the increased colonial effort to launch tractors, reapers, and threshers and their accessories in the interwar years. Some of this was spurred by war-induced shortages in food and rising costs. But largely there was a growing realization of stagnation in agriculture. This stagnation had come after decades of growth in aggregate yield on account of the expansion of cultivable areas. In the years preceding World War I no easy expansion of agriculture was possible using available means. Only a major infusion of a radically new input like irrigation or technology could materially change the situation. Thus, B. C. Burt, engaged in the steam-plowing operation at Kheri, noted that in recent times more and more zamindars approached the Agriculture Department asking them to clear land using tractors. The zamindars were more than willing to pay a fixed sum per acre for steam plowing. Writing in 1923, H. R. Stewart, professor of agriculture at the College of Agriculture in Lyallpur, and D. P. Johnston, a colonial official, noted that "cultivation . . . by mechanical means has been largely engaging the attention of agriculturists since the war." This realization led Stewart and Johnston to write an article on the availability of different models of tractors in the colony and their suitability for different tracts and different agricultural operations.[20]

This trend toward mechanization through adoption of machines and accessories continued through the 1920s. The use of tractors of English and American companies, as well as of accessories produced by such tractor firms, was popularized. Sam Higginbottom, a missionary, referred to the utility of "our American Titan tractor with three American plows behind it" for clearing land of weeds at Allahabad. He was referring to the famous model launched by the International Harvester Company of Chicago in 1915, a tractor that seemingly appeared in the colony within a few years.[21] Soon after, breaking with the previous convention of using tractors for clearing land, tractors—in very specific zones no doubt, and by a select few—were deployed for actual cultivation. The government farm at Lyallpur started using tractors for cultivation in 1920. In 1930 D. P. Johnston, at the Agriculture Department in Punjab, could claim that the use of tractor accessories over the past decade had led to "ample" knowledge about those implements. His observations held true especially for the Canal Colonies.[22] Johnston also provides information that suggests

Name of implement	Manufacturer	Agent	Price
			Rs. A.
1. Three-furrow self-lift plough 120-A.	International Harvester Co., Chicago, U. S. A.	Messrs. Volkart Brothers, Lahore.	445 8
2. Two-furrow self-lift plough .	Ransomes, Sims & Jefferies, Ltd., Ipswich.	Messrs. Duncan, Stratton & Co., Lahore.	640 0
3. Two-furrow non-self-lift plough.	Ditto .	Ditto .	610 0
4. Two-furrow disc plough, non-self-lift.	Ditto .	Ditto .	400 0
5. Grand Detour 5-disc self-lift plough.	J. I. Case Threshing Machine Co., U. S. A.	Messrs. Greaves, Cotton & Co., Lahore.	1,100 0
6. Ransomes "Orwell" cultivator, 11-tined.	Ransomes, Sims & Jafferies, Ltd., Ipswich.	Messrs. Duncan, Stratton & Co., Lahore.	510 0
7. Tandem disc-harrow, 32 discs	International Harvester Co., Chicago, U. S. A.	Messrs. Volkart Brothers, Lahore.	556 12
8. Spring-tined harrow . .	....	Ditto .	200 0

FIGURE 4.2. Tractor accessories of English and American make in India.
Source: D. P. Johnston, "Tractor Implements Tried at the Lyallpur Experimental Farm," *Agricultural Journal of India* 25 (1930): 317–320, at 317.

the growing and wide prevalence of machines of English and American companies in colonial India (fig. 4.2; see above).

The Presbyterian Moment

A distinctly American moment in the launch of technic in colonial India came with the establishment of the Allahabad Agricultural Institute by American Presbyterian missionaries in 1912 (fig. 4.3). The founding of the Allahabad Agricultural Institute hinged on a growing global agricultural mission movement. At the core of the latter was the belief that the task of spreading the gospel must stem from a foundation of improved agriculture and well-being of the masses, a majority of whom were poor agriculturists. It was this belief that saw the conversion of the Ohio native Sam Higginbottom from a pure "evangelist" to a "missionary farmer" in India.[23] Higginbottom, who first came to India in 1903, founded the Allahabad Agricultural Institute and worked there until his retirement in 1945. Just two years before India's independence, Higginbottom relinquished his position of principal at the institute and returned to the United States. The institute continued its work after India's independence under the control of American missionaries.

As the Allahabad Agricultural Institute developed as a teaching institution, a demonstration farm, and an agricultural implement-manufacturing center, it inaugurated, in a microcosm, a distinct approach to agricultural improvement in the colony. It was first approved to impart intermediate-level education, and then subsequently, in 1932, it became a degree-granting institution affiliated with Allahabad University. Gradually it came to have separate teaching departments in agronomy, agricultural extension, animal husbandry, and home economics. It began a program for making small agricultural implements and launched an agricultural engineering degree course in 1942, the first of its kind in South Asia. The agricultural engineers at the institute started a forty-two-acre experimental "rain-fed" farm on the precincts, and in 1954 the agricultural implements factory was turned over to a separate entity specializing in the manufacture and sale of implements—the Agricultural Development Society.

Primarily staffed by incoming waves of American missionaries trained in agriculture, Allahabad Agricultural Institute was a conduit for the importation of American technic and ideals. The American

FIGURE 4.3. Allahabad Agricultural Institute.
Source: Frank H. Shuman, *Extension for the People of India* (Urbana: University of Illinois Press, 1957), 1.

staff at the institute successfully navigated the colonial framework and formed linkages with prominent nationalists, notably Gandhi. The Allahabad missionaries evidently approached agrarian problems from a slightly different position than the colonial state officials, deeming all, including India's agrarian poor, as judged by the same god and thus immanently "improvable." The aims of the missionaries at Allahabad, in a striking way, also differed from the Gandhian program. Meeting Mohandas K. Gandhi for the first time at the inauguration of the Banaras Hindu University in 1916, Sam Higginbottom began a sustained correspondence with him. Gandhi was quite impressed with Higginbottom's program for village improvement. But he differed with him in terms of priorities. To Gandhi, the development of village industries was important as a supplement to agriculture. The cottage industries, spinning and weaving of khadi (cotton cloth), were foundational to Gandhi's program. But to Higginbottom, agricultural improvement was a priority, a goal unto itself, and required new techniques and ideas. While mutual respect between Gandhi and Higginbottom remained, Gandhi later took the line that no real village improvement was possible under an alien government. Freedom was a prerequisite for any substantial project of agricultural improvement and required vast resources and a political commitment that only a national government could summon.[24]

In many ways, the Allahabad missionaries were also advocates of the American republican ideal of the self-reliant, Jeffersonian yeoman farmer. This was reflected, most importantly, in their advocacy of the interests of the "small farmer." The ideal farmer of missionary vision would be one owning his plot and implements and would work his farm with family labor. On the basis of this vision, the Americans critiqued the practices of the Indian peasantry in villages around Allahabad. Although extrapolated in universal rather than contingent terms through allusions to "human" questions, the quintessential small farmer of missionary vision was contrasted with the eviscerated Indian "others" in terms of both economic marginality and lack of "Christian" thrift. Such a vision stemmed from a moral outlook that attributed farmers' poverty to their own work ethic, often absolving the colonial state of any wrongdoing and fixing responsibility for improvement on the peasantry through a culture of "self-help."

This particular orientation toward the "small farmer" was nowhere more apparent than in the program of mechanization of agriculture that the Allahabad missionaries sought to develop and perfect. Sam Higginbottom's earliest writings allude to the poverty and illiteracy of the inhabitants of rural India, and of the farmers in the United Provinces specifically. The average holding of a tenant in the United Provinces, Higginbottom wrote, was three and a half acres and that of a landowner four and a half acres, with most having "little capital, very little equipment, and entirely insufficient food and clothing." It was this class of agriculturists that had to become the target of missionary-initiated improvement at Allahabad. The best way to do that was to "train the best and brightest in a central institution so that the ones so trained can go out to their own folk in the villages [and influence others]." Higginbottom also outlined his goal as inducement of better farming methods with machinery adapted to Indian conditions.[25]

Nobody represented the spirit of "small farmer mechanization" more fully than the Missouri native and Allahabad Agricultural Institute's agricultural engineer Mason Vaugh. Mason Vaugh was born in Farmington, Missouri, attended the University of Missouri's College of Agriculture, and received a degree in agricultural engineering in 1921. That same year, he applied to the Board of Foreign Missions of the Presbyterian Church to be placed as an agricultural missionary. He was assigned to India to work at the Allahabad Agricultural Institute. Arriving in Allahabad the same year, he would work at the institute uninterruptedly until 1958.[26]

Mason Vaugh's two core responsibilities at Allahabad were teaching courses in agricultural engineering and managing the institute's workshop dedicated to manufacturing simple implements. At the institute's workshop, Vaugh helped develop three plows—Shabash, Wah-Wah, and U.P. No. 1. The implements-manufacturing plan was such a success that it caught the attention of the provincial Agriculture Department. From 1938 to 1939 Vaugh was contracted to develop implements that were then distributed through government-owned farms throughout the United Provinces of Agra and Oudh. In 1943 the director of the Agriculture Department asked him to develop "improved" implements for the whole of the United Provinces.[27] In addition to the implements program, two further developments of the 1940s are relevant to the argument made here. One was the launch of a degree program in agricultural engineering at the Allahabad Agricultural Institute in 1942. The second was the beginning of an experimental rain-fed farm in 1944, which was put under the charge of the Agricultural Engineering Department. The farm was to conduct trials on agricultural implements and cultivation practices, what Vaugh often called "cultural practices," employed on farms that did not use controlled irrigation but were dependent on rainfall instead.

Three salient aspects seem to define the program at Allahabad Agricultural Institute, and they were interconnected through a justification of the "needs" of the small farmer.[28] The experimental rain-fed farm was justified in the name of the small farmer, a majority of whom normally did not have access to sources of artificial irrigation. In 1945 some 15 percent of the cultivated land in India was irrigated with canals and tube wells. Another 15 percent received some rudimentary form of irrigation from wells, tanks, and inundation canals. The remaining 70 percent of cultivated land was entirely dependent on rainfall for irrigation. To be relevant to the majority, agricultural experiments, the backers of the rain-fed farm asserted, must align with the needs of the majority. In so arguing, these American advocates were also building a critique of the colonial Agriculture Departments, which, they said, "tend to simply assume irrigation in anything they do or recommend, and . . . neglect the problems of the farmer who lacks irrigation." In contrast, they said that the efforts at the Allahabad experimental farm would develop methods of cultivation "which will enable the small farmer, particularly the small farmer not having irrigation facilities, to most fully occupy his time and to derive the greatest profit from his efforts."[29]

The Allahabad program also involved advocacy of a particular form of mechanization, which was, once again, delineated with reference to the persona of the "small farmer." In a paper delivered in 1953, in Manila, Philippines, titled "Mechanization for the Small Farmer," Mason Vaugh discussed what he thought was the type of mechanization suitable to postcolonial Indian agriculture. He counseled against mechanization of farm operations such as plowing, harrowing, seeding, and weeding. Vaugh clearly articulated his opposition to any effort toward mechanization of farm operations as implausible and impractical, given the small size of holdings, lack of capital, and a redundancy of labor that would result from such mechanization. Besides, such a measure would lead to a disruption in social organization. Vaugh and his cohorts at Allahabad elaborated on the relevance of simple, "improved" tools and the use of animal power and of machinery drawn by animals. But, separately from "farm operations," Vaugh expressed his approval for engine-driven mechanization of "barnyard" operations, those operations that could be performed at the village site or at any other central node. This latter suggestion was reminiscent of the prior interwar focus on threshers, for instance, in the colonial paradigm.[30]

In a related, third area, the Allahabad dissenters also went out on a limb to oppose the idea of "cooperative farms," an idea that was popular in the colonial program as well as in nationalist visions. The cooperative program entertained the idea of pooling individual landholdings and labor in order to introduce economies of scale in the operation of agricultural machinery. The cohort at Allahabad were opposed to any such initiative that would change the land structure or the management of farming operations. Vaugh spoke against the approach involving cooperative farming and allied mechanization. On the basis of his own calculations he came to the conclusion that the introduction of tractors on cooperative farms was "an uneconomic proposition." He called the backers of this approach "starry-eyed reformers" and pointed out that the plan was not workable on the ground and would not lead to an increase in production. He also alluded to the roadblocks in bringing cultivators together and in pooling resources.[31]

It might be prudent here to focus on the figure of the "progressive farmer" that emerges time and again in the Allahabad model or imaginary in their proposal of a system attuned to the needs of the "small farmer." This is the figure of the quintessential farmer going back to mid-nineteenth-century American agriculture.[32] These notions embraced farmers engaged in "scientific farming" and favoring innovation. In this view, traditionalists, in contrast to the progressives, were narrow-minded and headed for failure in the modern economy. In particular, these notions were very committed to landownership and "independent" family farms, very much within a capitalist orientation, and thus in alignment with the colonial, capitalist program.

It is thus interesting that Vaugh based his most stringent criticism of cooperative holdings and the plan to introduce tractors on them in the early 1940s on the anticipation that these would lead to the destruction of "family farms." According to him, there were, after all, two solemn objectives for agriculture: "the increase of production and the development of human personality, the attainment of the largest possible measure of human happiness." And both of these, he argued, could be ensured "by the continuance of the family farm, each unit, whether large or small, in the hands of and under the complete control of a family and worked, to the largest degree possible, by that family."[33] The archetypal figure of this "progressive farmer" emerges, explicitly, in later explanations of the logic of the implements development program at Allahabad as well as of cultivation practices at Allahabad's own rain-fed farm. To quote Mason Vaugh again, the Allahabad experiments were "made to simulate the conditions of a progressive farmer . . . wishing to try to improve his land with only the use of better implements and better cultural practices associated with them." This was a world of imaginaries and practices that would soon lose ground in the face of a new moment in India's postindependence agriculture. Modernization accommodated several diverse practices, as the transition from missionary experiments at Allahabad to the next stage of innovations at Allahabad showed.

The Land-Grant Moment

A "land-grant moment" awaited Indian agriculture when American land-grant institutions became involved with the program of agricultural modernization in India.[34] This engagement started after India's independence, in the early 1950s, and extended through the 1960s and beyond, until a sharp cutoff point was reached in 1971–1972 when, aggrieved over US support for Pakistan in the Indo-Pakistan War, the government of India abruptly stopped its continuing collaborative programs with Americans. But before that, American land-grant institutions participated in developing India's rural universities and their programs. In addition, American academic faculty—technical and social scientific experts—were present in the role of advisers to the central government: as planners of statewide and district-wide projects, as top-tier trainers of village-level workers for community development projects, and as technical and social scientific specialists who offered input on specific questions or executed impact studies (fig. 4.4). The Indian government officials at all levels actively sought their input. The push for fostering such institutional ties first came at the behest of the Ford Foundation in 1950, and subsequently the US State Department made a concerted push to establish such ties. In 1952 the United States signed a more specific technical cooperation agreement with India under Point Four programs. The 1950s and 1960s saw a continuous stream of university faculty from the United States and of personnel from federal agencies visiting India, along with a reverse flow of Indian agricultural faculty and technicians

FIGURE 4.4. Mrs. Shuman and Frank Shuman. Frank Shuman, a University of Illinois agronomist, was among the very first group of American university faculty experts sent to India under the Ford Foundation's and the US State Department's programs. He served at the Allahabad Agricultural Institute. Shuman earned a reputation among villagers around Allahabad for his insistence on the nitrogen deficiency of Indian soils, since he repeatedly explained signs of "nitrogen hunger" in plants. After spending four years in the area, as he boarded a train in Allahabad en route to the United States, a group of villagers put a wreath around his neck with the sign "Nitrogen Zindabad," literally "Long Live Nitrogen." *Source*: Frank H. Shuman, *Extension for the People of India* (Urbana: University of Illinois Press, 1957), 23.

who spent a considerable amount of time at land-grant colleges in the United States, completing formal degrees or attending short-term refresher courses. This was altogether an American technocratic attack on the perceived problem of the lack of modernization in Indian agriculture in the decolonizing era.

The emergence of an unparalleled faith in technology as an instrument for social improvement in the United States after World War II provided the foundation for the establishment of institutional ties with

India.[35] President Truman's declaration of Point Four policies of technical aid to the developing countries in 1949 first opened up the possibility of exporting agricultural ideas and expertise.[36] The first major tranche of $53 million of aid under Point Four flowed into India in 1952. The total amount of aid dropped to $45 million the next year before rising sharply to $231.5 million in 1954. This upward trend was maintained over the next several years.[37] In addition, using the university contract programs after 1951, the American State Department tapped into faculty resources of universities and deployed them in India (as also in other postcolonial nations).

American foundations partnered with the State Department in implementing programs of agricultural and rural uplift in India.[38] In the 1950s they, too, made an effort to establish long-term ties with institutions and policy-makers in India. On a Ford Foundation initiative, the Allahabad Agricultural Institute and the University of Illinois at Urbana-Champaign were made "sister institutions" with a long-term program for exchange of faculty between them and for placement of American agronomists, extension specialists, and experts in domestic science and in animal husbandry at Allahabad.[39] The involvement of land-grant institutions in India became much more robust subsequently. The International Cooperation Administration and, later, the US Agency for International Development enlarged their agricultural exchange program in India between 1955 and 1972. India was divided into five sections, and five universities—University of Illinois at Urbana-Champaign, Kansas State (replaced by Penn State in 1966), Ohio State, the University of Missouri, and the University of Tennessee—were linked with agricultural universities in separate zones in India. The State Department signed interinstitutional agreements to send "technicians" from these universities to India to establish research, teaching, and extension in the image of the American land-grant college. When this program ended in 1971–1972, a total of forty-eight agricultural universities, colleges, and institutes in India had become partners of this American involvement (fig. 4.5).

Here, as before, the prescription for "modernization" in this phase was heterogeneous, fiercely debated among both Americans and Indians as to its efficacy and appropriated at the Indian end in vastly different ways and on terms that were far removed from those assumed on the American end. Working within the framework of India's parliamentary system, the political parties on the left vehemently criticized American technical aid. Thus, what passes off as "Cold War" politics in many

FIGURE 4.5. Penn State faculty in Poona.
Source: Special Collections Library, University Archives, Pennsylvania State University, College of Agricultural Sciences Records, India Project: Reports and Pictures, undated, box 119 AX/CATO/PSUA/07199.

current framings of American aid was critiqued as a version of "Anglo-American imperialism" by a few significant forces in newly independent India that emphasized the utmost need for the country to maintain its sovereignty. Writing in *India Today*, the official journal of the Communist Party of India, P. C. Joshi, secretary of the party, chided Nehru's administration for signing the technical cooperation agreement with the United States in 1952. Joshi accused Nehru of "surrendering India's sovereignty—in economic matters, at least—to the American imperialists." He warned that India was in danger of "becom[ing] another Philippines." He called on the press and other democratic elements to defy the designs of "Anglo-American imperialists."[40] As Nehru's administration launched community development projects with the help of American technical experts, *India Today* interpreted the United States' real intent as the opening up of the Indian economy to American capital.[41] Even Gandhi's close associate J. C. Kumarappa described India's partnership with the United States in launching its community projects as "unworthy of any self-respecting independent country." Kumarappa also criticized the program for sidestepping the key question of land reforms.[42] At the very least, these democratic critiques, coming from the Communist Party as well as from the left wing of Nehru's own Congress Party, had an impact on his administration and tempered the nature of programs implemented under him.

"Land-grant modernization" in India, therefore, showed the signs of diversity of paths. The statist initiative toward village development in India postindependence was broadly marked by two parallel efforts: one pursued a rural growth model, and a second pursued straightforward productivist goals of aggregate yield improvement. Land-grant faculty based in India aligned with both. The first evolved from the Community Development Project, launched in 1952, and involved experiments in cooperative farming and joint farming and included within its ambit efforts to build participation in "planning" at the village level, particularly through *panchayati raj* (local government) initiatives. The second strand developed from fertilizer importation programs from the Grow More Food program to the Integrated Area Program and the Integrated Area Development Program and eventually emerged as the New Agricultural Policy between 1964 and 1966. In the latter time frame President Lyndon Johnson's administration adopted a policy of leveraging food aid to compel India to move toward the New Agricultural Policy.[43] The long-running land-grant programs in India responded to these new yield enhancement strategies by launching new Agriculture Production Projects in all their five zones of operation in India.

In a telling sign that the American land-grant operatives in India were split in their recommendations for agricultural intensification, Mason Vaugh, in 1957, took issue with a group of American advocates who pushed for the use of fertilizers in India in a major way. Vaugh published an article in the premier journal of the Indian Council of Agricultural Research, *Indian Farming*, under the title "Simple Ways to Raise Farm Production." It is likely that this article was prompted by India's recent launch of "package programmes" that identified specific areas where in a focused way the use of fertilizers, irrigation, and better seeds would be complemented by offers of credit and a guaranteed minimal purchase price to farmers. Apparently, Vaugh did not emphasize the use of fertilizers in this article. Such was the momentum toward fertilizer use in India at this time that Vaugh's lack of mention of fertilizers invited a rejoinder from the chief agronomist of the Fertilizer Association of India and eventually led to an official communication from the Indian Council of Agricultural Research to Vaugh, asking if he would like to respond and clarify his position on the use of fertilizers.[44] Vaugh responded that his omission of reference to fertilizers was "deliberate." He explained that his experience in India "has led . . . [him] to profoundly disagree with many advocates of fertilisers and with what appears to be some phases of Government policy." The farmers he had interacted with told him that the use of fertilizers led to an immediate rise in productivity but that in the following years productivity dropped. Vaugh surmised that this was on account of a lack of adequate knowledge in India of soil conditions and the impact of inappropriate fertilizers on them. Indeed, he believed that the persistent use of fertilizers was causing diminution in the organic matter in soil, leading to disastrous results. He was not against the use of fertilizers per se, but he felt that rampant application of chemical fertilizers "without the necessary related practices" was wrong, and he even implied that "vested interests" were promoting their use in India.[45]

It should be noted that Vaugh's was a lone voice in the wider scheme of things, as the Indian government was moving speedily toward agricultural intensification. An influential Ford Foundation report of 1959 noted that many of the previous official efforts undertaken through "community development" had failed in the primary task of raising yield. It pointed with anxiety toward the dismal food situation in India and advised that urgent measures be implemented to raise aggregate yield. It would seem that the Indian government's steps were quite in keeping with such recommendations.[46]

If it can be said, for the sake of simplicity, that India was going along the Ford Foundation's suggested path of agricultural yield improvement, one that was based on mechanization, irrigation, chemicals, and support prices, then Mason Vaugh was becoming an outlier in this scheme of things. His growing distance from the Ford Foundation became more evident. And one can only imagine his growing isolation at the Allahabad Agricultural Institute itself, which had started receiving funds from the Ford Foundation as early as 1950. The initial Ford funding at Allahabad was for experiments in extension activities and teaching. By the late 1950s and in the 1960s, however, Ford moved toward productivist strategies in Indian agriculture. Its initiatives at Allahabad started to encroach upon a field that was very dear to Mason Vaugh, that of agricultural implements. Seeking to supply implements to villages that were to be put under the Intensive Agricultural Districts Program, the foundation supported Allahabad's Agricultural Development Society (ADS), a separate agricultural implements manufactory branch of the Allahabad Agricultural Institute that Vaugh had helped set up in 1954. Vaugh did not hide his distress. In July 1960 he wrote to G. Wallace Giles, an agricultural engineer at North Carolina State University who had recently accepted a position as consultant with the Ford Foundation and was about to leave for India to participate in "package programmes." Vaugh made it clear to the new recruit, who had been asked to monitor the ADS, that "I have not always agreed with everything they have done."[47] He told both Giles and the Foundation that he did not like the way the implement development plan at ADS was proceeding, only to be rebuffed.[48] This only sharpened his disagreement with Giles, who was considered a strong voice in favor of agricultural mechanization in the developing countries in the 1960s. His response to the editor of an American Society of Agricultural Engineers journal, who had asked him to evaluate an article submitted by Giles, bordered on contempt. Vaugh turned down the request to referee the manuscript, pointing out to the editor that he "violently disagreed with almost every paragraph of the paper" and that he "doubt[ed] the value of [its] publication, at all."[49]

Current Framings and the Transnational Approach

Three modes of analyses can be deployed to the history of agricultural modernization in India. Timothy Mitchell's work on the role of foreign experts and their monopoly of technological and managerial imperatives is one way.[50] Discussing the work by the US Agency for International Development in Egypt in the middle decades of the twentieth century, Mitchell explains how the discourse of development erases and makes invisible the politics undergirding the discourse. Echoing Partha Chatterjee's formulations for the study of planning by the Indian state in India, Mitchell maintains that the discourse of development practices "self-deception," because the object of development is partly constituted by the discourse. In other words, the object of development is not "outside" the discourse. There is a variation in the way Mitchell, on the one hand, and Chatterjee, on the other, implicate the local "state" and processes of state formation in Egypt and India respectively. To a greater extent than Mitchell, Chatterjee marks out the processes of state formation in India and their centrality to projects of planned development. In Chatterjee's analysis, the internal differences of modernization projects and internal shifts in policy are folded into a political analysis of planning. And perhaps that is a major weakness of Chatterjee's framing for exploring the experience of modernization in India, in the sense that it glosses over the internal differences and shifts in policy angles.

A second group of scholars has looked at agrarian projects in India postindependence with reference to the surrounding classes and caste groups. These scholars tell us that the state's agrarian projects were appropriated by specific groups to advance their own interests. The overwhelming conclusion in these studies is that powerful caste groups and classes foiled the projects of agrarian developers and ensured that the majority of depressed classes would remain outside the ambit of the positive impact of such policies.[51]

A third trajectory of historiography implicitly argues that the path to agricultural "modernization" in India is inherently fraught because modernization and its primary tool of development operate within a logic of capital and that subaltern life and consciousness are incommensurable with the logic of capital.

That agricultural modernization was heterogeneous and variously experienced is a given. In such a context, a radically revisionist history of postcolonial agriculture would need to move not only beyond US State Department records of foreign relations but also beyond the accounts of Nehrus and Indira Gandhis. A focus on forces unleashed by the processes of local state formation is welcome. At the same time, a focus on local narratives can add nuances to the story we have known thus far. The narratives of experts on the ground along with the discursive ruptures in those narratives can provide entry to an even richer history of agrarian projects that captures the experiences of non-elites. Such a focus, if anything, prepares for an understanding and appreciation of forces that facilitated the turn toward the highly mechanized model of agriculture in 1960s India that goes by the name of the "green revolution." Looking back from the era of the green revolution, one easily falls prey to a certain homogeneous view of "development" or "modernization."[52] Alternatively, one may focus on diverse technics and technopolitics of large machines imported from abroad, mechanization through improved implements manufactured and distributed locally, and the establishment of agricultural engineering as a discipline along with the technologically "heavy" and resource-intensive approach implied by Point Four policies and encouraged by the Ford Foundation. Each of these technopolitical options expressed different forms of modernity, suspended specific forms of social relations, had different ideological agendas embedded within them, and engaged in different registers with the state, state formation, and capitalism. Placing these different technological "solutions" to the "problem" of agriculture at the core of the analysis draws attention to the disparate, conflicting forces involved in the transnational experience of agricultural modernization and to the diverse modes of expression of the knowledge/power nexus. A comprehensive study of the evolution of the agrarian modern in India requires considering South Asia not only as a Cold War destination but also as a regional formation, a postcolonial archipelago, a locality, and a locale.[53] The openness to treatment of the global dimension and the post–World War II global order dominated by the United States requires at the same time a look at genealogies going back to colonial times, the local, and the multiple in India.

Notes

1. The tractor was sourced from Richard Garrett and Sons Limited, an English firm based in Suffolk that supplied agricultural machinery in the colony

through their agent, Messrs. Burn and Company, based in Calcutta. It is possible to think of this tractor as an imperial or colonial machine. But it is the fundamental foreignness of the machine that is critical to the interpretive angle developed here.

2. Elizabeth M. Whitcombe, *Agrarian Conditions in North India* (Berkeley: University of California Press, 1972). By and large, the "Canal Colonies" to the west have received more scholarly attention by historians of South Asia, who have examined the technoenvironmental context of creating irrigation canals and aspects of state control and community formation around the enlarging colonial irrigation networks. On the colonial nature of irrigation networks in the Punjab and the frontier regions, see David Gilmartin, "Irrigation and the Baloch Frontier," in *Sufis, Sultans and Feudal Orders*, ed. Mansura Haidar (New Delhi: Manohar, 2004), 331–389; David Gilmartin, "The Irrigating Public: The State and Local Management in Colonial Irrigation," in *State, Society and the Environment in South Asia*, ed. Stig Toft Madsen (London: Curzon Press, 1999), 236–265. For environmental implications, see David Gilmartin, "Models of the Hydraulic Environment: Colonial Irrigation, State Power and Community in the Indus Basin," in *Nature, Culture and Imperialism: Essays on the Environmental History of South Asia*, ed. David Arnold and Ram Guha (Delhi: Oxford University Press, 1995), 210–236. For work performed by irrigation engineers, see David Gilmartin, "Scientific Empire and Imperial Science: Colonialism and Irrigation Technology in the Indus Basin," *Journal of Asian Studies* 53, no. 4 (Nov. 1994): 1127–1149. A recent description of Chenab irrigation dams and channels appears in Daniel Haines, *Building the Empire, Building the Nation: Development, Legitimacy and Hydro-politics in Sindh, 1919–1969* (Karachi: Oxford University Press, 2013). See also David Gilmartin, *Blood and Water: The Indus River Basin in Modern History* (Berkeley: University of California Press, 2015).

3. B. C. Burt, "Steam Ploughing Experiments in the Aira Estate, Kheri, United Provinces," *Agricultural Journal of India* 9 (1914): 1–6; Sirdar Jogendra Singh, "Experiments in Steam-Ploughing," *Agricultural Journal of India* 13 (1918): 47–53.

4. Singh, "Experiments in Steam-Ploughing," 49; Burt, "Steam Ploughing Experiments in the Aira Estate."

5. For an analysis of colonial liberalism and its working in rural India, see Andrew Sartori, *Liberalism in Empire: An Alternative History* (Berkeley: University of California Press, 2014), 96–129.

6. More than three decades of South Asian scholarship has critiqued Western social thought and identified the distinct consciousness of the "subaltern" classes on the Indian subcontinent, using the blanket term to identify the historical experience of the subordinated and the dispossessed. See Rosalind O'Hanlon, "Recovering the Subject: Subaltern Studies and Histories of Resistance in Colo-

nial South Asia," *Modern Asian Studies* 22, no. 1 (1988): 189–224, quotation on 190; Gyan Prakash, "Writing Post-Orientalist Histories of the Third World: Perspectives from Indian Historiography," *Comparative Studies in Society and History* 32, no. 2 (Apr. 1990): 383–408; Gyan Prakash, "Can the 'Subaltern' Ride? A Reply to O'Hanlon and Washbrook," *Comparative Studies in Society and History* 34, no. 1 (1992): 168–184; Gyan Prakash, "Orientalism Now," *History and Theory* 34, no. 3 (Oct. 1995): 199–212. Even as concerns expressed by the subaltern scholars and postcolonialists have become mainstream in South Asian historiography, historians have embraced two distinct approaches to examine the position of the subaltern, with some arguing for the complete autonomy of subaltern consciousness, especially vis-à-vis modernization, and others speaking of a more interactionist "field" in which vectors of modernization are engaged by the subordinates. For a discussion of related theoretical and empirical issues, see Akhil Gupta, *Postcolonial Developments: Agriculture in the Making of Modern India* (Durham, NC: Duke University Press, 1998). For a history of agricultural change among the Kallar community in Tamilnadu in a strictly postcolonial mold, see Anand Pandian, *Crooked Stalks: Cultivating Virtue in South India* (Durham, NC: Duke University Press, 2009). Pandian studies agriculture not through the framings of "development" but rather through "virtues of cultivation" among the Kallar community that drew upon colonial legacies as well as Tamil conceptions of virtue, self-governance, and community life. In other words, Pandian shows that the history of the concepts and practices of "development" among the Kallar cannot simply be reduced to the internalization of external discourses of colonialism or development. See also the important work of Gyan Prakash that discusses the place of science and technology in the "modernization" agenda: Gyan Prakash, *Another Reason: Science and the Imagination of Modern India* (Princeton, NJ: Princeton University Press, 1999).

7. Sanjay Subrahmanyam, "Connected Histories: Notes Towards a Reconfiguration of Early Modern Eurasia," *Modern Asian Studies* 31, no. 3 (1997): 735–762; Sanjay Subrahmanyam, *Explorations in Connected History: From the Tagus to the Ganges* (Delhi: Oxford University Press, 2005).

8. David Ludden, "Why Area Studies?," in *Localizing Knowledge in a Globalizing World: Recasting the Area Studies Debate*, ed. Ali Mirsepassi, Amrita Basu, and Frederick Weaver (Syracuse: Syracuse University Press, 2003), 131–136.

9. Mrinalini Sinha's works have stressed the necessity of implicating not just the imperial but also the global in the study of colonial discourses. See Mrinalini Sinha, *Specters of Mother India: The Global Restructuring of an Empire* (Durham, NC: Duke University Press, 2006); Mrinalini Sinha, "Premonitions of the Past," presidential address, *Journal of Asian Studies* 74, no. 4 (Nov. 2015): 821–841. Scholars have drawn attention to the "externalities" of the colonial context. See Sugata Bose and Kris Manjapra, eds., *Cosmopolitan Thought Zones:*

South Asia and the Global Circulation of Ideas (New York: Palgrave, 2010); Sugata Bose, *A Hundred Horizons: The Indian Ocean in the Age of Global Empire* (Cambridge, MA: Harvard University Press, 2009). Recent works have emphasized the connectedness of South Asia history with the world beyond through a focus on diasporas and networks. See Pedro Machado, *Ocean of Trade: South Asian Merchants, Africa and the Indian Ocean, c. 1750–1850* (Cambridge: Cambridge University Press, 2014); Sana Aiyar, *Indians in Kenya: The Politics of Diaspora* (Cambridge, MA: Harvard University Press, 2015); Johan Mathew, *Margins of the Market: Trafficking and Capitalism across the Arabian Sea* (Berkeley: University of California Press, 2016).

10. Scholars have shown two distinct periodizations of American modernization as entirely possible. One historiographical trajectory pivots on the coalescing of ideas in the social sciences as reflected in the emergence of modernization "theory" in the late 1950s and its subsequent impact on US foreign policy. The other, alternative periodization builds on tracking the emergence of modernization as an "ideology" going back to the New Deal era. References to mid-twentieth-century "modernization" as a tool of American foreign policy appear in Michael Latham, *The Right Kind of Revolution: Modernization, Development, and U.S. Foreign Policy from the Cold War to the Present* (Ithaca, NY: Cornell University Press, 2010); David Ekbladh, *The Great American Mission: Modernization and the Construction of an American World Order* (Princeton, NJ: Princeton University Press, 2011); Daniel Immerwahr, "Modernization and Development in US Foreign Relations," *Passport* 43 (Sept. 2012): 22–25; Nils Gilman, *Mandarins of the Future: Modernization Theory in Cold War America* (Baltimore: Johns Hopkins University Press, 2003).

11. Michel Foucault, "Governmentality," in *The Foucault Effect: Studies in Governmentality*, ed. Graham Burchell, Collin Gordon, and Peter Miller (Chicago: University of Chicago Press, 1991), 87–104; David Scott, "Colonial Governmentality," *Social Text* 43 (Autumn 1995): 191–220.

12. Prakash (*Another Reason*, 159–60) uses Martin Heidegger's notion of technology as a device for enframing all beings as resources.

13. Partha Chatterjee, *The Politics of the Governed: Reflections on Popular Politics in Most of the World* (New York: Columbia University Press, 2006), 34; also see his "Two Poets and Death: On Civil and Political Society in the Non-Christian World," in *Questions of Modernity*, ed. Timothy Mitchell (Minneapolis: University of Minnesota Press, 2000), 35–48.

14. Singh, "Experiments in Steam-Ploughing," 50.

15. This is precisely the point made by postcolonial historians—that is, that the projects of modernity did not and could not connect with subjectivities and interests of the subaltern class.

16. H. C. Young and B. C. Burt, "Experiments with a Light Motor Tractor in the Oel Estate, Kheri," *Agricultural Journal of India* 9 (1920): 375–380.

17. S. Milligan, "Reaping Machines for Wheat in the Punjab," *Agricultural Journal of India* 3 (1908): 327–332.

18. R. Shearer, "Steam Threshing in India," *Agricultural Journal of India* 2 (1907): 246–251.

19. B. C. Burt, "Some Experiments with Steam Threshing Machinery at Cawnpore," *Agricultural Journal of India* 8 (1913): 346–354.

20. Burt, "Steam Ploughing Experiments in the Aira Estate"; H. R. Stewart and D. P. Johnston, "Tractor Cultivation at Lyallpur, Punjab," *Agricultural Journal of India* 18 (1923): 23–39, at 23. For one recent argument about stagnation in agriculture in the years leading up to World War I, see Tirthankar Roy, "Roots of Interwar Crisis in Interwar India: Retrieving a Narrative," *Economic and Political Weekly* 41, no. 52 (Dec. 30, 2006–Jan. 5, 2007): 5389–5400.

21. Sam Higginbottom, *The Gospel and the Plow* (New York: Macmillan, 1921), 61–62.

22. D. P. Johnston, "Tractor Implements Tried at the Lyallpur Experimental Farm," *Agricultural Journal of India* 25 (1930): 317–320.

23. Higginbottom, *The Gospel and the Plow*, 124; *Sam Higginbottom, Farmer: An Autobiography* (New York: Charles Scribner's Sons, 1949), 102. The most detailed account of Higginbottom's contribution to the establishment and expansion of the Allahabad Agricultural Institute appears in Gary Hess, *Sam Higginbottom of Allahabad: The Pioneer of Point Four to India* (Charlottesville: University of Virginia Press, 1967).

24. Hess, *Sam Higginbottom of Allahabad*, 62–65.

25. Higginbottom, *The Gospel and the Plow*, 13–14, 31, 51.

26. Mason Vaugh, "An Agricultural Engineer as a Missionary," record 3130, undated, typed, the State Historical Society of Missouri (hereafter MHS), Columbia.

27. Mason Vaugh, "Recent Activities and Interests," record 3130, undated, typed, MHS.

28. The use of "need" as a trope for defining scientific and technological practices is influenced by Dana Simmons's work on French history. See Dana Simmons, *Vital Minimum: Need, Science, and Politics in Modern France* (Chicago: University of Chicago Press, 2015).

29. Mason Vaugh, "A New Experimental Farm," *Allahabad Farmer* 19, no. 3 (May 1945): 1–6, quotations on 1, 6.

30. Mason Vaugh, "Mechanization for the Small Farmer," speech given at the Eighth Pacific Science Conference, Manila, Philippines, Nov. 14–28, 1953, C 2639, folder 140, Mason Vaugh Papers, MHS.

31. Mason Vaugh, "The Co-operative Farm: Is It the Solution of India's Agricultural Problem?," *Allahabad Farmer* 20, no. 3 (May 1946): 1–4.

32. This notion of "progressive" farmer should not be elided with the notions of rural America in the Progressive Era. If anything, the Country Life movement, a Progressive Era reform effort, viewed rural people in very limiting ways,

and mostly as disadvantaged and backward. But the notion of "progressive" farmer remained alive within the developing "agricultural establishment" of the land-grant colleges within their research and extension arm and in the US Department of Agriculture. I owe this insight to conversations with Sally McMurry of Pennsylvania State University.

33. Vaugh, "The Co-operative Farm," 3.

34. Land-grant institutions were set up under the Morrill acts (1862 and 1890) and were where the "cooperative extension program," launched first by the Smith-Lever act of 1914, later developed. The so-called extension program matured with the support of the collaborative institutional ties between the land-grant colleges and the US Department of Agriculture in the first quarter of the twentieth century.

35. Michael Adas, *Dominance by Design: Technological Imperatives and America's Civilizing Mission* (Cambridge, MA: MIT Press, 2006); Nick Cullather, *The Hungry World: America's Cold War Battle against Poverty in Asia* (Cambridge, MA: Harvard University Press, 2011).

36. For the early beginnings and initial shape of the Point Four program, see Stephen Macekura, "The Point Four Program and the U.S. International Development Policy," *Political Science Quarterly* 128, no. 1 (Spring 2013): 127–160.

37. RG 286, Records of the Agency for International Development and Predecessor Agencies, India Branch, 1951–54, box 1, multiple folders, National Archives and Records Administration, College Park, MD.

38. For the analysis of ideological convergence between the State Department and foundations such as the Rockefeller and Ford Foundations, see Inderjeet Parmar, *Foundations of the American Century: The Ford, Carnegie, and Rockefeller Foundations in the Rise of American Power* (New York: Columbia University Press, 2015). For the Ford Foundation's work in India, see "Foundation in the Field: The Ford Foundation New Delhi Office and the Construction of Development Knowledge, 1951–1970," in *American Foundations and the Coproduction of World Order in the Twentieth Century*, ed. John Krige and Helke Rausch (Göttingen: Vandenhoeck und Ruprecht, 2012), 232–260.

39. "Programmes in India Receiving Assistance from the Ford Foundation," "Foundation Activities in India," folder "US Projects in India, 1952–53," Special Collections, University of Illinois at Urbana-Champaign.

40. P. C. Joshi, "Nehru Mortgages India to America," *India Today*, Jan. 1952, 22–28, at 26, available at the P. C. Joshi Archives in New Delhi.

41. O. P. Sangal, "Our Community Projects," *India Today*, June 1952, 13–18, available at the P. C. Joshi Archives in New Delhi.

42. J. C. Kumarappa, "Community Projects," *India Today*, Sept. 1952, 14, 22, available at the P. C. Joshi Archives in New Delhi.

43. Kristin Ahlberg, *Transplanting the Great Society: Lyndon Johnson and Food for Peace* (Columbia: University of Missouri Press, 2008), 106–146.

44. Letter from M. G. Kamath, editor, *Indian Farming*, to Mason Vaugh, Jan. 2, 1958; copy of letter addressed to the editor, Dec. 24, 1957; letter from Mason Vaugh to M. G. Kamath, undated; all in folder 16, 3130, Mason Vaugh Papers, MHS.

45. Letter from Vaugh to M. G. Kamath, undated but probably following the Jan. 1958 letter by Kamath to him, folder 16, 3130, Mason Vaugh Papers, MHS.

46. Ford Foundation, *Report on India's Food Crisis and Steps to Meet It* (New Delhi: Government of India, Ministry of Food and Agriculture, 1959).

47. Letter from Vaugh to G. Wallace Giles, July 25, 1960, folder 10, 3130, Mason Vaugh Papers, MHS.

48. Ford Foundation letter from New Delhi office to Vaugh, Mar. 24, 1962, folder 10, 3130, Mason Vaugh Papers, MHS.

49. Letter from Vaugh to James A. Bassalman, May 13, 1964, folder 5, 3130, Mason Vaugh Papers, MHS. Vaugh's distance from Giles, especially given Vaugh's conviction of the utility of small-farm implements and animal power for the farm, is understandable. Giles was a strong supporter of heavy-machinery farm mechanization, and he authored a report in 1967 that was consistently cited by those who favored farm mechanization in many parts of the world. G. W. Giles, "Agricultural Power and Equipment," in *The World Food Problems*, vol. 3, *A Report of the President's Advisory Committee* (Washington, DC: Superintendent of Documents, Government Printing Office, 1967), 175–216.

50. Timothy Mitchell, *Rule of Experts: Egypt, Techno-politics, Modernity* (Berkeley: University of California Press, 2002), 233; Partha Chatterjee, *The Nation and Its Fragments* (Princeton, NJ: Princeton University Press, 1999), 200–219.

51. Francine Frankel, *India's Political Economy, 1947–2004: The Gradual Revolution* (Princeton, NJ: Princeton University Press, 1978; repr., New Delhi: Oxford University Press, 2005),; Francine Frankel, *India's Green Revolution: Economic Gains and Political Costs* (Princeton, NJ: Princeton University Press, 1971); Atul Kohli, *The State and Poverty in India: The Politics of Reform* (Cambridge: Cambridge University Press, 1987); Ashutosh Varshney, "Ideas, Interest and Institutions in Policy Change: Transformation of India's Agricultural Strategy in the Mid-1960s," *Policy Sciences* 22, nos. 3/4 (1989): 289–323.

52. Daniel Immerwahr, "Modernization and Development in U.S. Foreign Relations," *Passport* 43 (Sept. 2012): 24.

53. Arjun Appadurai, "The Production of Locality," in *Counterwork*, ed. R. Fardon (London: Routledge, 1995), 204–225.

CHAPTER FIVE

Transnational Knowledge, American Hegemony

Social Scientists in US-Occupied Japan

Miriam Kingsberg Kadia

After 1945 the United States solidified its status as a global hegemon in part through the "soft power" strategy of underwriting the intellectual reconstruction of its aligned nations. Recent scholarship has examined American support for the natural sciences and engineering in both devastated allies and erstwhile enemies. The result was a transnational network of knowledge production centered on the United States and supporting US geopolitical aspirations.[1] Less well studied but no less important is the contribution of social science. On the eve of the Cold War, American social scientists used research to advance "progress" according to the ideals of the then-popular ideology of modernization. Modernization presented the putatively American values of democracy, capitalism, and peace as a universal endpoint. Modernizers, in other words, believed that all societies were capable of progressing toward peaceful, liberal democratic capitalism. During the Cold War, they extended US assistance toward this end to recruit nations to the American fold in opposition to the competing allure of communism offered by the Soviet Union.

Japan, the site of America's longest peaceful postwar occupation (August 1945 to April 1952), was regarded as both a test case and a showcase of modernization theory.[2] As such, the nation poses a particularly fruitful lens for examining the geopolitical significance of social science in the sphere of US hegemony. Western-style social science reached Japan in the late nineteenth century and was grafted on to local intellectual traditions. By the early twentieth century, the nation had established itself within the burgeoning community of knowledge-producing states. However, the outbreak of World War II isolated Japanese scholars. The defeat of Japan in 1945 offered the United States an opportunity to reshape Japanese research practices according to American cultural values and to restructure the transnational intellectual network of the prewar years into a US-dominated entity that served national political ambitions in the Cold War era.

Understanding the importance of face-to-face interactions in accomplishing these goals, the American Occupation bureaucracy dispatched scores of scholars to Japan. Social scientists did not come, as they so often did among Native Americans and colonized peoples, to study "primitivity" or Otherness. Instead, as both Japanese and American scholars later claimed, their relationships often exemplified the teacher-student bond used as a largely positive metaphor for the Occupation itself. Through texts, lectures, and, most important, collaborative fieldwork, American social scientists modeled and promoted the ideals of modernization. Meanwhile, their Japanese counterparts capitalized on their unprecedented position and influence over the government and public to enshrine these values within an instrumental national identity as a US ally.

In contrast to Europe, to which the United States acknowledged a cultural similarity and historical debt that implicitly obligated the nation to restore local stability and prosperity, most early postwar Americans felt little sense of identification with Japan. On the contrary, World War II represented the culmination of decades of anti-Japanese sentiment. Virulent hate literature abounded, depicting the Japanese as a pathologically and incorrigibly inferior race. Racism was also at the heart of the US government policy of interning Japanese immigrants and their descendants, including many American citizens.[3] While the end of the war blunted the most visceral aversion, postwar American sentiments continued to reflect a sense of superiority and paternalism toward the former enemy, famously compared to a boy of twelve by Occupation chief General Douglas MacArthur.[4] These attitudes did not disappear in the years after 1945 and continue to mark the American stance toward Japan to some degree even to this day. The postwar embrace of mutual values could not override all existing prejudices. Nonetheless, in the establishment of the new order, faith in modernization helped to supplant the absence of a shared past by enabling the imagination of a collective future. By building a common circuit of knowledge production cohered by shared values, American and Japanese scholars jointly reimagined the world under US hegemony.

New Knowledge, Old Producers

Although Japan had a long and distinguished scholastic tradition, it was not until the mid-nineteenth century that Western-style social science and its accompanying institutions, including universities, museums, and research organizations, became entrenched in the national landscape. The earliest generation of Japanese social scientists largely mastered the disciplines through study abroad or from European or American tutors at home. As in much of the world, German social theories and methods were particularly influential. Within only a few decades, Japanese scholars had proven themselves capable not only of mastering but also of adding to social science. During the interwar years, the rate of Japanese participation in international conferences and publication in foreign languages soared.[5]

As the first non-Western power to gain recognition as a producer of legitimate, original knowledge, Japan gave the international intellectual community some status as a transnational rather than merely Euro-American entity. All too soon, however, the outbreak of war in the 1930s introduced new political, ideological, and even physical boundaries to the participation of Japanese scholars. Rather than continuing to cultivate intellectual linkages with the great powers, Japan came to focus on the development of academic networks in its burgeoning empire. In pursuit of professional status and resources, researchers sought evidence of the superiority of the Japanese people and their consequent right, even obligation, to impose the rule of the allegedly divine emperor over the inferior but confraternal peoples of Asia and Oceania. While surveying the bodies and behaviors of local peoples, fieldworkers provided information intended to facilitate the exploitation of human and natural resources, the pacification and administration of conquered territories, and the assimilation of populations within the Japanese Empire.[6] To maximize safety and efficiency in hostile and remote territories, they often ventured to the field in groups. Ranging from a handful to hundreds of participants, these team expeditions built a sense of professional solidarity among social scientists while spreading complicity with imperialism and war through the intellectual ranks.

Beginning in 1940, Japan's invasion of Southeast Asia set in motion a chain of events that ultimately prompted the United States to enter World War II. Against the Axis enemy, the Allies rallied under the banner of representative government, the free market, and peace. As the then-popular ideology of modernization taught, these values were the endpoint of development and universally accessible to all societies (with US tutelage). The looming victory of the Allies appeared to be "objective" confirmation of this belief.

Traditionally, the transnational community of social scientists upheld objectivity, thought of as universal "truth" free from proclivity or bias, as the defining value of legitimate scholarship. As historians have shown, however, in practice objectivity has often functioned as a rhetoric of legitimacy for various ideological positions.[7] Shocked and horrified by the devastation and atrocities of World War II, Allied social scientists asserted a paramount responsibility for creating knowledge that would not simply describe the human condition but also advance the modernization telos. "This is apparently the first time in world history when the people of many lands have officially turned to the social scientist to seek his aid in man's quest for enduring peace," enthused a multinational group of prominent scholars.[8] By the end of the war, the practice of objective research was identified with the pursuit of democracy, capitalism, and peace.

With the goal of understanding the enemy and preparing for peacetime reconstruction, American social scientists intensified their study of Japanese culture and society. The most influential research on this topic was renowned anthropologist Ruth Benedict's monograph *The Chrysanthemum and the Sword* (1946). Beginning her research on Japan in mid-1944 but writing largely after the war had ended, Benedict sought to set forth an understanding of Japan that would facilitate the transition to peace and promote a collaborative rather than punitive occupation. Benedict's work followed in the tradition of "national character" studies, which anthropomorphized and homogenized nations as individuals defined by personality traits rooted in cultural indoctrination, particularly during early childhood. Many of Benedict's colleagues viewed the Japanese national character as pathologically deviant, defined by aggressiveness, group-mindedness, authoritarianism, rigidity, and fear of dishonor. They viewed these predispositions as an explanation for the protracted, doomed struggle and atrocities of World War II.[9] Benedict, by contrast, rejected the idea of an incorrigible, abnormal national character. Instead, she attributed the nation's wartime course to a small militarist coterie that had taken the Japanese down a "wrong path." Removing power from authoritarian leaders, she implied, would free mass society to transition to US-style democracy. *Chrysanthemum*, American anthropologist Clifford Geertz later remarked, represented the Japanese as "the most reasonable enemy we have ever conquered."[10]

Studied by nearly all Occupation personnel in preparation for service overseas, *Chrysanthemum* lent the credibility of professional social science to an instrumental conclusion: that Japan might be quickly and effectively reconstituted as an American ally in East Asia and the Pacific Rim. US policy-makers were eager to use Japan (like West Germany) to counter the growing threat of Soviet power. The prewar emperor system had to be discarded as quickly as possible to clear the way for a new ideological orientation toward democracy, capitalism, and peace. Rather than painstakingly sifting through the general population for proponents of militarism, fascism, and imperialism, Benedict suggested that the United States might simply charge responsibility for war crimes and crimes against humanity "to specific individuals and institutions . . . identified and isolated from the mainstream."[11] Ultimately, tribunals of the late 1940s purged some two hundred thousand individuals, representing a mere 0.29 percent of the total population. (By contrast, some 2.5 percent of Germans under American occupation were legally excluded from public life.)[12] Victims in Japanese academia numbered fewer than one hundred, or about 0.3 percent of active professors.[13] Evidence suggests that the Occupation was more concerned with rooting out suspected communists than with prosecuting former advocates of empire and war.[14]

The American academic establishment endorsed this whitewashing of the past activities of Japan's intellectuals, faulting the irresistible domination of the military and bureaucracy for "imposing destabilizing restraints" that reduced the social scientist "to the level of a special pleader and propagandist."[15] Given their own contributions to the war effort—for example, by one estimate up to three-quarters of professional anthropologists in the United States worked at least part-time on applied research in the early 1940s—American scholars were inclined to understand their Japanese counterparts as "no more than normally patriotic for a period of nationalism."[16] Moreover, the consequences of social science research in the Japanese Empire were borne mostly by colonial subjects—a population that the United States and its allies largely overlooked in the postwar pursuit of justice.[17]

Japanese scholars embraced this justification of their wartime record. Exoneration incentivized them to prove the Occupation narrative of wartime oppression true by showing themselves as enthusiastic proponents of democracy, capitalism, and peace. With the defeat and consequent discrediting of the Japanese state and military as arbiters of national identity, they asserted unprecedented influence over public life. Nanbara Shigeru, president of the University of Tokyo (Japan's leading institution of higher learning), set the tone for this stance in a November 1945 speech declaring scholars' "especial obligation for rebuilding the nation . . . on a new foundation of truth and freedom."[18] Nanbara's successor, the distinguished economist Yanaihara Tadao, likewise emphasized social scientists' responsibility to cultivate the vaunted values of the United States: "If we are hereafter to make our knowledge really an active force, it is of primary importance that . . . knowledge be spread widely and freely among the people and thus intensive interest in peace be aroused in every aspect of their lives. In practice, we scientists can achieve something only when we place our trust in the people and walk in step with them."[19] In this way, Japanese social scientists sought to rebuild academia, fashion a positive national identity, and join the intellectual community associated with the geopolitical hegemony of the United States.[20]

Establishing the Pillars of Postwar Research in Japan

To American observers of the Occupation era, Japanese social scientists appeared "an extremely interesting group," "all very bright and effective people." The war, however, had cut them off from the transnational intellectual community, resulting in a "period of isolation which for Japanese scholars and scientists was as strict as that which had preceded the opening of Japan in the last century [following over 250 years of self-imposed withdrawal from contact with most foreign societies]."[21] Moreover, one American scholar observed, "What has been decidedly unfortunate is an overlong persistence of the influence of German social science."[22] Although American scholars themselves owed an incalculable debt to Germanic theories and methods, Germany's status as a defeated nation and the horrific misuse of science by the Nazis decisively disgraced its intellectual legacy. In consultation with over eighty Japanese academics, Harvard University anthropology professor Clyde Kluckhohn (who had studied in Austria during the 1930s) charged that Germanic logic, philosophy, and ideas of law and the state had molded prewar Japanese research into "a means of promoting autocracy within and aggression without," "not something based on free inquiry resulting in universal good."[23]

The Supreme Commander for the Allied Powers, or SCAP (an acronym typically used for the entire Occupation bureaucracy), sought to replace the "un-democratic" and "fascist" mentalities attributed to the former Axis powers with American ideals of democracy, capitalism, and peace. These values were the basis of a reconstituted transnational network of knowledge production. In contrast to the prewar intellectual community, which was dominated by Europeans, this network was to center on the United States and support the maintenance and extension of American hegemony against the much vaunted threat posed by the Soviet Union and its allies.[24]

The Constitution of Japan, drafted in 1946 and imposed in 1947 by SCAP, laid the foundations of Japanese participation in this network by guaranteeing academic freedom as well as freedom of thought and conscience.[25] The Fulbright Scholars and Government Aid and Relief in Occupied Areas programs sent a handful of Japanese social scientists abroad to train in US institutions. However, given limited funding and restrictions on travel for Japanese citizens during the Occupation, most training took place locally under the supervision of the Civil Information and Education Unit (CIE), established by SCAP in September 1945. CIE staff included numerous proficient speakers of Japanese who had lived in the country (often in missionary families) or received training at military language schools during World War II. Recruits ranged from renowned scholars at elite institutions to untested ABDs (all-but-dissertation graduate students) in search of professional opportunity and adventure.[26] David L. Sills, a sociology student at Yale University who joined the CIE in August 1947, described himself bluntly as "a pure mercenary," recalling, "I came to the occupation of Japan to make money so I could pursue my graduate work."[27]

As a starting point for their attempt to restructure knowledge production, CIE staff sought to build a library network through which Japanese colleagues might enjoy access to foreign scholarship. From the perspective of American policy-makers, the library was "a potent engine of democracy . . . mak[ing] available to all what would otherwise be reserved to the few." Although Japan had maintained a modern library system since the late nineteenth century, the war had interrupted foreign acquisitions and domestic publishing. By one estimate, firebombing claimed half of Japan's book resources in the early 1940s, leaving no more than five million volumes in the entire nation at the time of defeat.[28] The scarcity of recent literature was particularly acute. A concerned American scholar wrote, "I gather . . . that [Japanese social scientists] subscribe to few, if any, journals and that the students and faculty do not therefore have access to the many crucially important articles and monographs that have been published in the past."[29] Shortage was opportunity: "The field is open to the far-sighted nation that restocks the sources of supply for Japan's book-reading public," predicted one anthropologist.[30] Wrote another, "the U.S. task is to insure that an adequate quantity and a wide variety of information on democracy are made available to Japan's information-starved intelligentsia and highly literate masses. Only such information can provide the background needed for forming attitudes favorable to a democratic order."[31] Implicit in this exhortation was the fear that the Soviet Union might flood Japan with propaganda and win the hearts and minds of the people to communism.

Responding to geopolitical pressures, one CIE employee entreated a prospective colleague to "bring everything you can get your hands on which deals with public opinion, social psychology, social research, methods, etc. Write for permission pronto to get 300 pounds extra hold baggage, for books and papers. . . . We are especially desirous of monographs on actual research projects, as well as text and instructional material."[32] He and others petitioned their home institutions for surplus copies of important recent monographs and journals.[33] Additional donations poured in from charitable and scholarly foundations and societies, government agencies, publishers, and concerned citizens. By the midpoint of the Occupation, nearly 1.25 million English-language books had reached Japan.[34]

Distribution bottlenecks trapped some donations in warehouses for up to a year while the CIE worked to create a network of libraries throughout the Japanese archipelago. The flagship facility in downtown Tokyo housed some thirteen thousand books and five hundred periodicals. In the reading room, postwar leaders of Japanese social science studied the works of Benedict and others for the first time. Ultimately, twenty-three CIE libraries came to offer access to not just texts but also lectures, concerts, discussion groups, English language classes, documentary film screenings, and exhibits to as many as two million patrons annually.[35] SCAP also introduced legislation to create a public library network by expanding and rebuilding existing facilities and resources, implementing modern cataloging methods, and abolishing fees for patrons. By the end of the 1950s nearly every prefecture and over half of Japan's cities, as well as some towns and villages, operated public libraries.[36]

Beyond bringing English-language books to Japan, the CIE also supported Japanese translations of selected social science works, including Ruth Benedict's *The Chrysanthemum and the Sword*. Much to the disgust of American officials who saw no value in popular fare, translators also rendered manuals on baseball and housekeeping, Margaret Mitchell's 1936 bestseller *Gone with the Wind*, and children's literature. The program expanded quickly as SCAP rushed to counteract the perceived threat of communism embodied by a spate of Japanese-language editions of Soviet Russian works. By the midpoint of the Occupation, the CIE had sponsored some 150 translations and licensed 200 others.[37] By making available the classics of US civilization for Japanese consumption, SCAP sought to instill putatively American mentalities of democracy, capitalism, and international cooperation.

Public lectures offered an alternative mode of transmitting these values. Speakers not only conveyed information but also built personal relationships with audience members. Reflected one CIE anthropologist, "I do my best job around here in communicating new ideas to our Japanese, in showing our younger people how to organize a project, in introducing American methods and knowledge to them. . . . I teach all the time. I'd rather do it than anything else."[38] A thank-you note from a Japanese sociologist read: "We have learned so much from your lectures delivered from quite a different angle than ours. . . . In the near future, I hope, we will show you better sociology and contribute more to the social science of the world."[39] Through face-to-face interactions, American scholars recruited Japanese colleagues as partners in the establishment of a transnational knowledge network that supported the hegemony of the United States.

In 1950 a more systematic training program was inaugurated in the form of the American Studies Seminar (Amerika kenkyū seminā), jointly hosted by the University of Tokyo and Stanford University and funded by the Rockefeller Foundation. In the early postwar era, private foundations such as Rockefeller, Ford, and Carnegie provided critical support for the development of a global intellectual network friendly to US geopolitical ambitions.[40] Modeled on a similar endeavor in Salzburg, Austria, the American Studies Seminar aspired to "imbu[e] the defeated nation of Japan with the spirit of American democracy and . . . promot[e] intellectual, scholarly exchange between the United States and Japan."[41] As one proponent exhorted, "Democratic institutions exemplified in American life should become better known in Japan, and training in the history of American traditions should become part of the normal university curriculum in the new age. Successive generations of Japanese students should be encouraged to study American affairs so that they may carry to their leadership in public life a better comprehension of our country."[42] By teaching an interdisciplinary social science program according to American methods, the seminar sought to diffuse US values among Japanese intellectuals.

Over the course of five weeks in the summer of 1950, five well-known American senior professors, each representing a different discipline, lectured and held small roundtables for two hours each weekday afternoon for a total of nearly 125 Japanese participants ranging in age from twenty-three to fifty-four.[43] Owing to the varied levels of English-language proficiency among participants, seminar leaders relied on name cards, predistributed outlines, and nearly two dozen interpreters to facilitate communication. These teaching aids helped to achieve the facilitators' goal of free and uninhibited intellectual exchange, viewed as the essence of American-style democracy. As one applauded, "The give and take of the seminar method was established during the first week. The quality of the discussion was high, and absolute frankness between Japanese and Americans was achieved. The reputation of the seminars was well established among academic circles in Tokyo before the end of the first week."[44]

Adjudged "an outstanding success despite the unrelenting heat and the long sessions," the program was repeated annually through 1956, four years after the termination of the Occupation.[45] The seminar ultimately reached nearly six hundred professors and students (both graduate and undergraduate). Moreover, beginning in 1952 the University of Kyoto and Dōshisha University (a historically Christian college) inaugurated a parallel Kyoto American Studies Seminar that ran every summer (with the exception of 1953) through 1976. Heavy and rising competition for access ensured highly qualified and motivated classes. Participants were selected from throughout the archipelago in the hope that they would bring knowledge back to their home prefectures. Beyond the classroom, students and professors met during office hours, field trips, cultural events, and publicity opportunities with the national media.[46] The seminar created a library of assigned readings; by 1953 it included over one thousand books.[47] It also spawned a fellowship program that brought two Japanese scholars to the United States annually for a year of study and a public lecture series on topics such as "Japanese acceptance of and resistance to American democracy" and "appraisals of American influence on thought, religion, art, and way of life upon the Japanese."[48]

Given the "depressingly small" compensation and relative lack of amenities (one professor was advised to bring his own refrigerator), most visiting Americans were motivated by volunteer spirit.[49] One reflected, "My stay in Japan has been one of the happiest periods in my life. I know that I have *received* in abundance; if I have *given* something in return, if I have made some slight contribution to the thinking and teaching of my Japanese colleagues, I shall be satisfied" (emphasis in original).[50] Others lauded the surprising intellectual benefits they themselves derived from the program: "We are convinced that it is valuable for American scholars to confront the Japanese interpretations of American traditions and culture. Although we frequently found ourselves in disagreement with these interpretations, we ourselves derived many penetrating insights from the Japanese perspective. . . . These served as a constant stimulus for the discussion and reconsideration of our assumptions."[51] This generally humble attitude of American facilitators made a favorable impression on Japanese participants. Inaugurating the fourth seminar in 1953, Yanaihara Tadao captured the collaborative mood: "The American professors are our guests and at the same time they are our colleagues. They did not come here to make American propaganda nor did they come to diagnose Japanese feeling toward Americans. We stand on the equal ground of academic learning and are colleagues striving toward the common goal in search of scientific truth."[52] Japanese participants expressed their appreciation in similar terms. One wrote to his facilitator in gratitude, "I thank you heartily for your coming again to Japan to enlighten us young (spiritually) lovers of wisdom. Your zeal for education touches me deeply."[53] Through the reconstruction of social science, American scholars recruited their Japanese counterparts not as subordinates but as colleagues and partners in the entrenchment of shared values and the establishment of a transnational knowledge network that supported the hegemony of the United States.

Realizing American Ideals through Fieldwork in Japan

To a greater degree than training in the library and classroom, fieldwork came to symbolize and advance Japan's transformation into a peaceful, capitalist, democratic society. Through the collection of empirical data, Japanese and American social scientists sought "objectively ascertained facts" as the basis for democratic and inclusive policy-making. Meanwhile, relations among researchers modeled the collaborative, egalitarian spirit they hoped to cultivate in society at large.

By the time of the Occupation, Japanese social scientists had a long tradition of field research both at home and in the empire.[54] From the outset of the Occupation, interest in continuing and improving upon field practice was apparent to SCAP. In 1947 one survey reported that Japanese scholars throughout the archipelago were "exceedingly eager for field work" and that "many college administrators pay at least lip service to the idea of empirical social research."[55] One revealing indication was the proliferation of translations of works by Bronisław Malinowski (1884–1942), the Polish-born British social anthropologist often represented as the architect of methodological guidelines for "objective" fieldwork. Texts translated during the early postwar years included Malinowski's *Crime and Custom in Savage Society* (orig. 1926), *The Sexual Life of Savages in North-Western Melanesia* (orig. 1929), and *A Scientific Theory of Culture* (orig. 1944).[56]

Prior to the Occupation the sole English-language academic field study of Japan was *Suye Mura*, a 1936 village ethnography by University of Chicago sociologist John F. Embree. "Every anthropologist who went to Japan in the 1950s knew Embree's book well," recalled one CIE employee.[57] Writing a decade after Embree, Ruth Benedict incorporated some of his conclusions into *Chrysanthemum* but was prevented by the war from visiting Japan personally.[58] Instead, she worked according to a method known as "research at a distance." With the help of a second-generation Japanese American informant, Benedict interviewed and conducted psychological tests on interned Japanese emigrants and their descendants and analyzed Japanese texts, images, and films. To many

American anthropologists who read *Chrysanthemum* in preparation for service in occupied Japan, fieldwork represented an opportunity to confirm her findings in the field.

SCAP's Public Opinion and Sociological Research Division (PO&SR) coordinated the first field studies of the Occupation era. The PO&SR was created in early 1946 as a subcommittee of the CIE to train Japanese social scientists in American theories and methods and to supply research on the national mood. To SCAP policy-makers, "The democratic atmosphere created by the Occupation has resulted in a widespread feeling among both government officials and the people at large that knowledge of public opinion is important for democratic government."[59] Public opinion research, therefore, emerged as both an agent and a result of popular participation in politics.

Herbert Passin (1916–2003), an ABD in sociology from the University of Chicago, served as the deputy director of the PO&SR. Passin was an experienced survey researcher who developed an interest in Japan through his work with former inmates of internment camps for Japanese and Japanese Americans. He was fluent in written and spoken Japanese, having studied at the Army Language School at the University of Michigan during the early 1940s. Following the departure of the PO&SR's original director, Passin recruited his former classmate John W. Bennett (1915–2005), an assistant professor of anthropology at Ohio State University, for the position. The PO&SR also hired a handful of other American researchers, including Japanese American veterans of studies of wartime internment camps in the US West.[60] They were outnumbered by more than a dozen Japanese social scientists (as well as over thirty temporary and secretarial employees).

On financial grounds alone, employment at the PO&SR was highly desirable. In the desperate years of the late 1940s, academic positions in Japanese universities provided little economic stability. Advisers to SCAP described the "terrible fight" professors faced to "keep themselves alive":[61] "Totally inadequate university salaries do not give the individual scholars even a minimum living wage, with the result that time which would otherwise be devoted to research is instead devoted to supplementing the family income through repeating lectures in other universities and schools, through hack writing, and through other activities even further removed from research and the scholarly life."[62] By providing work, Bennett concluded, "the Division saved the professional lives of a number of Japanese sociologists, anthropologists and social psychologists."[63]

The relative seniority and reputation of many Japanese employees, compared with the youth and inexperience of Bennett, Passin, and their American coworkers, helped to forestall anticipated hierarchies of victor and vanquished. Bennett described his Japanese colleagues as "the top ranking social scientists of the country, fully comparable in skill and intelligence to the best in the States—better in fact." He wrote to his wife, "It is a strange feeling . . . to have around one's desk the minister of communications in the Jap[anese] gov[ernment], the chairman of the sociology dept. at the largest university, and the top social psychologist in Japan, all bowing and honoring me!"[64] Bennett's respect for the knowledge and experience of Japanese scholars, coupled with reciprocal Japanese interest in US methodologies and humility toward the victorious Allies, generally facilitated productive working relationships.

Among the PO&SR's earliest and most influential studies was an assessment of the impact of land reform, carried out in 1947–1948. One year earlier, the Occupation had mandated the breakup and redistribution of large estates, seeking to "replace traditional agrarian feudalism with a democratic way of life" by creating a nation of independent yeoman farmers.[65] To evaluate the resulting social and economic changes in villages, SCAP called upon Arthur F. Raper, a renowned sociologist at the US Department of Agriculture. Working with the PO&SR, Raper selected thirteen allegedly representative, geographically dispersed communities for study. When published, Raper's report was hailed as a follow-up to Embree's classic study and applauded as "a completely unbiased, uninfluenced account" of Japanese village life.[66]

Raper worked with four American and fifteen Japanese social scientists at the PO&SR over the course of three stints of fieldwork totaling nearly seven months between 1947 and 1949. His group expedition methodology set the tone for early postwar research. For Japanese scholars, teamwork was a familiar practice from the age of empire, when the dangers and expense of in situ investigation necessitated collaboration. Although independent fieldwork was the rule in American academia, US social scientists, too, had come to view cooperation favorably. No less a spokesperson than Margaret Mead extolled the personal and practical benefits of intellectual complementarity.[67] In addition to these advantages, Raper's decision to work with a team reflected conditions particular to occupied Japan. Whereas American scholars initially expected to train Japanese colleagues, the latter soon emerged as a critical source of expertise. Passin, a key contributor to the study, recalled, "When I started on this research, I drew upon my recent sociological research in southern Illinois during my graduate student days, my knowledge of black sharecroppers in the American South, my experience with Mexican peasants, and my general reading in the fields of anthropology and rural sociology. . . . But I did not even have a vocabulary to describe the new phenomena that came to my attention."[68]

Beyond language and cultural barriers, American researchers also confronted the distrust of their informants. Many rural communities, associating public opinion research with the wartime military police, mistrusted the foreigners in their midst. To alleviate suspicion, Japanese social scientists took the lead in the field, arranging for village officials to distribute questionnaires and conducting intensive interviews with local informants. Ultimately, they gathered up to 95 percent of the PO&SR data.[69]

In addition to serving as the locus of knowledge production, the field was also a space for direct tutelage in democracy, including the sharing of opinions, the expression of dissent, and the cultivation of consensus. Passin described his typical on-site routine: "At the end of each day of interviewing . . . we then sat around in a group and discussed the interviewing problems, the meaning of the results, and compared local results with those obtained in the Tokyo phase of the study. Suggestions for the recording of further verbatim materials, election and political records were outlined."[70] A Japanese social scientist later recalled the excitement of debating survey techniques for up to two nights straight.[71]

Collaboration in the field drew Japanese and American social scientists together in lasting personal bonds. Raper recalled his team positively, though not without reference to certain stereotypes of national character: "I was tremendously impressed with the capability of the people. They were very thoroughly regimented. I came back very convinced that if our civilization turned on learning calculus and theirs turned on learning calculus, they would survive and we wouldn't—because if they needed to learn calculus, they'd all learn calculus in one year, because they have it fixed up so they could operate in that kind of fashion."[72] Meanwhile, Japanese social scientists appreciated the hands-on training they received under Raper, though they chafed against the demand for speed. The second stint in the field was particularly rushed. Traveling by train and jeep, testing the goodwill of local officials, and working "almost without rest," researchers visited five villages in a mere forty-five days, returned to Tokyo for twenty-four hours, and then departed for the next six sites. Factoring in travel time, they spent one or at most two nights in each location. Concerning the challenge "to do a month's research in a day," one recalled, "there was a lot of complaining."[73]

Following the conclusion of the land reform study, the PO&SR undertook field and public opinion research on such topics as traditional fishery rights, neighborhood associations, family and household composition, the labor boss system (*oyabun-kobun*), problems of urban workers and consumers, the changing status of women, the reform of the zaibatsu, and literacy and language education.[74] Highlighting the advance of democracy, capitalism, and cooperation, such research established a convergence of values between postwar Japan and the United States.

Legacies

Their ambitions notwithstanding, Bennett and his coworkers ultimately failed to exert much impact on SCAP policy. In part, the understaffing of the PO&SR was to blame: one employee observed that twenty to thirty social scientists would have been needed to adequately discharge the workload assigned to two or three.[75] More cripplingly, SCAP paid no more than lip service to the importance of research, except through what the PO&SR regarded as obstructionist or interfering management. The application of research findings to decision-making was further hamstrung by the generally ill-defined aims that characterized the Occupation itself. In the words of a disgruntled Bennett, "Nobody has any idea of what policy really means in an Occupation, nor have they any concrete program. Just pass along from one small problem to another, solving each one as they go, with a total lack of vision or purpose other than the vague one of doing everything the American way."[76]

Yet the reformulation of Japanese academia under the Occupation left an enduring impact on postwar social science in the United States, Japan, and beyond. As a result of their experiences in the field and work with Japanese colleagues, American scholars in Japan came to question earlier assumptions about the Japanese national character. Instead of the homogeneity that Benedict had led them to expect, they confronted "a strong local divergence in type." Noting that "every group I meet seems different; every individual I meet is an individual," Bennett concluded that the assumption of homogeneity "is dangerous to use in the sense of conferring an ability upon the investigator to make generalizations about Japanese culture of an order comparable to those made for primitive societies."[77] *Chrysanthemum* was quietly removed from the PO&SR library, and many Occupation scholars came to dismiss the study of national character as "a highly elaborate structure built on flimsy and suspect evidence."[78]

The PO&SR was dissolved in June 1951 in anticipation of the termination of the Occupation. Only three months later Japan and the United States signed the San Francisco Treaty, marking the end of World War II, and the Treaty of Mutual Cooperation and Security (Sōgō kyoryoku oyobi anzen hoshō jōyaku, often referred to as Anpo), which set forth the terms of the post-Occupation relationship between the two nations. The agreement sheltered Japan, which was constitutionally banned from maintaining armed forces and waging war, beneath the American military and nuclear umbrella. It also itemized the ways in which Japan was called upon to support the hegemon's geopolitical agenda in Asia, including the maintenance of permanent bases for US air, land, and sea forces. As the treaties stipulated, Japan regained independent sovereignty in April 1952.

Repatriated American social scientists capitalized on their experiences under the Occupation to spearhead the study of Japan in the United States. The previously marginal field of Japanese studies blossomed in the years after 1952. It served as a cornerstone of Cold War area studies, the primary intellectual approach to developing nations during the 1950s and 1960s. Area studies, a multidisciplinary endeavor, sought to advance both theoretical and empirical knowledge of states and regions through intensive language preparation, on-the-ground research, and the incorporation of local viewpoints and interpretations. Critics today often understand area studies as an attempt to perpetuate the power structures of imperialism in the Cold War, replacing overt political control with indirect attempts to foster putatively American values in unaligned developing nations. Research established a hierarchy of "students" generously supported by their home governments and "subjects" dominated by knowledge thus produced. The trajectory of Japanese studies, however, did not conform to this pattern. By the 1960s Japan had transitioned from a developing nation to one of the world's largest economies. Social scientists accordingly took up the task of extrapolating structural, cultural, and psychological lessons for other nations under the sway of US geopolitical dominance.[79]

The Occupation also transformed Japanese academia. In February

1952, mere weeks before the departure of SCAP, twenty leaders of Japanese social science united for a long-planned roundtable on the state of research in the postwar nation. Participants reflected on the Occupation as "a bridge toward the reconstruction of Japanese scholarship" and the starting point of genuinely objective intellectual inquiry.[80] The shared conviction that the values of democracy, capitalism, and peace undergirded knowledge production facilitated cooperation between the United States and Japan and mutual satisfaction in both study and reform.

The intellectual partnership arising from common ideals was perhaps the most enduring legacy of SCAP's overhaul of Japanese knowledge production. The struggle to secure a job in postwar Japanese academia was intensely competitive, pitting venerable graybeards against new graduates and Japan-based faculty against repatriating scholars from universities in the former empire. In the post-Occupation years, connections with US social scientists came to function as a critical credential and source of contacts for obtaining a faculty position. Put simply, virtually all Japanese employed in academia in the 1950s had some experience working under SCAP, and hence exposure to American culture, friendship, and values. The CIE and PO&SR took a particular interest in the post-Occupation fate of their Japanese affiliates, helping many to secure university jobs. In cases where such employment was not available, SCAP helped locate positions in libraries, museums, newspaper and journal editorial boards, and independent research organizations. A few scholars even received fellowships to study in the United States.[81]

Long after the Occupation, Japanese scholars continued to undertake research that buttressed the geopolitical hegemony of the United States by indexing and furthering modernization in their nation and beyond. To be sure, receptivity to American values did not preclude the possibility of dissent with the United States. With the reappearance of far-Right nationalism in the 1950s, some social scientists promoted the rearming of Japan and the "restoration" of direct rule by the emperor. More common were leftist denunciations of Japan's complicity with the American geopolitical agenda, animating discourse in every branch of social science. Tsurumi Shunsuke (1922–2015), a historian and philosopher, gave voice to popular pacifism in his journal *Shisō no kagaku* (The science of thought). In art theory and practice, Okamoto Tarō (1911–1996) decried the hegemony of Western aesthetics, calling for Japan to break free of Euro-American domination by seeking inspiration from its "primitive" past and by cooperating with artists in the nonaligned Third World. Maruyama Masao (1917–1996), a political scientist today remembered as Japan's most prominent spokesman for early postwar liberalism, produced a stream of books and articles on prewar fascism and the need for an active citizenry capable of withstanding foreign pressures on its democracy.[82]

The increasingly critical stance of Japanese intellectuals did not pass unremarked by their American counterparts. Former Occupation attaché Edward Seidensticker (1921–2007), who studied at the University of Tokyo in the late 1950s, recalled,

> I was surrounded by very, very, intelligent boys, it was clear. That was simply beyond denying. . . . But they were unfriendly and they were opinionated, exceedingly opinionated, exceedingly doctrinaire . . . their view of the world which held America responsible for all of the mischief, all of the ails and all of the sufferings of the world, it just wasn't acceptable. . . . Their view of the world made me mad, but I think I also felt rather contemptuous of them. It seemed to me that they were misusing their undeniable talents. . . . I mean, this wasn't a view of the world which was worthy of a first-rate mind.[83]

Today, Seidensticker is widely considered to be one of the finest twentieth-century historians and translators of Japanese literature and a writer of extraordinary sensitivity and grace. In 1975 he received the Order of the Rising Sun, the highest medal awarded to cultural contributors by the Japanese government. Such words from a figure of this repute indicate the pervasiveness and durability of American paternalism and even racism toward Japan. However much the United States might rely on Japan as an ally, it remained locked in the hierarchical mentality of the immediate postwar period. The student could not challenge or supersede the teacher.

In 1960 prominent Japanese intellectuals did dispute Japan's ongoing relationship with the US military by leading mass protests against the renewal of the Anpo Treaty. In their enthusiasm for democracy, capitalism, and peace, Japanese intellectuals strongly objected to armed American engagement in Asia, including the maintenance of bases on national soil, the ongoing occupation of Japan's southernmost prefecture, Okinawa, and, by the end of the decade, the Vietnam War. The ratification of the treaty, forced by Prime Minister Kishi Nobusuke over the strenuous objections of the Japanese Diet, further provoked opposition to a political system that allowed a strongman to prevail over the will of elected representatives. At this moment, in the eyes of many Japanese social scientists, their nation's relationship with the United States appeared not to exemplify but rather to betray the values they had embraced under American tutelage.

Mobilizing millions of citizens representing a broad cross section of society, the protests, the largest in Japan's history, indicated the entrenchment and maturation of putatively American values in the Japanese national consciousness. And yet, perhaps for this very reason, opposition to Anpo petered out without producing substantive change. Fifteen years after a brutal war, the nation had too much at stake—politically, economically, and intellectually—to seriously contest its relationship with the United States.[84] Spearheaded by social scientists, the common ideals of democracy, capitalism, and peace had knit together a resilient network of knowledge production undergirding American hegemony.

Notes

1. E.g., see James R. Bartholomew, *The Formation of Science in Japan: Building a Research Tradition* (New Haven, CT: Yale University Press, 1989); Nakayama Shigeru, *Science, Technology, and Society in Postwar Japan* (New York: Routledge, 1991); John Krige, *American Hegemony and the Postwar Reconstruction of Science in Europe* (Cambridge, MA: MIT Press, 2006); John Krige, *Sharing Knowledge, Shaping Europe: U.S. Technological Collaboration and Nonproliferation* (Cambridge, MA: MIT Press, 2016).

2. Sebastian Conrad, "'The Colonial Ties Are Liquidated': Modernization Theory, Japan and the Cold War," *Past and Present* 216 (2012): 181–214.

3. John W. Dower, *War without Mercy: Race and Power in the Pacific War* (New York: Pantheon Books, 1986).

4. John W. Dower, *Embracing Defeat: Japan in the Wake of World War II* (New York: W. W. Norton, 2000), 556.

5. Miriam Kingsberg, "Legitimating Empire, Legitimating Nation: The Scientific Study of Opium Addiction in Japanese Manchuria," *Journal of Japanese Studies* 38, no. 2 (2012): 325–351.

6. Kawamura Minato, *"Dai Tōa minzoku" no kyojitsu* (Tokyo: Kodansha, 1996); Nakao Katsumi, ed., *Shokuminchi jinruigaku no tenbō* (Tokyo: Fūkyōsha, 2000); Sakano Tōru, *Teikoku Nihon to jinruigakusha: 1884–1952-nen* (Tokyo: Keisō Shobō, 2008).

7. On objectivity in the social sciences, see, e.g., Peter Novick, *That Noble*

Dream: The "Objectivity Question" and the American Historical Profession (New York: Cambridge University Press, 1988); Thomas L. Haskell, *The Emergence of Professional Social Science: The American Social Science Association and the Nineteenth-Century Crisis of Authority* (Urbana: University of Illinois Press, 1988); Thomas L. Haskell, *Objectivity Is Not Neutrality: Explanatory Schemes in History* (Baltimore: Johns Hopkins University Press, 1998); Lorraine Daston and Peter Galison, *Objectivity* (New York: Zone Books, 2007).

8. "Common Statement," in *Tensions That Cause Wars*, ed. Hadley Cantril (Urbana: University of Illinois Press, 1950), 17.

9. E.g., Geoffrey Gorer, *Japanese Character Structure* (New York: Institute for Intercultural Studies, 1942); Douglas Haring, *Blood on the Rising Sun* (Philadelphia: Macrae Smith, 1943); Arnold Meadow, *An Analysis of the Japanese Character Structure Based on Japanese Film Plots and Thematic Apperception Tests on Japanese Americans* (New York: Institute for Intercultural Studies, 1944).

10. Quoted in Pauline Kent, "Misconceived Configurations of Ruth Benedict: The Debate in Japan over *The Chrysanthemum*," in *Reading Benedict / Reading Mead: Feminism, Race, and Imperial Visions*, ed. Dolores Janiewski and Lois W. Banner (Baltimore: Johns Hopkins University Press, 2004), 179–190, at 189.

11. Herbert Passin, *The Legacy of the Occupation of Japan*, Occasional Papers of the East Asian Institute (New York: East Asian Institute, Columbia University, 1968), 4–5.

12. John D. Montgomery, *Forced to Be Free: The Artificial Revolution in Germany and Japan* (Chicago: University of Chicago Press, 1957), 26.

13. Sebastian Conrad, *The Quest for the Lost Nation: Writing History in Germany and Japan in the American Century*, trans. Alan Nothnagle (Berkeley: University of California Press, 2010), 82.

14. In one ironic case, a prominent Japanese social scientist was cleared of communist leanings owing to his wartime collaboration with Austro-German colleagues. See "Application for Employment: Ishida Eiichirō," Record Group 331, Records of the Allied Operational and Occupation Headquarters, World War II, box 5870, file "Ishida Eiichirō," National Archives and Records Administration, College Park, MD.

15. US Cultural Science Mission to Japan, *Report of the United States Cultural Science Mission to Japan* (Seattle: University of Washington Institute for International Affairs, 1949), 14.

16. John C. Pelzel, "Japanese Ethnological and Sociological Research," *American Anthropologist* 50, no. 1 (1948): 72.

17. Yuma Totani, *The Tokyo War Crimes Trial: The Pursuit of Justice in the Wake of World War II* (Cambridge, MA: Harvard University Asia Center, 2009).

18. Nanbara Shigeru, *Bunka to kokka* (Tokyo: Tokyo Daigaku Shuppankai, 1968), 339, 346.

19. "Sensō to heiwa ni kansuru Nihon no kagakusha no shōmyō," *Sekai* 39 (1949): 9.

20. Ishida Takeshi, *Nihon no shakai kagaku* (Tokyo: Tokyo Daigaku Shuppankai, 1984), 223; Laura Hein, *Reasonable Men, Powerful Words: Political Culture and Expertise in Twentieth Century Japan* (Berkeley: University of California Press, 2004), 2–3; Andrew E. Barshay, *The Social Sciences in Modern Japan: The Marxian and Modernist Traditions* (Berkeley: University of California Press, 2004).

21. Joseph C. Trainor, *Educational Reform in Occupied Japan* (Tokyo: Meisei University Press, 1983), 224.

22. John W. Bennett, "Some Comments on Japanese Social Science," Record Group 331, Records of the Allied Operational and Occupation Headquarters, World War II, box 5915, file "Comments on Japanese Social Science," National Archives and Records Administration.

23. US Cultural Science Mission to Japan, *Report of the United States Cultural Science Mission to Japan*, 9, 1.

24. Letter from John W. Bennett to Kathryn G. Bennett, Mar. 24, 1949, John W. Bennett Papers (hereafter JWB Papers), box 2A, file 38U, Rare Books and Manuscripts: Collections, Rare Books and Manuscripts Library, Ohio State University.

25. The Constitution of Japan (1946), accessed Oct. 21, 2016, http://japan.kantei.go.jp/constitution_and_government_of_japan/constitution_e.html.

26. Merle Fainsod, "Military Government and the Occupation of Japan," in *Japan's Prospect*, ed. Carl J. Friedrich (Cambridge, MA: Harvard University Press, 1946), 287–304, at 294.

27. Interview, David L. Sills, Hastings-on-Hudson, NY, Apr. 14, 1979, 12, Marlene J. Mayo Oral Histories with Americans Who Served in the Allied Occupation of Japan, Gordon M. Prange Collection, University of Maryland.

28. Theodore F. Welch, *Libraries and Librarianship in Japan* (Westport, CT: Greenwood, 1997), 17.

29. Letter from Julian H. Steward to Fred Eggan, Dec. 29, 1955, Fred Eggan Papers, box 23, file 8, Special Collections Research Center, University of Chicago.

30. Douglas G. Haring, "The Challenge of Japanese Ideology," in Friedrich, *Japan's Prospect*, 259–286, at 280.

31. Robert B. Textor, *Failure in Japan* (New York: John Day, 1951), 149.

32. Letter from John W. Bennett to Richard Morris, Apr. 12, 1949, JWB Papers, box 24, file 215.

33. Ishida Mikinosuke, "Tōhō minzokugaku kankei Ōbun kincho (ichi)," *Minzokugaku kenkyū* 13, no. 1 (1948): 80–85.

34. Textor, *Failure in Japan*.

35. Reorientation Branch Office for Occupied Areas, *Semi-annual Re-*

port of Stateside Activities Supporting the Reorientation Program in Japan and the Ryukyu Islands (Washington, DC: Office of the Secretary of the Army, 1951), 19.

36. Japanese National Commission for UNESCO, *Japan: Its Land, People and Culture* (Tokyo: Ministry of Education, 1958), 546.

37. Textor, *Failure in Japan.*

38. Letter from John W. Bennett to Kathryn G. Bennett, Sept. 4, 1949, JWB Papers, box 2A, file 38YYY.

39. Letter from Monkichi Nanba to John W. Bennett, July 6, 1950, JWB Papers, box 24, file 215.

40. Inderjeet Parmar, *Foundations of the American Century: The Ford, Carnegie, and Rockefeller Foundations in the Rise of American Power* (New York: Columbia University Press, 2012).

41. Quoted in Wada Jun, "American Philanthropy in Postwar Japan: An Analysis of Grants to Japanese Institutions and Individuals," in *Philanthropy and Reconciliation: Rebuilding Postwar U.S.-Japan Relations*, ed. Yamamoto Tadashi, Iriye Akira, and Iokibe Makoto (New York: Japan Center for International Exchange, 2006), 135–184, at 163.

42. George H. Kerr, "An Institution for American Studies in Japan, 1948–1958: A Prospectus for a Ten-Year Project," 5, American Studies Seminar in Japan, Records, 1950–1981, box 1, file 2, Special Collections and University Archives, Stanford University.

43. James Gannon, "Promoting the Study of the United States in Japan," in Yamamoto, Iriye, and Iokibe, *Philanthropy and Reconciliation*, 189–194; "Tōdai no Amerika kenkyū kōkai kōgi," *Yomiuri shinbun*, July 7, 1954, 6.

44. "Proposal to the Rockefeller Foundation concerning the Seminar in American Studies," 1953, American Studies Seminar in Japan, Records, 1950–1981, box 1, file 1, Special Collections and University Archives, Stanford University.

45. Julian H. Steward, "Report of the Director, Kyoto American Studies Seminar, 1956," 3, Julian H. Steward Papers, box 20, Special Collections and University Archives, University of Illinois at Urbana-Champaign.

46. "Daigakusei no sanka mo yurusu: Amerika kenkyū seminā," *Yomiuri shinbun*, Apr. 15, 1952, 3; "Tōdai no Amerika kenkyū seminā kōkai kōgi," *Yomiuri shinbun*, July 11, 1953, 6.

47. Yanaihara Tadao, "The Committee for the Seminar in American Studies Report, 1953," American Studies Seminar in Japan, Records, 1950–1981, box 1, file 7, Special Collections and University Archives, Stanford University.

48. "Round Table Conferences," American Studies Seminar in Japan, Records, 1950–1981, box 1, file 5, Special Collections and University Archives, Stanford University.

49. Letter from Virgil C. Aldrich to Julian H. Steward, Dec. 26, 1955, and let-

ter from Matsui Shichirō to Julian H. Steward, Jan. 5, 1956; both in Julian H. Steward Papers, box 20, Special Collections and University Archives, University of Illinois at Urbana-Champaign.

50. Fritz Machlup, "Report on My Activities at the Kyoto American Studies Seminar," Aug. 22, 1955, 3, Julian H. Steward Papers, box 20, Special Collections and University Archives, University of Illinois at Urbana-Champaign.

51. Joseph S. Davis, Claude A. Buss, John D. Goheen, George R. Knoles, and James T. Watkins, "American Studies in Japan, 1950: Report of the Stanford Professors," American Studies Seminar in Japan, Records, 1950–1981, box 1, file 1, Special Collections and University Archives, Stanford University.

52. Yanaihara Tadao, "Opening Address," July 13, 1953, American Studies Seminar in Japan, Records, 1950–1981, box 2, file 1, Special Collections and University Archives, Stanford University.

53. Letter from Kanamatsu Kenryo to John D. Goheen, July 31, 1951, Julian H. Steward Papers, box 1, file 6, American Studies Seminar in Japan 1951, Rare Books and Manuscripts Library, University of Illinois.

54. Kawamura, *"Dai Tōa minzoku" no kyojitsu*; Nakao, ed., *Shokuminchi jinruigaku no tenbō*; Sakano, *Teikoku Nihon to jinruigakusha*.

55. Clyde Kluckhohn and Raymond Bowers, "Report on Field Trip to Southern Honshu and Kyushu, 8–23 January 1947," JWB Papers, box 1, file 15.

56. Marinousukī, *Mikai shakai ni okeru hanzai to shūkan*, trans. Aoyama Michio (Tokyo: Nihon Hyōron Shinsha, 1955); B. Marinofusukī, *Bunka no kagakuteki riron*, trans. Himeoka Tsutomu and Kamiko Takeji (Tokyo: Iwanami Shoten, 1958); B. Marinofusukī, *Mikaijin no sei seikatsu*, trans. Izumi Seiichi, Gamō Masao, and Shima Kiyoshi (Tokyo: Kawade Shobō, 1957).

57. Robert J. Smith, "Time and Ethnology: Long-Term Field Research," in *Doing Fieldwork in Japan*, ed. Theodore C. Bestor, Patricia G. Steinhoff, and Victoria Lynn Bestor (Honolulu: University of Hawai'i Press, 2003), 252–366, at 354.

58. Ruth Benedict, *The Chrysanthemum and the Sword: Patterns of Japanese Culture* (Boston: Houghton Mifflin, 1946), 5.

59. Herbert Passin, "The Development of Public Opinion Research in Japan," *International Journal of Opinion and Attitude Research* 5 (1951): 21.

60. John W. Bennett, "Summary of Major Research Problems of the Public Opinion and Sociological Research Division, CIE," JWB Papers, box 1, file 4.

61. Hugh Borton, "The Reminiscences of Hugh Borton" (New York: Oral History Research Office, Columbia University, 1958), 48.

62. US Cultural Science Mission to Japan, *Report of the United States Cultural Science Mission to Japan*, 15.

63. John W. Bennett, "Social Research in the Japanese Occupation," JWB Papers, box 1, file 1.

64. Letter from John W. Bennett to Kathryn G. Bennett, Mar. 9, 1949, JWB Papers, box 2A, file 38M.

65. Arthur F. Raper, *The Japanese Village in Transition* (Tokyo: General Headquarters, Supreme Commander for the Allied Powers, 1950), 12.

66. Ibid., i–ii; John W. Bennett, "Community Research in the Japanese Occupation," *Clearinghouse Bulletin of Research in Human Organization* 1, no. 3 (1951): 5.

67. Margaret Mead, "The Organization of Group Research," in *The Study of Culture at a Distance*, ed. Margaret Mead and Rhoda Métraux (New York: Harper Row, 1953), 85–87.

68. Herbert Passin, *Encounter with Japan* (New York: Kodansha International, 1982), 143.

69. Bennett, "Summary of Major Research Problems of the Public Opinion and Sociological Research Division, CIE."

70. Herbert Passin, "Report of Field Trip to Yuzurihara," Herbert Passin Collection, "Field Trip to Yuzurihara Report," Special Collections, DuBois Library, University of Massachusetts at Amherst.

71. "CIE ni okeru shakai chōsa no tenkai," *Minzokugaku kenkyū* 17, no. 1 (1953): 68–80.

72. Arthur F. Raper, "The Reminiscences of Dr. Arthur F. Raper" (New York: Oral History Research Office, Columbia University, 1971), 149.

73. "CIE ni okeru shakai chōsa no tenkai," *Minzokugaku kenkyū* 17, no. 1 (1952): 73.

74. "Summary of Major Research Problems of the PO&SR Division, CIE," JWB Papers, box 1, file 4.

75. Textor, *Failure in Japan*, 174.

76. Letter from John W. Bennett to Kathryn G. Bennett, Aug. 8, 1949, JWB Papers, box 2A, file 38NNN.

77. Letter from John W. Bennett to Kathryn G. Bennett, Apr. 26, 1949, JWB Papers, box 1, file 1; John W. Bennett, "Social and Attitudinal Research in Japan: The Work of SCAP's Public Opinion and Sociological Research Division," 18, JWB Papers, box 1, file 12.

78. Fred N. Kerlinger, "Behavior and Personality in Japan: A Critique of Three Studies of Japanese Personality," *Social Forces* 31, no. 3 (1953): 257; "Inventory of Books," Record Group 331, Records of Allied Operational and Occupation Headquarters, World War II, box 5873, file "PO&SR," National Archives and Records Administration.

79. David L. Szanton, "The Origin, Nature, and Challenges of Area Studies in the United States," 1–33; Alan Tansman, "Japanese Studies: The Intangible Art of Translation," 184–216; both in *The Politics of Knowledge: Area Studies and the Disciplines*, ed. David L. Szanton (Berkeley: University of California Press, 2004).

80. "CIE ni okeru shakai chōsa no tenkai," *Minzokugaku kenkyū* 17, no. 1 (1953): 68–80; Japan Society for Ethnology, *Ethnology in Japan: Historical Review* (Tokyo: K. Shibusawa Memorial Foundation for Ethnology, 1968), 4.

81. Letter from Ishino Iwao to John W. Bennett, June 5, 1951, JWB Papers, box 20, file 197.

82. Each of these figures is the subject of a rich biographical literature. For recent works on Tsurumi, see Adam Bronson, *One Hundred Million Philosophers: Science of Thought and the Culture of Democracy in Postwar Japan* (Honolulu: University of Hawai'i Press, 2016); Kurokawa Sō, *Kangaeru hito Tsurumi Shunsuke* (Fukuoka: Genshobō, 2013); Kimura Tsuneyuki, *Tsurumi Shunsuke no susume: Puragumateizumu to minshushugi* (Tokyo: Shinsensha, 2005). On Okamoto, see Hirano Akiomi, *Okamoto Tarō no shigoto ron* (Tokyo: Nihon Keizai Shinbun Shuppansha, 2011); Kawagiri Nobuhiko, *Okamoto Tarō: Geijutsu wa bakuhatsu ka* (Tokyo: Chūsekisha, 2000). On Maruyama, see Yanagisawa Katsuo, *Maruyama Masao to Yoshimoto Takaaki: Kaisōfū shisōron* (Tokyo: Sōeisha/Sanseidō Shoten, 2014); Yoshida Masatoshi, *Maruyama Masao to sengo shisō* (Tokyo: Ōtsuki Shoten, 2013); Fumiko Sasaki, *Nationalism, Political Realism and Democracy in Japan: The Thought of Maruyama Masao* (New York: Routledge, 2012); Andrew Barshay, *The Social Sciences in Modern Japan: The Marxian and Modernist Traditions* (Berkeley: University of California Press, 2004).

83. Interview with Edward Seidensticker, Columbia University, Nov. 10, 1978, 71, Marlene J. Mayo Oral Histories with Americans Who Served in the Allied Occupation of Japan, Gordon M. Prange Collection, University of Maryland.

84. For recent scholarship on the Anpo protest movement, see Hosoya Yūichi, *Anpo ronsō* (Tokyo: Chikuma Shobō, 2016); Kobayashi Tetsuo, *Shinia sayoku to wa nani ka: Han Anpo hosei, han genpatsu undo de shutsugen* (Tokyo: Asahi Shinbun Shuppan, 2016); Nikhil Kapur, "The 1960 U.S.-Japan Security Treaty Crisis and the Origins of Contemporary Japan" (PhD diss., Harvard University, 2011).

CHAPTER SIX

Dispersed Sites

San Marco and the Launch from Kenya

Asif Siddiqi

This is a history of transnational science and technology whose unit of analysis is the *site*.[1] I use here the example of the Cold War era Italian American San Marco project off the coast of Kenya to advance some preliminary insights into how the notion of a *site* can be deployed to reveal aspects of transnational science that might be less evident with older, diffusionist narratives about the "spread of knowledge" or with more recent frameworks concerned with encounter and the coproduction of knowledge, often framed in terms of "circulation." In the case of the latter, we can now make use of a relatively large body of work, with historians and science, technology, and society scholars variously interested in how knowledge was not only circulated but also regulated, impeded, reshaped, and reconstituted in new settings.[2] Drawing particularly from postcolonial science studies, this work subverted the received language of dichotomies by revealing the unequal logic of encounter and circulation and, through them, the atomized and contingent vicissitudes of the production of knowledge at the level of communities, laboratories, and nations.[3]

This chapter draws on some of these insights into the circulation of knowledge, from historians of modern and colonial science, but redirects attention from knowledge, objects, and communities to the heuristic of "site," by which I mean, at its most reductive level, an engagement with not only the *why* and the *what* but more specifically the *where*. "Site" can be a spatial concept, a laboratory, a facility, a network, a nation-state, or an international society; it can also be an ontological concept, identifying the location of knowledge production, transcending geographic contours. Scale is also a factor in considering the notion of site, since the contours of any particular site, as one might expect, are contingent upon the scale of activity under study. We can imagine that the same phenomenon, for example, might appear to us as "local" and "global" depending on where we as scholars choose to demarcate our stories.[4] In the present example, I am less interested in distinguishing the "local" from the "global" than in illuminating the epistemic and material dislocations that result from the intersections between the "local" and the "global." Exploring these dislocations leads to further questions: What is a site in a transnational project, especially one involving a postcolonial state? Where are its boundaries? What happens in a site that makes it part of the site? What resists definition into the site? And finally, what kinds of social dislocations and violence are revealed as a result of the ontological collapse of the fixed notion of a site?

In his introduction to this volume John Krige has persuasively highlighted how we must not abandon the "national" in our meditations on the "transnational," that the nation-state always seeks—sometimes successfully and sometimes with less success—to regulate the flow of knowledge across borders, even in cases that seemingly involve knowledge outside the realm of formalized control. At a physical site, such as an international research facility at a university, for example, the regulation that state authorities might introduce can take many forms, including such obvious tools as visa restrictions, export regulations, and security protocols.[5] One can imagine the consequences of such regulatory regimes not only for the circulation of information during the period of the project but also for the way a particular transnational project might be articulated in the public imagination. Using Krige's observations as a jumping-off point, I use the San Marco project, a "successful" Cold War science project whose history has had no place for Kenya despite being literally based there, to investigate two phenomena that may arise from the nation-state's imperative to regulate the flow of information.

First, such measures to regulate rarely produce "airtight" regimes; either by design or by failure, knowledge (and objects and communities) of particular projects—and not just purely scientific and technical knowledge—frequently skirts or breaks through the "borders" proscribed for them. Regulatory measures also produce regimes of resistance and friction in the form of counterphenomena, which can take the form of political action, scientific boycotts, or opposing claims to the knowledge produced. Second, one might also see the process of choosing a particular site to produce collaborative science and technology itself as an act of regulation, a process that is embodied in two contradictory forces: on the one hand, one party cedes a claim to control by agreeing to locate the project in the other party's (geographic) control; or, from the other party's perspective, the process affirms its position of control. From either perspective, the choice is designed to destabilize the "airtightness" of the site.

Of course, as Gabrielle Hecht, Peter Redfield, and others have shown, when European scientific projects were located in postcolonial spaces, these led to further overlapping claims, conflicts, and frictions rooted in inequality, in terms of both resources and also claims about who "owns" the knowledge.[6] Inequality surfaces not only in power relations within the project but also in the language and frames used by both actors and historians to speak about it. I am particularly interested here in how Kenyans (represented by a number of different voices) made meaning out of San Marco that was entirely at odds with how Italians or Americans saw the project and how within the Kenyan perspective itself there was frequent disagreement about its import.[7] Through an examination of the contesting claims imprinted on the postcolonial space of San Marco—defined geographically, ontologically, and historically—my chapter offers some thoughts on the notion of the site; I introduce the idea of a scientific site as *dispersed* into *fragments*, where traces of knowledge, objects, and communities extend beyond formal borders and, depending on what is "seen" and who is doing the "seeing," reshape the ontological contours of the site itself. The result is often a deliberate occlusion of certain "fragments," especially ones characterized by intense social dislocations, as was evident in the case of San Marco.

San Marco: The Basic Architecture of the Story

The fundamental contours of the history of San Marco are well known.[8] To the extent that it has been considered by historians of science and technology at all, the San Marco project, which involved several launches of scientific satellites, has been framed as a highly successful cooperative effort between a Western European nation, Italy (represented by constituencies in academia and the Italian government), and the United

States (NASA, but also including scientists from academia). Here, geopolitical and scientific frames provide effective entry points for interrogating the vectors of knowledge circulation during the Cold War and serve as an example of what John Krige (via Charles Meier) has called "consensual hegemony" driven by American interests.[9] The original idea, proposed by Luigi Broglio, a professor at the University of Rome and an Air Force officer, was to establish a foothold for Italian science and industry in space research. Italy responded to a NASA offer for cooperation in space but with a twist: Broglio proposed that Italy could use NASA's Scout rocket to launch small Italian scientific satellites into equatorial orbit around the earth. To do this, Broglio suggested the use of a mobile seaborne platform that could be installed in equatorial waters at any point on the globe. A key element of the plan was that Italians, having been trained by NASA personnel, would carry out the satellite launches without any contribution from the Americans although the United States would provide key material and technical support. Broglio, representing the Commissione per le Ricerche Spaziali (Space Research Commission), and Huge Dryden, representing NASA, agreed to these stipulations in May 1962.[10]

In 1963 and 1964 several dozen Italian engineers traveled to the United States. At various locations, under the supervision of engineers from NASA and the Texas-based LTV Aerospace Corporation / Vought Missiles and Space Company (the prime contractor for the four-stage Scout rocket), they trained to build satellites, maintain a launch range, and launch a rocket. Meanwhile, the Centro di Ricerche Aerospaziali (Aerospace Research Center), based at the Urbe Airport near Rome and under the jurisdiction of the University of Rome, mastered the assembly of the Scout launch vehicle.[11] Most crucially, under Broglio's direction, the Italians procured two floating launch platforms, one called Santa Rita, and the other, San Marco. The former was used for initial test launches but was converted to hold facilities for range control, a blockhouse, and telemetry equipment. The latter was a rectangular steel barge with full facilities to launch a rocket capable of delivering a payload into earth orbit. The two platforms, located about 550 meters apart, were stationed just off the coast of Kenya, near the equator.[12] After a series of test rocket launches, on April 26, 1967, the Italians launched a Scout rocket from the San Marco platform and delivered a satellite into equatorial orbit around the earth, the first time an object had reached such an orbit. The satellite, known as *San Marco 2*, was built by the University of Rome and carried two scientific experiments, one to directly measure air density below 350 kilometers (using the spherical shape of the satellite) and the other, an ionospheric beacon, to observe electron content between the earth and the satellite. Both operated with great success, obtaining important data on the equatorial bands of the earth's upper atmosphere, a region hitherto largely unstudied. The satellite remained operational until August 5, 1967, and eventually decayed from orbit on October 19.[13]

The Italians launched several more satellites from the San Marco platform through the late 1980s, including a number of NASA satellites. The most notable of these was the so-called *X-Ray Explorer*, launched in 1970, the first ever satellite launched for the explicit goal of research on x-ray astronomy.[14] In orbit, the satellite carried out the first comprehensive survey of the entire sky for x-ray sources, and it is remembered as one of the great scientific successes of the early space age. The data returned from the satellite helped compile the first comprehensive x-ray catalog. *X-Ray Explorer* also found one of the first strong candidates for identification as a black hole, Cygnus X-1.[15] Although its main instrument was built in Cambridge, Massachusetts, the *X-Ray Explorer*'s principal scientific investigator was Riccardo Giacconi, the famed Italian scientist who later (in 2002) won the Nobel Prize in Physics "for pioneering contributions to astrophysics, which have led to the discovery of cosmic X-ray sources."[16] His involvement in the entire project and the extraordinarily valuable results from the *X-Ray Explorer* undoubtedly brought the Italian American project much visibility and accolades in the broader world of space-based astrophysics. On the basis of that measure alone, the San Marco experiment was undoubtedly an unqualified scientific success.

In the San Marco project, the imaginary of the "site" can best be described as dispersed, extending to multiple global locations. We can include the various facilities in Rome (such as the University of Rome, laboratories at the Urbe Airport run by the Centro di Ricerche Aerospaziali), offices in the United States (Wallops Island, Goddard Space Flight Center, NASA Headquarters in Washington, DC, East Texas, etc.), and the actual equatorial orbit of the earth where the Italian and American scientific instruments collected data. Dispersion is, of course, the hallmark of *all* large-scale scientific endeavors that draw on multiple sources of knowledge, infrastructure, people, and funding. What distinguished the San Marco project—and similar transnational projects situated in postcolonial spaces—was that the choice of the core locale, on and off the coast of Kenya, was primarily dictated by a confluence of geographic, scientific, and political considerations that had nothing much to do with Kenya itself. But once the site was chosen, the primary architects of this project, the Italians, sought to limit Kenyan involvement; their goal was cleanly to overlay the San Marco *project* onto the San Marco *site*, with the fewest incongruities possible. This clean mapping, however, became increasingly more difficult over time to sustain or to imagine, as *fragments* of the dispersed site came into violent friction with the entrenched human geographies and political imperatives of a newly independent and powerful postcolonial state, Kenya.

Italy and the United States

Italian aspirations to launch a domestically constructed satellite were inextricably tied to NASA's strategy of actively inviting friendly countries to cooperate in space exploration. When NASA was established by congressional fiat in 1958, policy-makers included in the National Aeronautics and Space Act the explicit requirement that one of its principal goals be "[c]ooperation by the United States with other nations and groups of nations," which was to be done "under the foreign policy guidance of the President."[17] As John Logsdon has noted, "the new space agency interpreted this provision as giving it authority to take the initiative in international space dealings," and "within six months, NASA began to develop a program of international cooperation in space that over the following three decades has resulted in agreements with more than 100 countries."[18] The earliest invitation was made at the second annual meeting of the Committee of Space Research in March 1959 at The Hague, when NASA, through a representative of the US National Academy of Sciences, announced that the United States hoped that the committee "could serve as an avenue through which the capabilities of satellite launching nations and the scientific potential of other nations may be brought together."[19] The United Kingdom was the first to seize this opportunity and initiated NASA's first international cooperative program, leading to the launch of the British satellite *Ariel 1* on an American rocket in 1962.[20] Likewise, Luigi Broglio welcomed the American offer, and by early 1961 had formulated a distinctly different idea than that of the British: his proposal was to use a NASA rocket to launch an all-Italian satellite into orbit. Originally, the plan was to launch from Sardinia, but then, in May 1961, Broglio suggested something unique: to launch the satellite into equatorial orbit off a mobile, ocean-based platform that would be adapted from an offshore oil-drilling rig. The requirement to launch into *equatorial* orbit, implying a launch from the equatorial region of the earth, was driven, apparently, by NASA and Italian scientists who saw this as an excellent opportunity to study phenomena that other satellites, launched from relatively northern latitudes, could not.[21]

The first and most important question for the Italians was: where on the earth's equator? The first idea, in 1961, was to build the "mobile launching platform to be anchored close to the equator, offshore from the coast of Somalia."[22] Anecdotal evidence suggests that Broglio was considering a platform "in extra-territorial waters" about twenty-four kilometers from Kismayo, a port city in southern Somalia that was close to Somalia's border with Kenya and was equipped with large docks for moving equipment.[23] The choice of Somalia was, in many ways, an obvious one, since it was the only former Italian colonial holding with a landmass and a coast through which the equator passed.[24] A contemporaneous media account of the charismatic Broglio vividly describes how geography and politics firmly constrained the site of this project: "The professor pulled out a world map and started scanning the equatorial regions. The site had to be coastal, with offshore waters sufficiently shallow for anchoring the launch platforms. It also had to be situated in a country with a government stable enough to survive the duration of the project."[25]

For reasons that are not entirely clear, the Somalia option did not endure long, and subsequently, the actual NASA-Italy memorandum of understanding from May 1962 makes no mention of a particular location, pointing enigmatically merely to "equatorial waters."[26] Broglio's attentions soon moved to Kenya, just south of Somalia.[27] In December 1962 with the help of Vought Missiles, the builder of the Scout rocket, the Italians initiated a study of "Formosa Bay" off the coast of Kenya as a more plausible location.[28] The deliberations behind the switch from Somalia to Kenya remain unknown, but one San Marco veteran recalls that there was concern among the Italians that the newly independent Somali government of Aden Abdullah Osman Daar, being far too friendly with the Soviets, might make it difficult to solicit help from the United States.[29] We may also speculate that, besides a "stable" government, one of the driving motivations was that the Italians wanted the two launch platforms to be as close to the coast as possible—to provide logistics support and ensure that personnel could be housed—but far enough away to be in *international* waters so that the principal activity would not have to be regulated by any national entity beyond Italy. In the case of Somalia, such a location would have been too far from the coast; Kenya, however, had recently declared that its coastal waters would extend no more than three nautical miles (5.56 kilometers).[30] This was perfect for the San Marco project.

The provenance of these two platforms, rarely considered, extends the story of San Marco into unexpected locales within the United States. The Santa Rita platform was originally built by an American company based in Longview, Texas, R. G. LeTourneau, in 1959 as an offshore oil-drilling platform, known as Mobile Tender-Assisted Platform Number 9.[31] The company was founded by Robert G. LeTourneau (1888–1969), a fundamentalist Christian and self-styled inventor whose products, largely earthmoving machinery, were so ubiquitous in the mid-twentieth century that, by some accounts, machinery from his company accounted for nearly 70 percent of the earthmoving equipment and engineering vehicles used by Allied forces during World War II.[32] LeTourneau's other achievement was to found LeTourneau University, a private Christian university based in Longview, Texas. In his autobiography, LeTourneau mentions, among his many works devoted to God, his evangelical and "colonization" projects in Liberia and Peru, eerily anticipating the role that one of his rigs would play in Kenya.[33] The platform in question was purchased originally by an Italian oil company, SAIPEM, and renamed Scarabeo before being purchased in turn by the Centro di Ricerche Aerospaziali in early 1963 for the San Marco project.[34] It was then renamed once again, this time to Santa Rita, and remodeled to fit the requirements of the satellite project and transported to the coast of Kenya. Meanwhile, the actual launch platform, San Marco, was procured from US Army surplus storage, which had in its inventory floatable steel barges typically used for rapid docking. One of these barges, with a deck area of about 27.4 by 91.4 meters, was located at a naval facility where it had been mothballed since 1957. Eventually, NASA signed an agreement with the US Army to lease the barge, which was then towed by an Italian tugboat in May 1965 from Charleston, South Carolina, to La Spezia, Italy, to be remodeled into the San Marco platform.[35]

Italy and Kenya

Kenya, despite its centrality in the San Marco story, remains almost entirely absent from the literature on the San Marco project. This is striking not only because San Marco operations relied on the Kenyan government for support but also because a significant portion of the project's physical, logistical, and technical operations were located *on* Kenyan territory. On the other hand, its effacement from the standard narrative is simultaneously unsurprising given that Kenya is an example of a fragmented site, one that doesn't fit neatly into a narrative constructed around a "project" that exhibits a discrete beginning, end, contours, and geographies; Kenyans, after all, were not involved in the definition of the project, nor did they receive any of the scientific data collected by satellites launched from the San Marco platform.

There is one more curious aspect of Kenya's involvement: the original Italian interest in Kenya (after the Somalia option was abandoned) emerged, in 1962, when Kenya was still a British colony. The original details of the agreement were laid out by administrators from the University of Rome, representing the Italian government, and by the University of East Africa Royal College of Nairobi (now known as the University of Nairobi), representing the University of London. At the time of signing, September 18, 1963, the official head of the Kenyan state was still Queen Elizabeth II, who, as one Kenyan parliamentary representative, John Mbadi, later pointed out, "was the one who negotiated all this."[36] It is probably not a coincidence that this agreement was signed just two months *prior* to Kenyan independence, after which the new Kenyan government was obligated to abide by the San Marco agreement (along with many others pushed through hurriedly by the British prior to independence).[37] In other words, the original agreement to allow the Italians into Kenya may not have been negotiated by the Kenyans at all.

The 1963 agreement was followed by another signed after independence, between the Royal College of Nairobi and the Italian Space Commission, allowing participation of the former under the direction of Professor A. N. Hunter, a British physicist based in Nairobi. A group of four staff members of the college (plus one technician) were permitted to receive data from two scientific experiments from the initial sounding rockets, and this involvement was vigorously advertised by both Broglio and Hunter in press conferences.[38] Broglio, for example, repeatedly and disingenuously called San Marco "a joint program between the Royal College and the Italian Space Commission." Similarly, Hunter noted that the project would put "Kenya on the science 'map'" with "[w]orld scientists welcom[ing the] Kenya[n] space project."[39] US State Department documents confirm that bringing in the Royal College of Nairobi was a cynical move, "useful from several angles including [the] fact that any irresponsible or radical attack [from Kenyans] on [the] project would run against [the] University and its scientific interests both of which [are] sacrosanct in the public eye."[40] Unsurprisingly, beyond the two initial experiments on sounding rockets, there is no record of Kenyan involvement in any of the science collected by the satellites, an issue that, for some Kenyans, became a point of contention in later years.

In terms of operations to support the Italian project, in a postindependence agreement the government of Kenya granted the Italians a number of rights to operate at Mombasa, including at its airport and port.[41] According to American State Department documents, the Kenyan "Prime Minister's Office [was] basically sympathetic to [the] Italian project" but was nervous about the visibility of any American personnel in Kenya, "fearing criticism from leftist elements in Kenya . . . during a period of substantial political danger."[42] Through all this, there was an understanding that Kenyans would have no claims to the project, its operations, or its results. No Kenyans were, in fact, ever involved in operations on the two platforms, the San Marco and the Santa Rita, which were technically in international waters, and none were involved in defining scientific goals or had access to the scientific results from any of the satellites.

A significant portion of the San Marco operation was, however, routed through Kenyan territory. The casings of the Scout rocket, as well as the actual Italian satellite, were flown into Nairobi from Washington, DC, and Rome, respectively. At Nairobi airport, these sensitive technological artifacts in sealed containers were then transferred to large trucks with the help of Kenyan airport personnel under the supervision of Italians. From Nairobi, the trucks drove first to Mombasa and then about two and a half hours (145 kilometers) further up the coast to the Malindi area, the base of operations for the San Marco and Santa Rita platforms. Access to Nairobi airport required special licenses and permissions from the Kenyan government but also assurances to NASA and the Italians that Kenyan airport staff would not have access to any sensitive technology. Mombasa, meanwhile, was not merely a way-stop for the trucks—its harbor was also the principal location for preparation and repair of the San Marco and Santa Rita platforms, work performed with the help of local Kenyans. The settlement of Malindi meanwhile provided support in the form of a helicopter (taking ninety minutes) and motorboats (taking three and a half hours) to take personnel to the platforms.[43]

The most obvious physical manifestation of Kenyan involvement in San Marco was the so-called "base camp," created as "a staging area for platform operations." The base camp was located near the village of Ngomeni, about 21 kilometers north of Malindi and 113 kilometers north of Mombasa. A two-acre plot of land (75 by 126 meters) was acquired by the Italians for this purpose "on long term leases from the property owners . . . sanctioned by the government of Kenya." The entire operation, including four tents, one open-sided building, barracks, kitchen, dining hall, radio station, and boat dock, ensured full support to the two platforms, which were about five kilometers from shore in Ungwana Bay (known as Formosa Bay by the Italians). Italians staying at base camp would make the twenty-five-minute trip to the two platforms on two fifteen-meter-long motorboats. Hazardous materials that were deemed unsafe to deliver by air, such as solid propellants, rocket motors, and pyrotechnics, were delivered directly to the platforms by ocean-going freighter ship after a monthlong trip from Virginia. These materials were transferred using cranes to the San Marco platform for full assembly of the rocket and payload. While the San Marco platform served as the assembly and launch site for the rocket, the Santa Rita platform, about 550 meters away, was the staging ground for command and control. No Kenyans were allowed on the two platforms, located as they were on international waters.[44]

Contemporary accounts of the San Marco project hint at the seeds of conflicting claims about the site. A glowing *Life* magazine piece from 1967, on the occasion of the first satellite launch, mentions "the cheers of a bunch of Bajuni tribesmen" when the rocket lifted off from the San Marco platform. Implicit in the account is that the scientific project disrupted no local life: the *Life* magazine reporter helpfully notes (undoubtedly quoting Broglio) that the area was suitable for the project because it "was inhabited only by bands of wild baboons and a handful of Bajuni tribesmen who eked out a marginal living by fishing and selling wood," thus inadvertently linking the Bajuni with wild baboons. *Life* magazine reassured its readers that because of the San Marco project "the Bajuni tribal village of Ngomeni was inundated by prosperity. The Italians hired 60 natives to help in the base camp, paying four times the going wage. Seven new Bajuni huts of sticks and mud were completed and four more were abuilding when the satellite went up."[45] Although the Italians apparently faced difficulties with the local food (mostly English fare, a legacy of the British colonial presence), "[t]hey solved it by teaching the natives how to cook spaghetti [and thus, the base camp] quickly earned a reputation for the best cuisine in Kenya." Diet apart, the Bajuni, the principal population at Ngomeni and a vulnerable ethnic minority in Kenya and Somalia who speak a Swahili dialect known as Kibajuni, had an extremely tortured and conflicted relationship with the Kenyan government because of property rights. Like the fortunes of many other communities along Kenya's coast, the history of this community, who made a livelihood as sailors or fishermen, was marred by ethnic conflict rooted in and often enabled by British colonial rule.

Malindi: The Uneasy Decline

While the San Marco and Santa Rita platforms served as launch points for the Italians and while Ngomeni played host to the base camp, it was the city of Malindi, twenty-one kilometers south of base camp, that Italians considered their home away from home. Famous for having been visited by Vasco da Gama in 1498, Malindi was part of a contested sixteen-kilometer strip on the coast of Kenya populated largely by a heterogeneous mix of "Swahili" people and migrant "Arabs."[46] In 1861 Sultan Majid of Zanzibar brought it under his rule, but in 1895 he signed a treaty to "lease" it, as a legal "Protectorate" of the island state of Zanzibar, to Great Britain. Through the twentieth century Malindi was a somewhat unremarkable tourist destination in colonial Kenya, but as Kenyan independence approached, a British review board, the Robertson Commission, which was formed to determine the precise administrative divisions of coastal Kenya, recommended that Malindi and its neighboring towns be folded into newly independent Kenya (rather than Tanzania, as was the case with Zanzibar). Local Arab and Swahili residents (including the Bajuni), afraid of political domination by Kenyan Africans, launched a movement to retain autonomy within newly independent Kenya. The deep ethnic and racial enmity between Africans, on the one side, and Swahili and Arab residents, on the other, threatened to explode into violence (as it did in Zanzibar in the early 1960s), exacerbated by the growing number of African squatters in and around Malindi who claimed property rights.[47]

The uneasy postindependence status quo was marred by sporadic interventions from the government and then the entry of the Italians. Soon after independence, Kenyan president Jomo Kenyatta reneged on an agreement to allow land tenure for the Bajuni, a move that effectively opened up the land around Ngomeni to outside prospectors, including the Italians.[48] Italian intervention into this conflict, with their long-term lease for the base camp (which is still in effect in 2017), undoubtedly exacerbated frictions between the Bajuni (and other Swahili peoples) and the Kenyan authorities, further disenfranchising the former.

The Italians who came for the San Marco project arrived in larger and larger numbers, slowly transforming Malindi by adding to its mix of Swahili, Arabs, and Africans. Increasing numbers decided to stay on permanently. In the 1970s waves of new Italians moved to Malindi, starting tourist businesses, opening restaurants, and creating a go-to destination, especially for Italians keen to experience "the tropics." In his study of growing European tourism to Malindi in the 1970s, anthropologist Robert Peake imagines them as a kind of postcolonial European "leisure class" (echoing Thorstein Veblen): "Tourism represents the West's power over the underdeveloped world," given that, while on holiday in the Third World, the "tourist enjoys facilities and activities usually beyond the means of most hosts."[49]

Beginning in the 1970s chartered flights directly from Rome brought in waves of tourists. As a journalist recently wrote, "In the 1980s, Malindi's reputation as a haven for Italian fugitives and pensioners with mafia links had become a staple in the Italian media. Malindi was the sex-and-drugs destination of the Italian bad boy, helped by the absence of an extradition treaty between Kenya and Italy."[50] Everything about the city took on an Italian flavor—the food, the clothes, the language, the street signs, and the hotels. The city became associated with a kind of tourism that was gleefully lowbrow with tacky affectations of glamour. Italian tycoon Flavio Briatore set up a string of ventures based in Malindi before a series of fraud convictions forced him to downscale. In 2014 he was back again, setting up the Billionaire Resort, whose opening was attended by former Italian prime minister Silvio Berlusconi, Formula One race car driver Fernando Alonso, and even Kenyan president Uhuru Kenyatta. Yet the resort also caused controversy after accusations that it had encroached upon and damaged protected Kenyan marine lands. The clientele for Malindi's tacky nouveau riche culture was not only Italians and other European tourists but also many local Italians who had permanently immigrated with their families and who still remain in Malindi. They include people like former satellite engineer Franco Esposito, who arrived in Malindi in 1964 as part of the San Marco team but then never left. In 2017, now no longer involved in anything to do with space, he was a Kenyan citizen, running for office for a place in the national parliament.[51] By 2015 Malindi had one thousand full-time Italian residents, thousands of Italian homeowners, and as many as fifty thousand annual visitors.[52]

Despite all the Italian involvement, Malindi has remained an extremely poor area. One journalist noted, "After three decades of [Italian] mafia money and a rampant sex trade Malindi's tourist income is drying up."[53] The lack of tourist money, the tourist industry's negative effect on local industries, and general disinterest from the Kenyan government have ensured that Malindi is now one of the poorest cities in all of Kenya, where 78 percent of the population live at or below the poverty threshold.[54] The poverty rate undoubtedly contributed to the most unsavory legacy of the Italian presence: a booming sex trade in underage girls that has drawn the attention of feminist and anti-sex-trafficking activists from both Kenya and Italy, a campaign that fortunately appears to have at least reduced the horror of this activity, although some speculate that the trafficking has merely gone underground.

Kenya and Its Claims

The Kenyan government, once sympathetic to the San Marco project, has also begun to weigh in with censure. The original agreement from 1963 had been renewed in 1964, 1987, and 1995. In between, in 1969, the Italians almost had to stop their work on the satellite project when the site itself was subjected to a counterclaim when its status transformed, not in any physical way, but in a legal capacity. That year, President Jomo Kenyatta issued a proclamation declaring that Kenyan "internal waters" now extended beyond three nautical miles to twelve nautical miles (22.24 kilometers). The immediate implication was that the two San Marco project platforms in Ungwana Bay were now rather abruptly and directly under Kenyan sovereignty.[55] The Italians scrambled to placate the Kenyan government, and the latter, remarkably, decided not to renegotiate a territorial claim on the platforms, allowing the Italians an exemption to operate as before. The next Italian launch, of NASA's *X-Ray Explorer*, fortuitously was scheduled for Kenya's independence day in 1970. American State Department officials spotted an "unusual public relations opportunity" and "suggest[ed] naming the rocket in Swahili, possibly 'Uhuru,'" the Swahili word for "independence."[56] As one journalist noted, launching "Uhuru [into space] was . . . a statement of recognition [on the part of the Italians] that the space center was [now] within Kenya's borders."[57]

After twenty-seven rocket launches, the Italians effectively abandoned the San Marco satellite launch facility after a final firing in 1988. The causes were many and included internal struggle among different Italian scientific and industrial factions as well as Broglio's inability to expand the offshore platform to a land-based (on the Kenyan coast) launch site, which had long been the project's ultimate goal.[58] Yet although the twin platforms were abandoned (and now lie derelict and damaged in the ocean), Italy (through the Italian Space Agency, since 2003) retained "ownership" of the old base camp, which was upgraded and expanded to 3.5 hectares (about 8.6 acres) and renamed the Luigi Broglio Space Center. After a major tripartite agreement signed in 1995 between Kenya, the Italians, and the European Space Agency, the site continues to be a major facility for tracking satellites and data acquisition from orbit, mostly from remote-sensing satellites.[59] More important, the Italians began to contract out the center's facilities to third parties, such as China and the European Space Agency, who require tracking and data reception in the equatorial region.

The new surge of activity at the space center, as well as the impending expiration of the original agreement with the Italians, led Kenyan government authorities to review the whole arrangement. A burst of media attention followed. A study was commissioned in 2012 jointly by the Kenyan government's Energy and Education Committees and, among other things, found that the Italians were delinquent on a number of accounts: they had refused to share revenue gained from the center with Kenya; they had failed to get permission from Kenya to host third-party countries; they had declined to share with Kenyan authorities data collected at the center by third parties; and they had trained in space technology operations far fewer than the forty Kenyans they had originally promised. All these stipulations had apparently been enumerated in an agreement between the Italians and the Kenyans in 1995.[60] During the media attention in 2012, the Italian space center—and the few Kenyan government representatives who advocated for its presence—were subjected to intense criticism in the Kenyan parliament. The complaints were varied. Many were upset that no Kenyans were employed by the Italians. Some objected on the basis of sovereignty issues. It was learned that the Italian government, as a gesture of conciliation, had disbursed 240 million Kenyan shillings (about $2.88 million in 2012 dollars) through the Kenyan Coast Development Authority for community development in wider Malindi. This, the first direct funds to be shared by the Italians since the project began in the early 1960s, did little to placate some Kenyans, as this money was widely seen as promotional money for the unseemly Malindi tourist industry or as far too insignificant after nearly five decades of Italian presence. One angry member of Parliament, Danson Mungatana, stood up and noted:

> for the Assistant Minister to stand here and tell us that since 1963 up to now, the San Marco Project has only given [$2.88 million] to community projects is a joke. What we want to ask the Assistant Minister is: What benefits are there that you can show for the people of Malindi and Coast Province in general from that project? . . . We need this thing to be taken seriously by the Ministry of State for Defence. It is very unfair for the Government to stand here and tell us that they are making nothing.[61]

Later, the minister for forestry and wildlife, Dr. Noah Wekesa, weighed in:

> I think the agreement that we, as Kenya, made with the Italians is overdue [for renegotiation]. But like all these agreements that were made at the time of independence, the Kenya Government was being duped by a lot of people. In this particular case, there is the need to review the agreement. What I found out, as Minister, is that a lot of things that the station does are not known by us. They are very secretive.[62]

The brewing animus of some Kenyans toward the Italian presence was resolved, at least temporarily, by the work of a Joint Steering Committee, representing both Italians and Kenyans, which first met in September 2014.[63] After over two years of work, in October 2016, Italy and Kenya formally signed a renewal of the agreement on the Luigi Broglio Space Center. In a statement, the Italian Ministry of Foreign Affairs noted how the center was "an instrument of scientific and technological dialogue with the entire continent of Africa."[64] For the Kenyan side, Italy now offered support to Kenya's burgeoning space agency, help with a regional center for Earth observation, and access to data collected at the Broglio center. Most critically, the Italians would provide training and education for locals and help with telemedicine. The agreement was reached not without much conflict and dissension. Earlier in the year Member of Parliament Dr. Wibur Ottichillo, a highly respected Kenyan space scientist and one of the strongest opponents of the Italy-Kenya deal, warned that Kenya would lose a significant amount of revenue from various tracking and communications experiments conducted at Malindi unless the terms of any agreement were made more equitable. Ottichillo and many others in Parliament had long lobbied for the government to not renew the agreement with Italy, suggesting that the final agreement in 2016 was made despite significant objections of a sizable domestic Kenyan constituency that included parliamentarians, academics, and scientists.[65]

Conclusions

In this chapter, I use "site" as a heuristic and the San Marco project as a case study to interrogate assumptions about transnational science in the postcolonial setting during the Cold War. Politics, science, and geography were obvious determinants of the San Marco project, but a deeper investigation of "where" (in addition to "what") reveals the ambivalences, blurred boundaries, and inequalities in the contacts between key actors. Focusing on the site of Malindi (in addition to project or community or scientific goals, for example) reveals a much more heterogeneous coterie of principal actors who were constantly in contact, often in consonance but frequently in friction, with each other. Such an approach allows us to subvert the received history of the San Marco project, which not only completely excluded Kenya from the story but also framed it as an unqualified success. Here, defining it as a success depended critically on making the site coterminous with the project, thus excluding actors—such as the Malindi area population, Kenyan parliamentarians, Italian immigrants and tourists, offshore laborers at Mombasa, and factory workers in Texas who built the Santa Rita—all of whom contributed to its implementation and ultimately to its ambiguous outcome. In particular, it opens up the possibility of reconciling counterclaims, such as the one by a Kenyan parliamentarian who noted in 2012, after forty years of an Italian presence, that San Marco was in fact *not* successful given that the "venture [has] not [been] an income generating project."[66]

I make three broad arguments here, each derived from looking at the *site* of San Marco in addition to the more obvious frame of reference, the *project* of San Marco. First, a deeper exploration of the multiple histories of San Marco invites us to think of the site not as a singular and discrete quantity but as one that was dispersed into fragments. This dispersion was manifested in a global scattering of actors, actions, and areas involved in and affected by an ostensibly transnational scientific project orchestrated by Europeans and Americans and requiring the geographical advantages proferred by a specific locale, one defined locally but, in actuality, dispersed globally. Fragments of this dispersed site came into conflict with non-*project* phenomena, provoking conflicting claims and resulting in a cluster of dislocations, such as the disposession of an ethnic minority in Kenya, the rise of a vocal Kenyan opposition against the project, and, most unsettlingly, the neoliberal reinvention of Malindi that has resulted in sexual exploitation of a vulnerable population. The postcolonial setting of San Marco provides potent language to make sense of the violence of this social dislocation, offering an alternative frame to historicize San Marco, completely at odds with the received narrative shaped by Italians and Americans. This postcolonial frame becomes legible when San Marco is viewed as a dispersed and fragmented site, with incursions into contested postcolonial spaces, still raw from the original violence of colonialism.

Second, I argue that *site dispersion* not only explains the physical dispersion of the site across Kenya and into global points but also highlights a kind of *ontological* dispersion where "what the project is" has different meanings for different actors. These very conflicting worldviews of the same project result from accounting for multiple, incommensurate, and conflicted geographies of the same project—each with its own history, imperatives, and demarcations. They engender deeply contested claims that are manifest in a particularly striking way in postcolonial contexts, where older colonial forms of violence endemic to knowledge production are reformulated in new language to enact new forms of control. The outcome, as in the case of San Marco, was highly ambiguous, characterized by the production of valuable scientific knowledge and concomitant social disenchantment and dislocation.

Finally, the San Marco case suggests that the choice of a particular site for transnational science itself represents a countervailing force against the regulation of knowledge, partly because, by definition, the choice to locate a project at a particular site creates and intensifies certain frictions. A site for doing transnational science and technology is a statement by one party that it has more control and an acknowledgment by another that it has ceded it. Sites for transnational science and technology are, thus, by definition, unstable. In that context, site dispersion, the physical and ontological spread of the sites required to produce a particular kind of scientific and technological knowledge, often works to resist the nation-state's campaign to regulate the flow of knowledge. Such transnational scientific and technology endeavors—which all to some degree or other involve dispersion of sites—have embedded within themselves, in the form of fragments, forces that inflict friction and violence to spaces outside the strict boundaries of the project. Such forces, and the phenomena they produce, belong outside the history of the project and are rendered legible only when we think of transnational sites as mutable, unfixed, and ultimately dispersed, encompassing all types of heterogeneous and nonprogrammatic phenomena. The San Marco project has always been understood as a success because it has been seen as physically contained, as a discrete project. But in fact, using the perspective of the site, and, particularly, the postcolonial site, with all its blurred edges and multiple claims, we see that its success was much more conditional and its legacy marred by deeply uncomfortable social realities.

Notes

1. For some prior meditations on the heuristic of "site" in the global history of science, see Asif Siddiqi, "Another Global History of Science: Making Space for India and China," *British Journal for the History of Science: Themes* 1 (2016): 115–143; Asif Siddiqi, "Another Space: Global Science and the Cosmic Detritus of the Cold War," in *Space Race Archaeologies: Photographs, Biographies, and Design*, ed. Pedro Igancio Alonso (Berlin: DOM, 2016), 21–37.

2. Naomi Oreskes and John Krige, eds., *Science and Technology in the Global Cold War* (Cambridge, MA: MIT Press, 2014); Grégoire Mallard, Catherine Paradeise, and Ashveen Peerbaye, eds., *Global Science and National Sover-*

eignty (New York: Routledge, 2009); Kapil Raj, "Beyond Postcoloniaism . . . and Postpositivism: Circulation and the Global History of Science," *Isis* 104 (2013): 337–347; Francesca Bray, "Only Connect: Comparative, National, and Global History as Frameworks for the History of Science and Technology in Asia," *East Asian Science, Technology and Society: An International Journal* 6 (2012): 233–241; Fa-ti Fan, "The Global Turn in the History of Science," *East Asian Science, Technology and Society: An International Journal* 6 (2012): 249–258; Lissa Roberts, "Situating Science in Global History: Local Exchanges and Networks of Circulation," *Itinerario* 33, no. 1 (2009): 9–30; Sujit Sivasundaram, "Science and the Global: On Methods, Questions, and Theory," *Isis* 101 (2010): 146–158; Simone Turchetti, Nestor Herran, and Soraya Boudia, "Introduction: Have We Ever Been "Transnational'? Towards a History of Science across and beyond Borders," *British Journal for the History of Science* 45, no. 3 (2012): 319–336.

3. For trenchant summaries of this work, see Warwick Anderson, "Introduction: Postcolonial Technoscience," *Social Studies of Science* 32, nos. 5/6 (2002): 643–658; David Arnold, "Europe, Technology, and Colonialism in the 20th Century," *History and Technology* 21, no. 1 (2005): 85–106; Roy M. MacLeod, introduction to *Nature and Empire: Science and the Colonial Enterprise*, ed. Roy M. MacLeod, Osiris, 2nd ser., vol. 15 (Chicago: University of Chicago Press, 2000), 1–13; Suzanne Moon, "Introduction: Place, Voice, Interdisciplinarity: Understanding Technology in the Colony and the Postcolony," *History and Technology* 26, no. 3 (2010): 189–201; Michael A. Osborne, introduction to *Science, Technology and Society: An International Journal Dedicated to the Developing World* 4, no. 2 (1999): 161–170; Suman Seth, "Putting Knowledge in Its Place: Science, Colonialism, and the Postcolonial," *Postcolonial Studies* 12, no. 4 (2009): 373–388; Asif A. Siddiqi, "Technology in the South Asian Imaginary," *History and Technology* 31, no. 4 (2015): 341–349.

4. I offer some preliminary thoughts on the issue of scale in my "Science, Geography, and Nation: The Global Creation of Thumba," *History and Technology* 31, no. 4 (2015): 420–451.

5. See chapters 1 and 2 in this volume

6. Gabrielle Hecht, *Being Nuclear: Africans and the Global Uranium Trade* (Cambridge, MA: MIT Press, 2012); Peter Redfield, *Space in the Tropics: From Convicts to Rockets in French Guiana* (Berkeley: University of California Press, 2000).

7. For useful insights into the relationship between language and power in the way "we" speak about science and technology in Africa, see particularly Clapperton Mavhunga, "A Plundering Tiger with Its Deadly Cubs? The USSR and China as Weapons in the Engineering of a 'Zimbabwean Nation,' 1945–2009," in *Entangled Geographies: Empire and Technologies in the Global Cold War*, ed. Gabrielle Hecht (Cambridge, MA: MIT Press, 2011), 231–266.

8. The only scholarly treatment, albeit a brief one, can be found in John

Krige, "Introduction and Historical Overview: NASA's International Relations in Space," in *NASA in the World: Fifty Years of International Collaboration in Space*, by John Krige, Angelina Long Callahan, and Ashok Maharaj (New York: Palgrave Macmillan, 2013), 31–33. A semiofficial account of the project also provides very useful details: Michelangelo De Maria and Lucia Orlando, *Italy in Space: In Search of a Strategy, 1957–1975* (Paris: Beauchesne, 2008), 77–106, 189–222. See also an earlier version of this work that covers a longer period of time: Michelangelo De Maria, Lucia Orlando, and Filippo Pigliacelli, *Italy in Space, 1946–1988* (Noordwijk: ESA Publications Division, 2003), 13–20.

9. John Krige, *American Hegemony and Postwar Reconstruction of Science in Europe* (Cambridge, MA: MIT Press, 2006).

10. The Commissione per le Ricerche Spaziali was the principal policy-making institution of the seedling Italian space program, although the commission was staffed with academics: Edoardo Amaldi (University of Rome), Nello Carrara (University of Florence), Corrado Casci (Polytechnic of Milan), Mario Boella (Polytechnic of Turin), Giampiero Puppi and Guglielmo Righini (University of Bologna). See De Maria and Orlando, *Italy in Space*, 57. For an early snapshot of the program, see Warren C. Metmore, "San Marco Satellite to Probe Air Density," *Aviation Week and Space Technology*, Aug. 26, 1963, 76–78.

11. The early Italian space effort was grouped under the Consiglio Nazionale delle Ricerche (National Center for Research), which included two subordinate bodies, the Commissione per le Ricerche Spaziali (Space Research Commission) and the Instituto per le Ricerche Spaziali (Institute for Space Research). The Commissione per le Ricerche Spaziali was basically a policy-making institution, while the Instituto per le Ricerche Spaziali defined scientific priorities. A third organization, the Centro di Ricerche Aerospaziali (Center for Aerospace Research), was primarily involved in R&D and production of satellites and scientific instruments.

12. The exact geographical coordinates of the San Marco platform were longitude 40°12′15″ east and latitude 2°56′18″ south. See *San Marco Range User's Manual*, Dec. 2, 1968, pp. I–11, NASA History Office Archives, Italy Space Program, San Marco folder.

13. Data collections from the two *San Marco 2* experiments can be accessed at NASA's Space Science Data Coordinated Archive at http://nssdc.gsfc.nasa.gov/nmc/datasetSearch.do?spacecraft=San%20Marco%202. The chief scientist for the set of experiments was not Broglio but Dr. Robert F. Fellows (1920–), who, from 1961, served as chief of the Chemistry Program of Geophysics and Astronomy at NASA. Through the 1960s and 1970s, Fellows oversaw the scientific programs of a number of NASA and international satellite projects.

14. In NASA parlance, the satellite was known as *Small Astronomy Satellite-A* (*SAS-A*) or *Explorer 42*.

15. N. Jagoda et al., "Uhuru X-Ray Instrument," *IEEE Transactions on Nu-*

clear Science 19, no. 1 (Feb. 1972): 579–591; R. Giavvoni et al., "An X-Ray Scan of the Galactic Plane from *Uhuru*," *Astrophysical Journal* 165 (Apr. 15, 1971): L27–L35; *SAS-A* press kit, Dec. 2, 1970, https://ntrs.nasa.gov/archive/nasa/casi.ntrs.nasa.gov/19710003482.pdf.

16. "The Noble Prize in Physics 2002," http://www.nobelprize.org/nobel_prizes/physics/laureates/2002/giacconi-facts.html.

17. National Aeronautics and Space Act of 1958, Pub. L. 85-568, 72 Stat. 426, signed by the president on July 29, 1958, reproduced as document II-17 in *Exploring the Unknown: Selected Documents in the History of the U.S. Civil Space Program*, vol. 1, *Organizing for Exploration*, ed. John M. Logsdon (Washington, DC: NASA, 1995), 334–345, esp. 335, 339.

18. John M. Logsdon, "The Development of International Space Cooperation," in *Exploring the Unknown: Selected Documents in the History of the U.S. Civilian Space Program*, vol. 2, *External Relationships*, ed. John M. Logsdon (Washington, DC: NASA, 1996), 1–15, at 1.

19. Porter to van de Hulst, Mar. 14, 1959, in Logsdon, *Exploring the Unknown*, 2:18–19.

20. Arnold W. Frutkin, *International Cooperation in Space* (Englewood Cliffs, NJ: Prentice-Hall, 1965), 42–43.

21. Broglio's first official communication to NASA about the equatorial satellite project dates to August 9, 1961, followed by a formal proposal submitted on September 26. See memo to Mr. Jesse Mitchell on Ad Hoc Committee Meeting on Italian satellite proposal, Nov. 21, 1961, NASA History Office Archives, Italy Space Program, San Marco folder.

22. De Maria and Orlando, *Italy in Space*, 79.

23. Freddie del Curatolo, "Franco Esposito, Fifty Years of Kenya from S. Marco to Parliament," *Malindikenya.net*, Mar. 6, 2017, http://www.malindikenya.net/en/articles/news/people/franco-esposito-fifty-years-of-kenya-from-smarco-to-parliament.html. See also American Embassy Rome to State Department, "Italian Space Program as Discussed with Congressman Anfuso," Oct. 10, 1961, folder 965.8011/6-2861, box 3076, Central Decimal File, 1960–1963, RG 59, General Records of the Department of State, National Archives and Records Administration (hereafter NARA), College Park, MD.

24. Italy colonized and ruled over most of current-day Somalia as "Italian Somaliland" from the 1880s to 1941 and then again, through a UN Trusteeship, from 1949 to 1960. See Robert L. Hess, *Italian Colonialism in Somalia* (Chicago: University of Chicago Press, 1966); Paolo Tripodi, *The Colonial Legacy in Somalia* (New York: St. Martin's Press, 1999).

25. Michael Durham, "Italy's African Space Triumph," *Life*, May 26, 1967, 101–105.

26. "Memorandum of Understanding between the Italian Space Commission of the National Council of Research and the United States National Aeronautics

and Space Administration," May 31, 1962, reproduced in H. N. Nesbitt, *History of the Italian San Marco Equatorial Mobile Range*, NASA CR-111987 (Washington, DC, 1971), appendix A, A-3 to A-5. See also "US-Italian Space Program Aims at Equatorial Launch," NASA News Release 62–192, Sept. 6, 1962, NASA History Office Archives, Italy Space Program, San Marco folder.

27. The Kenya option was being seriously considered as early as September 1962. See State Department to American Embassy Rome, "Reports on San Marco Meetings," Oct. 12, 1962, folder 965.802/1-1262, box 3076, Central Decimal File, 1960–1963, RG 59, General Records of the Department of State, NARA. See also memo "Comments on the Communication Section of the San Marco 'Project,'" Dec. 6, 1962, NASA History Office Archives, Italy Space Program, San Marco folder.

28. "San Marco Environmental Study Program, Preliminary Report, Vought Missiles and Space Company—Texas, 00.173," Feb. 22, 1963, cited in Nesbitt, *History of the Italian San Marco Equatorial Mobile Range*, 16, R-1.

29. Del Curatolo, "Franco Esposito."

30. Nesbitt, *History of the Italian San Marco Equatorial Mobile Range*, 14.

31. In 2012 the company was bought out by Joy Global, based in Milwaukee, Wisconsin, and is now known under the new name. See Mike Elswick, "Manufacturing Plant to Drop LeTourneau from Name," *Longview News-Journal*, May 18, 2012.

32. Dennis Karwatka, "Technology's Past: R. G. LeTourneau and His Massive Earth-Moving Equipment," *Tech Directions* 65, no. 10 (May 2006): 8.

33. R. G. LeTourneau, *Mover of Men and Mountains* (New York: Prentice-Hall, 1960), 256–260.

34. SAIPEM stood for Società Azionaria Italiana Perforazioni e Montagge. See Nesbitt, *History of the Italian San Marco Equatorial Mobile Range*, 17; "Platform No. 9, Scarabep [*sic*] (Santa Rita)," https://museum.letu.edu/items/show/335.

35. Nesbitt, *History of the Italian San Marco Equatorial Mobile Range*, 19–20. This was a "self-elevating pier barge" (or BPL). See *Army Water Transport Operations, Field Manual* (Headquarters, Department of the Army, Oct. 1976), 2–5.

36. Kenya National Assembly, Official Report (Hansard), Wednesday, Apr. 18, 2012, 24–25. In a meeting between NASA and Italian officials in September 1962, the Italians assured NASA that they were seeking UK approval to use a variety of resources located at Mombasa. See State Department to American Embassy Rome, "Reports on San Marco Meetings," Oct. 12, 1962.

37. For Kenyan independence, see Bethwell A. Ogot and William Robert Ochieng, *Decolonization and Independence in Kenya, 1940–93* (London: J. Currey, 1995); Daniel Branch, *Defeating Mau Mau, Creating Kenya* (Cambridge: Cambridge University Press, 2009).

38. For the agreement with the university, see "Memorandum of Understanding between the Italian Space Commission of the National Council of Research and the University of East Africa Royal College, Nairobi," Jan. 11, 1964, folder SP-Space & Astronautics IT-A 1/1/64, box 3076, Central Decimal File, 1960–1963, RG 59, General Records of the Department of State, NARA.

39. "Possible Answers by Prof. Broglio at the Press Conference on San Marco Project," Feb. 8, 1964, folder SP-Space & Astronautics IT-A, box 3076, Central Decimal File, 1960–1963, RG 59, General Records of the Department of State, NARA; "Space Probe May Help East Africa," *East African Standard*, Feb. 11, 1964; "Kenya on the Science 'Map,'" *Daily Nation*, Feb. 11, 1964; A. N. Hunter, "World Scientists Welcome Kenya Space Project," *East African Standard*, Feb. 14, 1964, 1.

40. American Embassy Nairobi to State Department, May 28, 1964, folder SP 12-4 Scientific Satellites IT/San Marco, box 3076, Central Decimal File, 1960–1963, RG 59, General Records of the Department of State, NARA.

41. The terms of the postindependence project are laid out in Italian Consul General to Robert J. Ouko, Government of Kenya, Jan. 10, 1964, folder SP-Space & Astronautics IT-A 1/1/64, box 3076, Central Decimal File, 1960–1963, RG 59, General Records of the Department of State, NARA.

42. First quotation from American Embassy Nairobi to State Department, May 28, 1964, folder SP 12-4 Scientific Satellites IT/San Marco, box 3076, Central Decimal File, 1960–1963, RG 59, General Records of the Department of State, NARA. Second quotation from American Embassy Rome to State Department, May 22, 1964, folder unnamed, box 3141, Central Foreign Policy Files, 1964–1966, RG 59, General Records of the Department of State, NARA.

43. The Scout and the satellite payload could not be flown directly to Mombasa or Malindi because the airports at these two cities were not equipped to handle such large and sensitive pieces of cargo. Thus, they were flown into Nairobi and then driven to Malindi.

44. Nesbitt, *History of the Italian San Marco Equatorial Mobile Range*, 25, 26.

45. Durham, "Italy's African Space Triumph," 104.

46. For Malindi, see E. R. Bradley, *The History of Malindi: A Geographical Analysis of an East African Coastal Town* (Nairobi: East African Literature Bureau, 1973). I use the terms "Arab" and "Swahili" as reductive terms that flatten much nuance and complexity. "Arab" here generally indicates ancestral origins in Arabia. "Swahili" here denotes communities that have adopted a coastal culture, including the use of the Swahili language and self-identification as Muslims. James R. Brennan notes that in the past century, "'Arab' and 'Swahili' [have] categorically differentiate[d] themselves from mainland Africans." See James R. Brennan, "Lowering the Sultan's Flag: Sovereignty and Decolonization in Coastal Kenya," *Comparative Studies in Society and History* 50, no. 4 (2008): 831–861, esp. 832, ref. 3.

47. Brennan, "Lowering the Sultan's Flag."

48. For the Bajuni in general, see James de Vere Allen, *Somali Origins: Swahili Culture and the Shungwaya Phenomenon* (Athens: Ohio University Press, 1993); Derek Nurse, "Bajuni Database General Document," http://www.ucs.mun.ca/%7Ednurse/bajuni_database/general_document.pdf. See also Inter-Agency, *Land, Property, and Housing in Somalia*, Norwegian Refugee Council / UN Habitat / UNHCR, July 2008, http://www.refworld.org/docid/496dfeb82.html.

49. Robert Peake, "Swahili Stratification and Tourism in Malindi Old Town, Kenya," *Africa: Journal of the International African Institute* 59, no. 2 (1989): 209–220, at 210.

50. Parselelo Kantai, "Kenya's Malindi, a Paradise Lost," *Africa Report*, Oct. 29, 2014, http://www.theafricareport.com/Society-and-Culture/kenyas-malindi-a-paradise-lost.html.

51. Del Curatolo, "Franco Esposito." For another permanently relocated Italian, Cleto Ancona, see Freddie del Curatolo, "Cleto, an Italian Istitution [*sic*] Man in Malindi," *Malindikenya.net*, Dec. 19, 2016, http://www.malindikenya.net/en/articles/news/people/cleto-an-italian-istitution-man-in-malindi.html.

52. Tristan McConnell, "Kenya's 'Little Italy,'" http://www.tristanmcconnell.co.uk/kenyas-little-italy/.

53. Kantai, "Kenya's Malindi, a Paradise Lost."

54. Ibid.

55. Republic of Kenya, *The National Assembly: Official Report (Hansard), Third Session, Tuesday, 21st March, 1972 to Friday, 12th May, 1972*, entry for Apr. 11, 1972, 586.

56. State Department to American Embassy Rome, Dec. 10, 1970, folder SP IT, box 3005, Central Foreign Policy Files, 1967–1969, RG 59, General Records of the Department of State, NARA.

57. "The Space Center Kenya Doesn't Own," *Owaahh*, Mar. 31, 2016, http://owaahh.com/space-center-kenya-doesnt/.

58. Michelangelo De Maria and Lucia Orlando, "Early Italian Space Activities—the San Marco and SIRIO Miracles," in *Proceedings of the Concluding Workshop: The Extended ESA History Project, 13–14 April 2005, ESA SP-609*, ed. B. Battrick and L. Controy, 55–65 (Paris: ESA Headquarters, 2005), 55–65.

59. The original formal name, the San Marco Equatorial Range (SMER), was changed to the Luigi Broglio Space Center in 2004. Broglio passed away in 2001. After the trilateral Italy–Kenya–European Space Agency Protocol of 1995 expired on March 14, 2010, it was extended to December 31, 2010, then to June 2012, and then to June 2013.

60. Consumer Federation of Kenya, http://www.cofek.co.ke/index.php/news-and-media/561-how-the-italian-government-has-duped-kenya-continues-to-make-a-kill-from-the-san-marco-space-application-centre-in-malindi-as-locals

-get-a-raw-deal?showall=; "Kenya 'Hasn't Gained from Space Center,'" *Daily Nation*, Sept. 17, 2012, http://www.nation.co.ke/news/Kenya-hasnt-gained-from-space-centre/-/1056/1510422/-/view/printVersion/-/m99tm7/-/index.html.

61. Kenya National Assembly, Official Report (Hansard), Wednesday, Apr. 18, 2012, 24–25, at 21.

62. Ibid. 25.

63. The Kenyan delegation was chaired by Kiritu Wamai, the secretary of administration of the Ministry of Defense, while the Italian was led by Marco Claudio Vozzi, the deputy director for sub-Saharan Africa in the Italian Ministry of Foreign Affairs. See "San Marco Project—Latest Developments," http://www.ambnairobi.esteri.it/ambasciata_nairobi/en/ambasciata/news/dall_ambasciata/2014/09/20140930sanmarco.html.

64. "Italy and Kenya Renew the 'Luigi Broglio' Space Centre Agreement (Trento, 24 October 2016)," Oct. 25, 2016, http://www.esteri.it/mae/en/sala_stampa/archivionotizie/approfondimenti/2016/10/diplomazia-e-hi-tech-italia-e-kenya.html; "ASI in Kenya for the First Joint Steering Committee," *Research Italy*, Jan. 27, 2017, https://www.researchitaly.it/en/news/asi-in-kenya-for-the-first-joint-steering-committee/.

65. John Ngirachu, "Don't Renew Space Deal, MPs Tell State," *Daily Nation*, June 15, 2014, http://mobile.nation.co.ke/news/Dont-renew-space-deal-MPs-tell-state/1950946-2349776-format-xhtml-5a7hp1z/index.html; "Ottichillo Warns Kenya Set to Lose Sh60m in Space Tests," *Business Daily*, Feb. 1, 2016, http://www.businessdailyafrica.com/Ottichillo-warns-Kenya-set-to-lose-Sh60m-in-space-tests/539546–3056562–15hwq0rz/index.html.

66. Kenya National Assembly, Official Report (Hansard), Wednesday, Apr. 18, 2012, 20.

CHAPTER SEVEN

Bringing the Environment Back In

A Transnational History of Landsat

Neil M. Maher

In the spring of 1976, as the US government made preparations to celebrate the country's bicentennial during the upcoming July Fourth holiday, the United Nations published the quarterly report of the Mekong Committee. Although the committee had originated back in 1957 to promote the development of Southeast Asia's lower Mekong River basin through large-scale dams and irrigation projects, this particular report publicized scientific data captured by orbiting satellites developed by the National Aeronautics and Space Administration (NASA). To make this scientific information more legible, the report included a full-page map of parts of Laos, Cambodia, Thailand, and Vietnam that was overlaid with ten orbital tracks of one of NASA's satellites (see fig. 7.1). "Landsat-II imagery is showing important new information," explained the report, adding that this particular Earth-observing satellite had collected more than 160 "frames" of data as it circled high above the 230,000-square-mile region between September 1975 and January of the following year. This valuable scientific work, assured the report's conclusion, "is continuing."[1]

The Mekong Committee's *Landsat* map also shows how technologies and the scientific knowledge they help to create, while often initiated nationally, in this case within the United States, almost always travel far beyond national borders. The map of the lower Mekong basin, for instance, illustrates not only the movement of space technology across the political boundaries of Southeast Asia but also the circulation of scientific

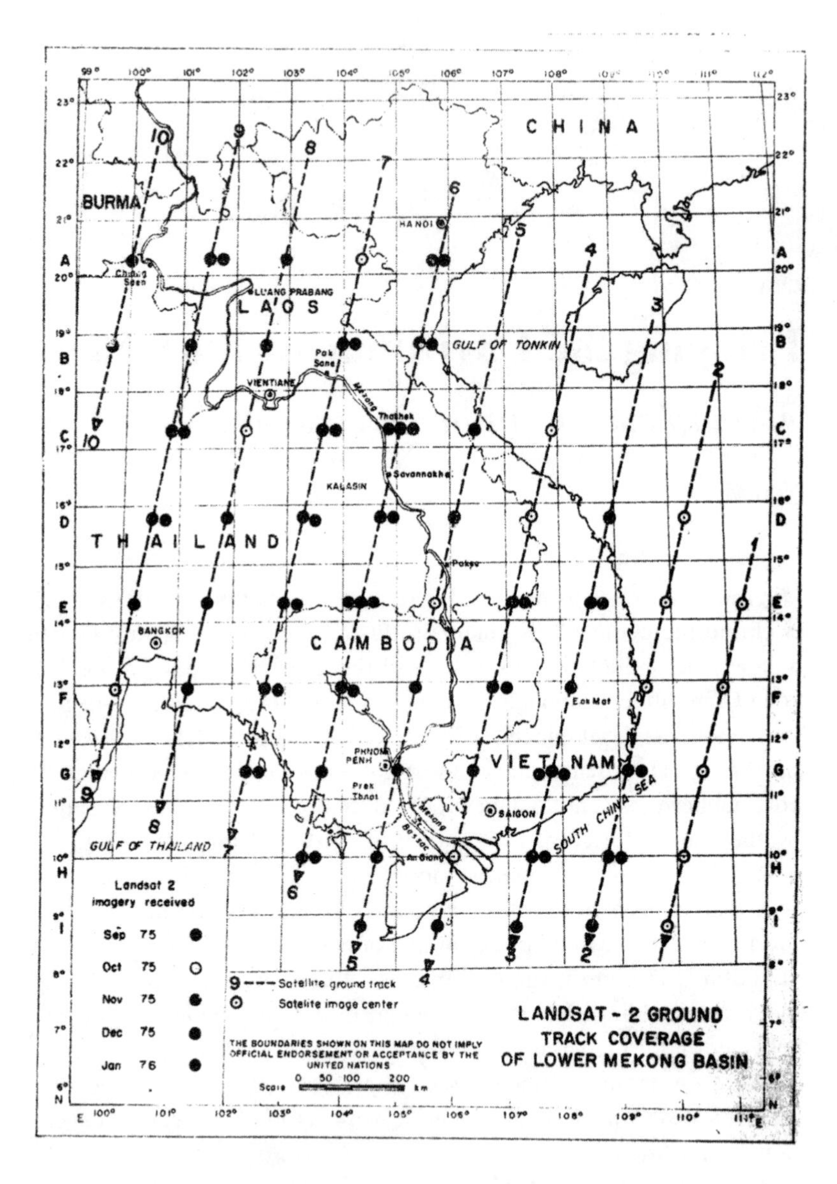

FIGURE 7.1. Map depicting the orbital track coverage of *Landsat 2* over the lower Mekong River basin from September 1975 through January 1976.
Source: Courtesy of National Aeronautics and Space Administration.

knowledge; the small, empty circles on the map represent "satellite image centers" that received and processed data on the ground in South Vietnam, Cambodia, Thailand, and Laos. This UN map thus neatly displays the novel framework of this collection of essays, which analyzes national governments and state agencies as actors, often quite powerful ones, functioning within dispersed international networks that both produce and circulate technologies and scientific knowledge. Rather than erasing the nation-state, the Mekong Committee's map, much like this collection, places it within its transnational context.

The satellite map and the Mekong Committee's overall report, however, also identify another important historical agent that is often missing from the current historiography on transnational technoscience. Far below *Landsat 2*, which orbited more than five hundred miles above Earth, flowed the Mekong River. While the UN map represents the waterway with a double line winding its way from Burma in the north to the southern tip of Vietnam, the body text of the report focuses entirely on the basin's natural environment. The publication explained that the *Landsat* data collected by the Mekong Committee would be used to analyze the basin's hydrology, especially with respect to flooding, to differentiate between different forest types, such as evergreen, deciduous, and mangrove, and to identify a wide variety of land use practices from rice farming to rubber tree plantations. "The main objective of the Mekong Committee investigations using Landsat data," explained the introduction to the report, was to collect scientific information that could be used to map "agricultural crops and land use, and for soil moisture monitoring."[2] As this full-page illustration suggests, technology and science do not operate alone on the transnational stage. Rather, they interact with, and most often seek information about, the natural world.[3]

Although historians of transnational science and technology have shied away from incorporating nature into their analyses, environmental historians have for decades been analyzing how nation-states explore and extract natural resources within their borders, as well as how federal governments regulate such land use and enact legislation to correct environmental problems.[4] This state-centered approach served early practitioners well. However, as environmental historian Donald Worster explained in his seminal 1982 essay "World without Borders: The Internationalizing of Environmental History," "the nation-state is no longer a suitable framework." The field's future success, Worster predicted, "will be found in research that moves easily across national borders." Environmental historians have followed Worster's advice, tracking various natures, whether they be flowing water, wafting pollution, or migrating animals, weeds, and diseases, across political boundaries.[5] This transnational approach, Worster concluded more than three decades ago, "calls for the reformulation of our research, so that when we find our Walden Pond to study we will also have found the River Ganges."[6]

The present chapter treats the Mekong River basin, as well as other natural environments beyond the United States, as important historical actors within the transnational history of science and technology during the twentieth century. I use the history of NASA's *Landsat* satellites as a case study to analyze how a technology developed within the United States became a hub that bound together a thick transnational network of space and ground communications systems, international and national agencies, American corporations and NASA, as well as indigenous engineers, technicians, and scientists attempting to better understand, and control, the natural environment. In the process, *Landsat* became a mechanism for both American hegemony and limited local control within the developing world.

I begin by examining not only the development of *Landsat* technology in America but also several of the impediments that limited the technology's success abroad. Installing ground stations in remote regions of foreign countries was sometimes dangerous; foreign researchers had to learn how to use the data and images provided by American satellites and computers; and government officials across the developing world were also concerned that the orbiting technology would infringe upon their countries' national sovereignty. To overcome these problems, the US government in the early 1970s began "selling" *Landsat* across Asia, Africa, and Latin America. In the end, although local knowledge about various natural environments situated within their own nation-states gave indigenous scientists and government officials a modicum of power regarding *Landsat*, the satellite's thick transnational network allowed the US government to maintain ultimate control over both the space technology and the scientific knowledge it produced. In other words, asymmetries in technology, scientific knowledge, and political power, although masked, were never fully effaced, even though NASA claimed otherwise.

Engineers and scientists at NASA could proudly claim, quite correctly, that their *Landsat* satellite was "made in America." This was because the technology, which the space agency first launched on July 23,

1972, was initially developed from both military hardware such as the *CORONA* spy satellite and civilian technology used clandestinely for war, including the *TIROS* and *ATS* satellites. Such top-secret origins precluded international cooperation on *Landsat*, which circled 560 miles above Earth in near-polar orbit taking 13,000-square-mile "snapshots" of the planet's surface.[7] During the next quarter century, six additional *Landsat* satellites gathered data for millions of images of planet Earth. By radioing back "pictures" of Earth from space, the *New York Times* explained in mid-January 1975, *Landsat* was "providing new insight into man's continuing effort to better manage earth's limited resources as well as aiding in the assessment and understanding of environmental changes."[8]

This thoroughly American technology included multispectral scanners that measured from space four different wavelengths of electromagnetic radiation reflecting off objects on the surface of Earth. Originally, *Landsat* satellites beamed these wavelength measurements back down to NASA's receiving stations in Fairbanks, Alaska, in Goldstone, California, and at the Goddard Space Flight Center in Greenbelt, Maryland. In each of these locations, technicians converted the raw data into visual maps by assigning coded false colors to Earth-bound objects with different wavelengths. *Landsat*, in other words, made the natural environment more legible by measuring the extremely slight temperature variations of the solar heat bouncing off rocks, trees, water, and even animals.[9] As *Science* magazine reported on the tenth anniversary of *Landsat 1*, the maps created from the Earth-observing satellite depicted "scarlet forests, red patchwork farms, blue city grids, brown crinkled mountains, and a delicate web of highways."[10]

Landsat's colorful maps quickly became scientific tools for analyzing natural resources, and NASA immediately began promoting such capabilities through easy-to-read pamphlets and booklets with appealing titles such as *Improving Our Environment*, *Ecological Surveys from Space*, and *Photography from Space to Help Solve Problems on Earth*. According to these publications, *Landsat* satellites aided agriculture and forestry by making possible the inventory of different types of crops and trees, the identification of early signs of plant diseases and insects, and the assessment of soil moisture to guide future land use practices. The space technology proved equally beneficial for the study of hydrological and atmospheric resources; *Landsat* data mapped fresh and salt water, forecast droughts and floods, and identified sources of both water and air pollution. It also provided geological measurements that located underground resources, including oil, natural gas, and mineral deposits, and even allowed biologists to track migratory wildlife both across the land and under the seas.[11]

By helping to manage natural resources, *Landsat* was also helping to manage NASA's public image, which during the early 1970s was suffering on the domestic front from a severe case of "NASA fatigue." As the *Los Angeles Times* put it in April 1972, "A long mental yawn will roll over America next Sunday when Apollo 16 spits fire from its tail and streaks skyward to the moon."[12] Partly because of such apathy, between the Moon landing of 1969 and the launch of Apollo-Soyuz in 1975 Congress drastically cut the space agency's funding. All told, during this six-year period beginning after Apollo 11, the federal government slashed NASA's budget by more than 40 percent, after accounting for inflation, to its lowest real-dollar level since 1962.[13]

In a conscious effort to reverse this trend, NASA administrators began publicizing to the American public *Landsat*'s role in scientifically assessing natural resources located within the United States. From 1972 to 1974 this entailed the development of the Large Area Crop Inventory Experiment, or LACIE. The joint venture by NASA, the Department of Agriculture, and the National Oceanic and Atmospheric Administration (NOAA) combined crop acreage measurements obtained from *Landsat* with meteorological information from NOAA satellites to forecast wheat production in an effort to stabilize the commodity's price for American consumers.[14] Such publicity efforts by NASA succeeded; not only did Congress authorize two additional *Landsat* satellites in 1975 and 1978, but it also increased the space agency's budget by more than 10 percent, after accounting for inflation, between 1975 and 1980.[15]

President Richard Nixon quickly realized that *Landsat* could do for the United States internationally what it had done for NASA domestically. Early on he understood *Landsat*'s promotional potential and announced in September 1969 to the UN's General Assembly that America's new Earth-observing satellites would "produce information not only for the U.S., but also for the world community."[16] Space agency officials were even more explicit, focusing many of their public comments concerning productive uses of *Landsat* data specifically on the natural resources of poorer countries. The new space technology would "assist both the developed and developing areas of the world alike in providing maps and other important resource inventory data," explained a

NASA position paper on remote sensing. In doing so, the report went on to argue, "the use of remote sensors in NASA spacecraft to aid developing countries thus represents an important way for the United States to enhance its world image."[17] By giving poor nations access to scientific data that could help them better manage their own natural resources, *Landsat* technology could raise the international standing of the United States by helping developing countries develop.

There were just two problems with this rosy scenario. First, at least initially, several developing nations openly resisted NASA's remote-sensing technology for fear that it would infringe upon their national sovereignty. While the Soviet Union was concerned that *Landsat* could be used for spying, countries across Latin America were more worried that developed countries would employ the technology to exploit natural resources located in the developing world; wealthier nations such as the United States could use satellite data not only to identify previously undiscovered resources, such as mineral and oil deposits, within poorer countries but also to forecast global crop production in an effort to manipulate agricultural commodity prices.[18] To protect against such actions, in 1975 several developing nations, including Argentina, Chile, Venezuela, and Mexico, cosponsored an unsuccessful UN proposal that would have prohibited any remote-sensing activity relating to natural resources under a country's national jurisdiction without prior consent from the nation being remotely sensed from space.[19]

The second problem hindering the US government's desire to promote *Landsat* globally was that scientists in developing countries were not trained in how to use the data being captured by NASA's satellites to assess their own country's resources. Such was the conclusion of an exasperated Verl Wilmarth, one of NASA's Earth observation managers, who during the summer of 1971 lamented the quality of proposals submitted by foreign scientists interested in participating in future *Landsat* experiments. The "poorly prepared proposals," he wrote, "indicate lack of knowledge of the program content and capabilities."[20] Administrators at NASA were equally concerned that even if foreign scientists did eventually understand *Landsat*'s capabilities, they would nevertheless continue to lack the technological and scientific expertise necessary to take full advantage of the new space technology. Of particular concern was the dearth in developing countries of trained photointerpreters both to analyze the images obtained from satellites and to extract from them the types of data with economic value.[21] To build a transnational network of knowledge producers and users, the space agency thus not only had to convince leaders of developing countries that *Landsat* did not pose a threat to their national sovereignty but also had to educate foreign scientists regarding the space technology's scientific and economic benefits for their own countries.

Government officials and NASA administrators started addressing such problems in the early 1970s. They began by inundating the international scientific community with press releases describing how *Landsat* technology worked. They also called for proposals from foreign scientists themselves that would improve natural resource management specifically in developing countries.[22] The space agency then augmented such efforts by teaming up with international institutions such as the UN, the World Bank, and the Inter-American Development Bank to sponsor conferences, symposiums, and workshops, some up to two weeks long, on the scientific uses of *Landsat* remote-sensing data.[23] Initially, the space agency invited foreign scientists, engineers, and politicians to such events held in the United States, both at academic institutions such as the University of Michigan and also at NASA's research facilities, including the Johnson Space Center in Houston, which conducted a week-long "Earth Resources Survey Symposium" during the summer of 1975. At the Houston *Landsat* conference some of NASA's heavy hitters, including Apollo astronaut Russell Schweickart, Marshall Space Flight Center director Wernher von Braun, and Johnson Space Center director Chris Kraft, addressed an audience of more than 1,200 scientists, engineers, politicians, and administrators from at least two dozen foreign countries on the practical applications of Earth-observing technology.[24]

During the mid-1970s NASA administrators and the US government also brought these educational training opportunities directly to foreign scientists and government leaders within developing nations. During the summer of 1975, for instance, the space agency conducted several three-day symposiums on Earth-observing technology in West Africa. They sought both to educate scientists and government officials in the region about the capabilities of *Landsat* technology and to encourage them to submit scientific proposals aimed at better managing their countries' scarce natural resources. The very first of these conferences, held in Ghana for English-speaking participants, was attended by scientists, engineers, and government leaders from that country as well as from several other nearby nations, including Nigeria, Liberia, and Togo. They listened, along with US ambassador to Ghana and former child movie star

Shirley Temple Black, as the keynote speaker implored those present to make use of "accelerating tools" such as *Landsat* in order to bridge the "technological gap" between underdeveloped and developed nations and propel the former along the arc of modernization (see fig. 7.2).[25] During the 1970s similar NASA conferences promoting the benefits of *Landsat* technology for developing countries took place in Asia and throughout Latin America.[26]

While NASA's conferences, workshops, and symposiums helped to educate participants from developing nations regarding *Landsat*'s scientific usefulness, the space agency simultaneously tried to alleviate con-

SYMPOSIUM — Dr. A. N. Tackie, Executive Chairman, Council for Scientific and Industrial Research, Ghana, addresses members of the Earth Resources symposium in Accra. Seated behind Tackie is the U.S. Ambassador Shirley Temple Black. To the right of Black is the U.S. Information Service Director, Ed Pancoast; and E. Lartyte, Director of the Ghana Institute of Industrial Research.

FIGURE 7.2. Dr. A. N. Tackie, executive chairman of Ghana's Council for Scientific and Industrial Research, addressing attendees of NASA's Earth resources symposium. Seated third from right is US ambassador to Ghana Shirley Temple Black.
Source: Courtesy of National Aeronautics and Space Administration.

cerns regarding the technology's encroachment on national sovereignty by training foreign scientists to collect, analyze, and interpret Earth observation data on their own. As with its *Landsat* conferences, such training took place both within the United States and abroad. In the early 1970s, for example, NASA expanded its international fellowship program to encourage foreign scientists to travel to American universities to take courses on the fundamentals of remote sensing.[27] The space agency also brought scientists from developing countries such as Brazil and Mexico to NASA centers, including the Johnson Space Center, to familiarize them with the acquisition, processing, and analysis of remote-sensing data.[28]

In an effort to institutionalize such training within these less developed nations, NASA, along with the US government, encouraged political leaders around the world to create their own remote-sensing departments, to train their own photointerpreters to assess remote-sensing data, and to establish their own national committees to determine for themselves the best applications and distribution of remote-sensing information.[29] Perhaps most important, the US government urged these developing nations to establish their own *Landsat* receiving stations to collect data on their country's natural resources. In South America this process began in 1974 when Brazil built its own receiving station, and continued three years later when Chile signed an agreement to build another and Venezuela formally expressed interest in doing the same. By early 1977 Egypt and Iran in the Middle East and Zaire in Africa had also established their own stations to receive and process *Landsat* data (see fig. 7.3). Each of these host countries funded, owned, and operated their *Landsat* ground stations, making their scientific experiments less dependent on the United States.[30]

Such efforts by NASA, both to educate the international scientific community about *Landsat* and to alleviate concerns of foreign government officials regarding the technology's impact on national sovereignty, proved enormously successful. The conferees at NASA's symposium in Ghana, according to one participant, were "openly receptive in their response to prospective remote sensing programs in their respective countries."[31] Participants across the developing world seemed to agree; by 1977 more than fifty countries worldwide were relying on *Landsat* data to better manage their natural resources.[32] "The benefits of this new capability promise to be particularly significant in the developing countries of the world," explained *Science* magazine in the mid-

GROUND STATION – This photograph shows the coverage which the Zaire ground station will have when it becomes operational. Within the circle, the ground station will be able to receive data from either LANDSAT I or II.

FIGURE 7.3. Map depicting the coverage area in Africa for the *Landsat* ground receiving station planned for Zaire.
Source: Courtesy of National Aeronautics and Space Administration.

1970s, because they "lack other means of surveying and assessing their resources."[33]

Across Asia many scientists used *Landsat* to map, for the very first time, the natural resources of their countries. In Burma, for instance, local scientists used NASA's multispectral scanners to delineate two-dozen categories of land types, such as wetland, grassland, and barren land, and also different land uses, such as agriculture and forestry. Scientists undertook similar studies in India and Bangladesh.[34] Throughout Africa such efforts tended to focus instead on improving the continent's food supplies. *Landsat* data allowed biologists from Sudan to inventory land, vegetation, and soil resources, game managers from Kenya to administer more efficiently rangeland for both domestic and wild animals, and hydrologists from Botswana to assess their country's only perennially flowing waterway, the Okavango River, for possible agricultural development.[35] Finally, in Latin America, local scientists from Bolivia, Venezuela, Colombia, Chile, and Argentina relied on remote-sensing data to locate mineral deposits, estimate water availability in arid regions, and produce the first accurate maps for large portions of the continent.[36]

While *Landsat* data helped scientists from the developing world to assess their local environment, nature on the ground in these countries was in turn influencing how *Landsat* data was being used. This was most evident during so-called "natural disasters" that struck several developing nations during the early to mid-1970s.[37] One such event was the severe and prolonged drought during the early 1970s that parched the Sahel region of Africa and caused widespread famine across the northern portion of the content. Scientists and government officials in Mali, one of the hardest-hit countries, responded by submitting a proposal to NASA to host a "Sahelian Zone Remote Sensing Seminar and Workshop," which was held in April 1973 and attended by more than thirty scientists and project managers from nine West African countries.[38] As a direct result of the training, local scientists used *Landsat* data to track the deteriorating impact of sand and dust storms on Sahelian plant communities and soil fertility as well as to determine range management techniques that could reverse the process of desertification.[39]

In Latin America the natural environment played a somewhat different, yet equally active role. This became obvious in July 1975 when an unexpected frost destroyed more than 80 percent of the trees in one of Brazil's most productive coffee-growing regions. In this case the extreme weather spurred local agronomists to lobby Brazil's space agency, the Instituto de Pesquisas Espaciais, to capture *Landsat* data for the region to study the frost's ecological effects.[40] Additional unexpected natural phenomena—from earthquakes in Nicaragua, to floods in Pakistan, to volcanic eruptions in Guatemala—also spurred local scientists in the developing world to submit proposals to NASA that resulted in novel uses of *Landsat* data.[41]

Although these natural disasters gave indigenous scientists and technicians some control over *Landsat* data, their experiences in the field also highlighted the significant social impediments to forging such transnational technological networks. In Brazil, for example, technicians analyzing satellite data constantly lacked supplies and replacement equipment, which had to come from the United States. They also had to overcome opposition from military officers who were concerned about the aerial surveillance of strategic sites, convince politicians to relax prohibitive laws that restricted natural resource exploration, and educate and train potential users of remote-sensing data from other government agencies in the art of photointerpretation. "It's hard to run a high technology effort in Brazil," admitted one scientist involved in the country's remote-sensing program. A reporter from *Science* agreed, noting in 1977 that such social obstacles on the ground in Brazil illustrate quite clearly "what is often involved in introducing a novel technology in a developing country."[42]

Which finally brings us back full circle to Southeast Asia and the Mekong Committee's quarterly report of 1976. During the committee's *Landsat* experiments, local scientists and government officials from the four countries straddling the Mekong River basin—Laos, Cambodia, Thailand, and South Vietnam—used NASA's satellite data to create three natural resource maps. The first, which was a land use map that differentiated between agricultural and forestlands as well as among different types of crops and tree species, was intended to help government officials from these developing countries better understand their current natural resource practices. The second map, which assessed the region's "land capabilities," was essentially a soil atlas aimed at improving planning for future natural resource management.[43] Together this pair of maps illustrate not only the physical movement of *Landsat* technology across national borders but also the complicated process of producing and circulating scientific knowledge through transnational networks composed of unequal partners.

The final map in the series demonstrates the often-forgotten role played by the natural environment in this transnational partnership. Back in 1966 the lower Mekong basin experienced the largest flood on record up to that time; 82 percent of cultivated land in the Vientiane plain in Laos became inundated, and some areas remained submerged under three to four meters of water for nearly one month. The Mekong Committee responded by investigating ways to address this environmental crisis. "The devastating flood of the Mekong River which occurred in September 1966," explained the committee's annual report, "served to emphasize the need for flood protection and control."[44] In the early 1970s *Landsat* offered a solution, and the committee encouraged local scientists and government officials from the basin to use remote-sensing data to map annual flood and drainage patterns in the Mekong lowlands. The result was the Mekong Committee's third map, a hydrological survey of basin flooding during different times of the year. Here in Southeast Asia it was flooding, rather than African drought or Brazilian frost, that influenced *Landsat* and its network.

Although all three of these *Landsat* maps were, as the Mekong Committee argued, "urgently needed in order to finalize a realistic post-war development program for the basin," NASA's remote-sensing technology was ultimately more of a mixed blessing for the inhabitants of the developing world, including those in Laos, Cambodia, Thailand, and South Vietnam.[45] On the one hand, *Landsat* measurements of natural resources from Botswana to Brazil to Burma depended on the cooperation of local scientists and politicians for success; biologists on the ground knew best which of their country's natural resources needed study from space, while native government officials had the political and economic capital to construct receiving stations and train photointerpreters. *Landsat*'s focus on local nature, in other words, left room for local control over *Landsat*'s scientific data.[46]

Yet the US government, in cooperation with NASA, still fabricated and launched *Landsat* satellites, decided when they should be "turned on" over what geographic regions, and determined which countries could and could not participate in the program. Administrators at NASA, sometimes guided by federal bureaus such as the Department of Defense, even had the power to demand that proposals by foreign scientists for *Landsat* experiments be "negotiated," or revised, before being officially approved.[47] As a result, while politicians and scientists from developing countries embraced Earth-observing programs in part because they could influence them from below, the American government ultimately controlled this modernizing project from above in ways that almost always supported its own foreign policy agenda and served the needs of military intelligence. When it came to the construction and maintenance of *Landsat*'s transnational network, in other words, the centralized political power of Washington, DC, trumped the peripheral influence of scientists in the developing world.

This foreign policy predicament for citizens of Asia, Africa, and Latin America had taken root soon after World War II, when American scientific and government elites worked together to rebuild research and development in war-ravaged Europe. While European technicians understandably welcomed such efforts, just as scientists and government leaders from developing countries welcomed *Landsat*, by sharing in this scientific diplomacy they ultimately helped to coproduce it and were thus less able to oppose more objectionable US foreign policy initiatives. *Landsat* functioned similarly by enhancing America's soft power across the developing world.[48]

Such was the case regarding NASA's involvement with the Mekong Committee in Southeast Asia, which began in 1973 as US troops starting leaving Vietnam. By enlisting *Landsat* to help the four countries straddling the basin to better manage their natural resources, the US government and the space agency ceded some control over the project to local government and scientific officials. To verify the accuracy of *Landsat* data, NASA technicians had to compare it both with aerial photographs provided by government administrators in Laos, Cambodia, Thailand, and South Vietnam and with field observations made by indigenous scientists from local forestry, agriculture, and other natural resources agencies. The "short term objectives" of the lower Mekong River basin *Landsat* experiment, explained NASA's Frederick Gordon, who oversaw the project from the Goddard Space Flight Center in Greenbelt, Maryland, were "supported by ground truth data and field surveys" and with "aerial photographs made available by the national departments" in the basin's four riparian countries.[49] The land use, soil, and flood maps of the basin created three years later from NASA's *Landsat* data were thus also coproduced, a joint effort by both the space agency and locals on the ground in Southeast Asia.

Yet this joint effort was not between equals. The overwhelming ability of NASA and the US government to direct the Mekong Committee's *Landsat* project in ways that supported American foreign policy was quite apparent in the quarterly report from April 1976, which the space agency coauthored. Although NASA officials completed the report more than six months after the fall of Saigon to communist forces, these adminstrators, perhaps wishfully, referred in the text to the nation of "South Viet-Nam" even though the country no longer functioned. American interests were likewise front and center in the full-page *Landsat* map accompanying the report (see fig. 7.1). The illustration by NASA, which superimposed the ten orbital tracks of the satellite over a political map of the region's national borders, refrained from identifying the soon-to-be reunited country by its official name, the Socialist Republic of Vietnam, and also included, quite prominently, a dotted line for the demilitarized zone that until quite recently had divided North from South Vietnam near the 17th parallel.[50] Additionally, while NASA did not "turn off" *Landsat* over Vietnam when the country became reunited in 1976, the US government's decision to ban assistance to the victorious communist government essentially halted the Mekong Committee's remote-sensing program.[51]

The history of *Landsat*'s promotion and use during the mid-1970s in developing countries, including those devastated by the war in Vietnam, illustrates important lessons regarding the transnational history of technology and science in the twentieth century. Too often historians have focused their sights solely on technology and science as it circulates both within nations and across the borders that divide them. In doing so they have ignored unruly nature as well as the enormous work involved in trying to tame it. The second lesson is that such work is almost always coproduced. Native scientists and engineers strove to weave together transnational networks, composed of government officials, local agencies, NASA, and the US government, that created, maintained, and circulated scientific knowledge on how to better manage their countries' natural resources. The space agency provided this scientific information from above while helping to train locals on the ground to process and use this data to assess their own crops, forests, deserts, and even their own national borders.

Indigenous nature situated within the developing world, whether it be infamous floods and frosts or more mundane crops, trees, and minerals, gave local scientists and government leaders the ability to influence, to a degree, *Landsat* technology and the scientific information it created. The result was reduced anxieties on the local level regarding *Landsat*'s potential to threaten national sovereignty. However, such national autonomy was always constrained by the asymmetrical power relationships that shaped this network, even though NASA constantly downplayed such inequalities. On the domestic front, an international community of *Landsat* users helped enhance NASA's public image and secure an increase in the agency's budget. Internationally, NASA was necessarily engaged in the US government's global ambitions and could build or break this technoscientific network at a moment's notice by denying access to *Landsat*.

Nature was the material that bound together this transnational community of *Landsat* users. It still does, just as the natural environment continues to connect networks of communities relying on technology and science in our twenty-first-century world. Our current climate change crisis is merely the most pressing example. Environmental history brings this natural world—which is often disorderly, fragile, and exploited—back into the history of science and technology, not as a passive stage on which this history plays out, but rather as an actor that shapes human behavior and the ultimate trajectory of social change. When studying such events beyond the national framework, therefore, historians must remember to place both technology and science in their environmental, as well as transnational, contexts.

Notes

1. Mekong Committee Secretariat, Mr. W. J. van der Oord and Mr. Frederick Gordon, Technical Monitor, Goddard Space Flight Center, "Agriculture/Forestry Hydrology," 5a, Quarterly Report, Dec. 1975–Feb. 1976, Mekong Committee Secretariat, c/o ESCAP Sala Santitham, Bangkok, Thailand, dated Apr. 1, 1976, NASA Technical Reports Server, Document ID 19760016569, Accession ID 76N23657, report no. E76-10330, NASA-CR-147211, REPT-2.

2. Ibid., 1.

3. The terms "nature" and "environment" are historically complex. On the difficulty of defining "nature," see Raymond Williams, *Keywords: A Vocabulary of Culture and Society* (New York: Oxford University Press, 1976), 184–189. The notion that material nature is embedded within culture and therefore also socially constructed is widely accepted within the field of environmental history and was initially analyzed in William Cronon, ed., *Uncommon Ground: Rethinking the Human Place in Nature* (New York: W. W. Norton, 1995). In this important volume, see esp. William Cronon, "The Trouble with Wilderness; or, Getting Back to the Wrong Nature," 69–90; and Jennifer Price, "Looking for Nature at the Mall: A Field Guide to the Nature Company," 186–203. Since *Uncommon Ground* appeared, examples of environmental histories that portray nature and culture as mutually constitutive abound. For examples, see Jennifer Price, *Flight Maps: Adventures with Nature in Modern America* (New York: Basic Books, 1999); Kathryn Morse, *The Nature of Gold: An Environmental History of the Klondike Gold Rush* (Seattle: University of Washington Press, 2003); Paul Sutter, *Driven Wild: How the Fight against Automobiles Launched the Modern Wilderness Movement* (Seattle: University of Washington Press, 2009); and the essays in Thomas Lekan and Thomas Zeller, eds., *Germany's Nature:*

Cultural Landscapes and Environmental History (New Brunswick, NJ: Rutgers University Press, 2005).

4. This research by environmental historians is vast. For overviews of the field, which include discussions of this sort of work, see Mart Stewart, "Environmental History: Profile of a Developing Field," *History Teacher* 31, no. 3 (May 1998): 351–368; J. R. McNeill, "The State of the Field of Environmental History," *Annual Review of Environment and Resources* 35 (Nov. 2010): 345–374.

5. For essays by environmental historians that call for this transnational approach, see Samuel Truett, "Neighbors by Nature: Rethinking Region, Nation, and Environmental History in the U.S.-MexicoBorderlands," *Environmental History* 2, no. 2 (Apr. 1997): 160–178; Paul Sutter, "Reflections: What Can U.S. Environmental Historians Learn from Non-U.S. Environmental Historiography?," *Environmental History* 18, no. 1 (Jan. 2003): 109–130; John McKenzie, "Empire and the Ecological Apocalypse: The Historiography of the Imperial Environment," in *Ecology and Empire: Environmental History of Settler Societies*, ed. Tom Griffiths and Libby Robin (Seattle: University of Washington Press, 1997), 215–228; Jeremy Adelman and Stephen Aron, "From Borderlands to Borders: Empires, Nation-States, and the Peoples in between in North American History," *American Historical Review* 104, no. 3 (June 1999): 814–841. For just a few examples of environmental historians tracking nature across national borders, see esp., Samuel Truett, *Fugitive Landscapes: The Forgotten History of the U.S.-Mexico Borderlands* (New Haven, CT: Yale University Press, 2008); Marc Cioc, *The Game of Conservation: International Treaties to Protect the World's Migratory Animals* (Athens: Ohio University Press, 2009); Mark Fiege, "The Weedy West: Mobile Nature, Boundaries, and Common Space in the Montana Landscape," *Western Historical Quarterly* 36 (Spring 2005): 22–47.

6. Donald Worster, "World without Borders: The Internationalizing of Environmental History," in "Papers from the First International Conference on Environmental History," special issue, *Environmental Review* 6, no. 2 (Autumn 1982): 8–13, at 13.

7. As the *Wall Street Journal* reported, "The earth resources program owes most of its technology to the highly classified military programs." William Burrows, "Sizing Up the Planet: Satellites Will Seek to Inventory Resources of Earth from Orbit," *Wall Street Journal*, June 8, 1970, 1. On the evolution of NASA technology from *TIROS* to *ATS* to *Landsat*, see Henry Hertzfeld and Ray Williamson, "The Social and Economic Impact of Earth Observing Satellites," in *Societal Impact of Spaceflight*, ed. Steven Dick and Roger Launius (Washington, DC: NASA Office of External Relations, History Division, 2007), 237–263. *Landsat* was originally called the *Earth Resources Technology Satellite* (*ERTS*).

8. United Press International, "NASA Satellite to Be Launched: 2d Earth Resources Craft to Relay Environmental Data," *New York Times*, Jan. 15, 1975, 53. See also *Photography from Space to Help Solve Problems on Earth: NASA*

Earth Resources Technology Satellite, 2, pamphlet published by NASA's Goddard Space Flight Center, ca. 1972, NASA Headquarters Archives, Washington, DC, NASA Historical Materials, folder 5745, ERTS Photos and Booklets.

9. M. Mitchel Waldrop, "Imaging the Earth (I): The Troubled First Decade of Landsat," *Science* 215 (Mar. 26, 1982): 1601.

10. Ibid.

11. "Ecological Surveys from Space," Office of Technological Utilization, NASA SP-230, 1970, NASA Headquarters Archives, NASA Historical Materials, folder 5754, Earth Resources Satellite, 1970; NASA, *Improving Our Environment* (Washington, DC: Government Printing Office, 1973); *Photography from Space to Help Solve Problems on Earth.*

12. Jeffrey St. John, "Space Effort: No Apologies Necessary," *Los Angeles Times*, Apr. 9, 1972, C3. See also "Space: Can NASA Keep Its Programs Aloft?," *BusinessWeek*, Feb. 13, 1971, 23.

13. NASA's budget during these years decreased from 3.9 to 3.2 billion dollars annually. For historical data on NASA's total budget for these years in both real and in 2008 inflation-adjusted dollars, see United States President, United States, and National Aeronautics and Space Council, *Aeronautics and Space Report of the President: Fiscal Year 2008 Activities* (Washington, DC: Government Printing Office, 2008), "Appendix D-1A: Space Activities of the U.S. Government, Historical Table of Budget Authority (in Millions of Real-Year Dollars)," 146, and "Appendix D-1B: Space Activities of the U.S. Government, Historical Table of Budget Authority (in Millions of Inflation-Adjusted FY 2008 Dollars)," 147.

14. On the economic benefits to American farmers of NASA's LACIE program, see Hertzfeld and Williamson, "Social and Economic Impact of Earth Observing Satellites," 240–241, 262. In the late 1970s the LACIE program was also used to forecast Soviet wheat supplies in an effort to avoid fluctuations in the international wheat market. On LACIE, see also "Landsat-2 Data Aid Research Management," *Bioscience* 25, no. 4 (Apr. 1975): 280. On the economic benefits for the United States of other satellite observations, see "Aerospace Research Profits Earth," *Roundup* (newspaper of the Johnson Space Center), Feb. 18, 1972, 2.

15. Between 1975 and 1980 Congress increased NASA's total budget from 3.22 to 5.24 billion dollars. When adjusted for inflation according to 2008 dollars, this represents an increase of 10 percent. For historical data on NASA's total budget in both real and inflation-adjusted dollars, see United States President, United States, and National Aeronautics and Space Council, *Aeronautics and Space Report of the President: Fiscal Year 2008 Activities*, 146, 147.

16. "Text of Address by President Nixon to General Assembly of the United Nations," *New York Times*, Sept. 19, 1969, 16.

17. "NASA Position Paper on the Remote Sensing of Planetary Surfaces

(Earth, Moon, Mars, Venus, etc.) from Orbital and Fly-by Spacecraft," paper attached to memorandum by NASA Advanced Missions Program Chief Peter Badgley, Oct. 8, 1965, box 075–14, series Apollo, Johnson Space Center History Collection, University of Houston at Clear Lake, Houston, TX.

18. For an example of this concern, see Edward Keating, "Hard Times: World Spy," *Ramparts* 9, no. 8 (Mar. 1971).

19. The proposal, which ultimately stalled in the UN's Scientific and Technical Sub-committee, was actually stricter than a similar UN proposal put forth by France and the Soviet Union the year before. On both these proposals, see Hamilton DeSassure, "Remote Sensing by Satellite: What Future for an International Regime?," *American Journal of International Law* 71, no. 4 (Oct. 1977): 714, 720. On *Landsat* possibly being used by developed countries to exploit natural resources in developing countries, see also John Hanessian, "International Aspects of Earth Resources Survey Satellite Programs," *Journal of the British Interplanetary Society* 23 (Spring 1970): 548.

20. "ERTS-EREP Proposal Review," memorandum by Verl Wilmarth to TA/Director of Science and Applications, July 7, 1971, record no. 14994, box 529, Johnson Space Center History Collection, University of Houston at Clear Lake.

21. On the lack of trained photointerpreters in the developing world, see Hanessian, "International Aspects of Earth Resources Survey Satellite Programs," 545. For additional concerns within NASA regarding a lack of skilled scientists in developing countries to take advantage of space technology, see "Practical Applications of Space Systems," 1975, NASA-CR-145434, National Academy of Sciences, Washington, DC.

22. For examples of these press releases by NASA, see "Earth Resources Experiments RFP," press release no. 70-117, July 14, 1970, NASA Headquarters Archives, NASA Historical Materials, folder 5754, Earth Resources Satellite, 1970; and "Skylab Experimenters Sought," press release no. 71-5, Jan. 19, 1971, record no. 142778, box 502, Johnson Space Center History Collection, University of Houston at Clear Lake.

23. In 1971 the UN created a Space Applications Program specifically to promote the use of Earth-observing remote-sensing data throughout the developing world. On the efforts of these international organizations to promote Earth-observing technology in developing countries, see V. Klemas and D. J. Leu, "Applicability of Spacecraft Remote Sensing to the Management of Food Resources in Developing Countries," Mar. 31, 1977, 31–48, Center for Remote Sensing, University of Delaware, report prepared for the School of Engineering and Applied Science, George Washington University, Washington, DC, and the Division of International Programs, National Science Foundation, Washington, DC.

24. On NASA's *Landsat* conference at the University of Michigan, see

"Earth Resources Survey Workshop," NASA press release no. 70-215, Dec. 23, 1970, NASA Headquarters Archives, NASA Historical Materials, folder 5754, Earth Resources Satellite, 1970. On the Johnson Space Center conference, see "All You Wanted to Know about Earth Resources," *Roundup* 14, no. 12 (June 6, 1975), 1. Lady Bird Johnson had NASA administrator Thomas Paine give a similar lecture on remote sensing of Earth resources at a cocktail party she threw for a group of foreign correspondents covering one of the early Apollo launches. On this cocktail party lecture, see oral interview of Dr. Thomas O. Paine by T. Harri Baker (tape 2), NASA Headquarters Archives, NASA Historical Materials, folder 4185, Paine Interviews Conducted by Baker, Lodsdon, Cohen, and Burke.

25. For a description of *Landsat* conferences in Ghana and Mali, see "Landsat May Help Bridge Technological Gap," *Roundup*, May 23, 1975, 2; "W. Africans Confer on Uses of Remote Sensing Data," *Roundup*, June 6, 1975, 2.

26. On similar *Landsat* conferences in the Philippines, see "Earth Resources Team Visits to the Philippines," Sept. 21, 1971, record no. 210333, report SRE, box 546, Johnson Space Center History Collection, University of Houston at Clear Lake. On efforts to promote *Landsat* across Latin America, see "Inter American Geodetic Survey Proposal for Multi-national ERTS (Earth Resources Technology Satellite) Cartographic Experiments," Apr. 7, 1972, record no. 213145, report IAGS-EROS, box 563, Johnson Space Center History Collection, University of Houston at Clear Lake.

27. On the expansion of NASA's international fellowship program to include study at US universities on *Landsat*, see John Hanessian Jr. and John Logsdon, "Earth Resources Technology Satellite: Securing International Participation," *Astronautics and Aeronautics*, Aug. 1970, 60.

28. On NASA's training of Brazilian and Mexican scientists and engineers at the Johnson Space Center, see Hanessian and Logsdon, "Earth Resources Technology Satellite," 59; Hanessian, "International Aspects of Earth Resources Survey Satellite Programs," 546. In this particular case the remote-sensing data was obtained from aircraft circling above those countries rather than from satellites.

29. "Landsat May Help Bridge Technological Gap," 2.

30. On NASA encouraging developing countries to build their own receiving stations, see Hertzfeld and Williamson, "Social and Economic Impact of Earth Observing Satellites," 239. On developing nations building *Landsat* receiving stations, see Klemas and Leu, "Applicability of Spacecraft Remote Sensing to the Management of Food Resources in Developing Countries," 42. On Zaire's *Landsat* receiving station in particular, see "Landsat May Help Bridge Technological Gap," 2.

31. "Landsat May Help Bridge Technological Gap," 2.

32. On this extensive use of *Landsat* data by fifty countries worldwide, see Klemas and Leu, "Applicability of Spacecraft Remote Sensing to the Management of Food Resources in Developing Countries," 41.

33. Allen L. Hammond, "Remote Sensing (I): Landsat Takes Hold in South America," *Science* 196, no. 4289 (Apr. 29, 1977): 511–512.

34. On material on NASA's use of *Landsat* in cooperation with the World Bank in Asia, see Klemas and Leu, "Applicability of Spacecraft Remote Sensing to the Management of Food Resources in Developing Countries," 35–36.

35. On *Landsat* data being applied to Sudan's Kordofan Province and Botswana's Okavango delta region, see Klemas and Leu, "Applicability of Spacecraft Remote Sensing to the Management of Food Resources in Developing Countries," 7, 33. On *Landsat* data being used for range management in Kenya, see "Landsat May Help Bridge Technological Gap," 2. For a list of African countries undertaking experiments with *Landsat* data, see Klemas and Leu, "Applicability of Spacecraft Remote Sensing to the Management of Food Resources in Developing Countries," 47.

36. For a description of *Landsat* data being used in Central and South America, see Klemas and Leu, "Applicability of Spacecraft Remote Sensing to the Management of Food Resources in Developing Countries," 31–48; Allen L. Hammond, "Remote Sensing (II): Brazil Explores Its Amazon Wilderness," *Science* 196, no. 4289 (Apr. 29, 1977): 513–515.

37. There is a rich literature on "natural" disasters within the field of environmental history. While the great majority of this scholarship accepts the unpredictability of these phenomena—from floods to wildfires to extreme weather such as hurricanes—it argues that the effects of such "natural" disasters are almost always dependent upon cultural and social practices. Here I am focusing on how these natural phenomena influenced both locals in the developing world (including scientists and government officials) and *Landsat*'s transnational technological network. For a sampling of this literature within environmental history, see Donald Worster, *Dust Bowl: The Southern Plains in the 1930s* (New York: Oxford University Press, 1979); Theodore Steinberg, *Acts of God: The Unnatural History of Natural Disasters in America* (New York: Oxford University Press, 2000); Christof Mauch and Christian Pfister, eds., *Natural Disasters, Cultural Responses: Case Studies toward a Global Environmental History* (New York: Lexington Books, 2009).

38. For a detailed discussion of the workshop hosted by the Malian government, see Brian Jirout, "One Satellite for the World: The American Landsat Earth Observation Satellite in Use, 1953–2008" (PhD diss., Georgia Institute of Technology, 2016), 139–143.

39. On NASA's *Landsat* program across the Sahel region of Africa, see Klemas and Leu, "Applicability of Spacecraft Remote Sensing to the Management of Food Resources in Developing Countries," 41; "Aid to W. Africa Aim of US Profs.," *Chicago Defender*, Sept. 8, 1973, 25.

40. Hammond, "Remote Sensing (I)," 512.

41. For a description of *Landsat* being used during these disasters, see

Charles J. Robinove, "Worldwide Disaster Warning and Assessment with Earth Resources Technology Satellites," in *Proceedings of the Tenth International Symposium on Remote Sensing of Environment*, Ann Arbor, MI, Oct. 6–10, 1975 (Ann Arbor: Environmental Research Institute of Michigan, University of Michigan, 1975), 2:811–820.

42. These quotations and a description of the various problems encountered by Brazilian technicians in creating the social networks necessary for *Landsat* to function properly in the country can be found in Hammond, "Remote Sensing (I)," 511–512. For a description of *Landsat* data being used in Central and South America, see Klemas and Leu, "Applicability of Spacecraft Remote Sensing to the Management of Food Resources in Developing Countries," 31–48; Hammond, "Remote Sensing (II)," 513–515.

43. These three maps are described in detail on pp. 1–6 of the Mekong Committee quarterly report cited in n. 1.

44. As quoted in Jeffrey W. Jacobs, "Mekong Committee History and Lessons for River Basin Development," *Geographical Journal* 161, no. 2 (July 1995): 142. Jacobs's article also describes the 1966 flood.

45. Mekong Committee Secretariat Willem J. van Liere, "Applications of Multispectral Photography to Water Resources Development Planning in the Lower Mekong Basin (Khmer Republic, Laos, Thailand and Viet-Nam," 76, Mar. 9, 1973, NASA Technical Report, Document ID 19730008739, Accession ID 73N17466, report no. E73-10257, PAPER-W3, NASA Headquarters Archives.

46. For a similar discussion of local indigenous people using *Landsat* data for their own purposes, see Karen Litfin, "The Gendered Eye in the Sky: A Feminist Perspective on Earth Observation Satellites," *Frontiers: A Journal of Women Studies*, Fall 1997, 41.

47. On NASA and the US government retaining ultimate control over *Landsat* experiments, see Hanessian, "International Aspects of Earth Resources Survey Satellite Programs," 552. In reviewing *Landsat* proposals from developing countries, NASA administrators were able to list an experiment as "N," meaning "Negotiation Required." Doing so gave NASA more control over the experiment being proposed. For an example of this process, see "Additional EREP Investigations," memorandum by NASA Associate Administrator for Applications Charles Mathews to Manned Spacecraft Center (Johnson Space Flight Center) Director Chris Kraft, Apr. 21, 1972, record no. 146924, box 535, Johnson Space Center Archives, University of Houston at Clear Lake.

48. My thinking here has been influenced by John Krige, *American Hegemony and the Postwar Reconstruction of Science in Europe* (Cambridge, MA: MIT Press, 2006), especially its introduction. For examples of the rich literature on coproduction within science studies, see Shelia Jasanoff, *States of Knowledge: The Co-production of Science and the Social Order* (London: Routledge, 2004).

49. This satellite data was compiled by both *Landsat 1* and *Landsat 2*. For

a description of this involvement by local scientists and government officials in corroborating *Landsat* data, see pp. 2–3 of "Annex II: Note on the Land Use Map of the Lower Mekong Basin" of the report cited in n. 1.

50. For references to "South Viet-Nam," see p. 2 of the report cited in n. 1. The *Landsat* map, which is titled "Landsat-2 Ground Track Coverage of Lower Mekong Basin," appears in the same publication on what is labeled by hand "5a," which is actually the eighth page of the report.

51. On the devastating impact of this ban on the Mekong Committee's *Landsat* work in Vietnam, see Jeffrey W. Jacobs, "The United States and the Mekong Project," *Water Policy* 1 (1998): 592.

PART III

Individual Identities in Flux

CHAPTER EIGHT

Manuel Sandoval Vallarta

The Rise and Fall of a Transnational Actor at the Crossroad of World War II Science Mobilization

Adriana Minor

"pressed" and "ordered"—but by whom? . . . by you and me, by each and all of us. And we do so precisely because of habits of thought and expression deeply rooted in us all; because of a narrow, exclusive, bigoted, simplistic attitude that reduces identity in all its many aspects to one single affiliation, and one that is proclaimed in anger. — Amin Maalouf, *In the Name of Identity*[1]

This chapter reflects on the conditions that make possible, and those that constrain, transnationalism in science, through an analysis of the circumstances in which Manuel Sandoval Vallarta (1899–1977), a Mexican-born MIT physics professor, decided to leave behind his life in the United States (where he had arrived in 1917) and return to Mexico in 1942. By the time he left, Sandoval Vallarta had achieved a prestigious scientific career in the United States. He had also developed a profile as a transnational actor by taking advantage of his ties to different national contexts. In this respect, he stood out by building professional links and promoting the movement of people, instruments, and practices between the scientific communities of the United States and Mexico and other Latin American countries. However, World War II altered his professional situation in the United States. During this period he headed a government-funded commission for the encouragement of

US-Latin American scientific relations, taking a leave of absence from his teaching duties. This type of science mobilization was an aspect of the cultural diplomacy that the US government undertook toward Latin American countries in this period, for which Sandoval Vallarta's transnationalism became instrumental. While leading this project, though, disputes over the definition of war effort priorities at MIT brought national sentiments to the surface at the expense of transnationalism, with critical implications for Sandoval Vallarta's life.

In this chapter, I use the term "transnational" as an adjective to refer to individuals and their capacity to establish connections across borders. "Transnationalism" is a related term I use to describe how individuals who live in a different place from where they were born maintain ties with their country of origin while developing identities and social relations in the country where they have settled.[2] The ability to establish cross-border relations emerges not as an automatic result of migration but as dependent on certain personal convictions and historical contingencies. Precisely, the case of Sandoval Vallarta shows how his active involvement in transcending frontiers and fostering connections among scientific communities and cultures fitted with the different goals of other scientists, private institutions, and governments. In his case, I emphasize the professional relations he maintained that contributed to creating circuits in which scientific knowledge traveled. As we will see, Sandoval Vallarta's professional trajectory was crucially defined by his transnationalism, a condition nourished by heterogeneous ways of belonging and mixed cultural and professional values.

Previous considerations of Sandoval Vallarta in historical scholarship reveal the limitations and contrasting interpretations of "methodological nationalism"—that is, the tendency to analyze events through the lens of a single nation as the predominant framework.[3] This is the reason his case has been interpreted in the historiography of science in Mexico mainly through his contributions to national science, without interrogating the long period he spent in the United States.[4] On the other hand, US histories of twentieth-century science and physics barely refer to Sandoval Vallarta, although he contributed in important ways to the formation of a physics community at MIT and belonged to what has been described as the "lucky generation of physicists" of the interwar period in the United States.[5] These remarkable biases arise from the preconception that historical actors have to be boxed in a national frame. The case of Sandoval Vallarta highlights precisely the contrary, because he belonged to, traveled in, and created connections among different national contexts. Sandoval Vallarta is understood differently through a transnational perspective, which allows us to see him as a representative example of a historical actor distinguished by his capacity to create connections across national borders that favored the mobilization of knowledge. In fact, as some scholars have suggested, focusing on individuals is the most basic scale at which transnational relations are built, along with the infrastructure that maintains them.[6]

In my analysis, I highlight the hybrid identity that Sandoval Vallarta built by developing a sense of belonging related to his ties to Mexico and the United States, and I show how this allowed him to connect different places together. This is an important feature that situates him as a transnational actor. It does not deny his national identity (no doubt, a sticky condition, as Michael Barany put it during the workshop). Indeed, he chose to maintain his formal national identity as Mexican even though he fulfilled the requirements to apply for US naturalization. For most of the time this was compatible with the fact that he belonged to and configured his scientific culture in association with the US community of physicists. That being said, his case illustrates how his condition as an alien who was not naturalized as a US citizen constrained the kind of things he was able to do in that country.[7] In other words, to characterize the parameters that shaped Sandoval Vallarta's intellectual trajectory, we historians have to consider the combination of all these personal, cultural, and professional ties. It was the exceptional situation of emergency in World War II that obliged Sandoval Vallarta to collapse his hybrid identity and redefine himself as belonging to a single nation. In this context, what was questioned was not his national identity but rather his transnationalism, considered then as a mark of disloyalty and of insufficient alignment with the national war effort. As the opening quotation in this chapter suggests, this singling out may be more common than we might expect.[8]

This chapter shows especially the extent to which different forms of science mobilization during World War II—research, teaching, and diplomacy—were accessible to the transnationalism characterizing Sandoval Vallarta. His transnationalism prevented his participation in the more notorious, and studied, engagement of science during the war—that is, research—and prompted his intervention in science diplomacy. The ensuing crisis contested Sandoval Vallarta's transnationalism and confined him to another form of science mobilization: the formation of scientific manpower. He was compelled to define himself in terms of a single nation, and he was obliged to make a personal and professional choice which determined not only on what side of the US-Mexico border he would thereafter work but also the type of scientist he would be remembered as. In this chapter, we will see how transnationalism is possible in the formation of a scientific career but that it is also a fragile condition to maintain.

A Transnational Actor to Encourage Inter-American Scientific Relations

The scientific career of Manuel Sandoval Vallarta was intimately linked with his capacity to transit across different nations, particularly between Mexico and the United States. He migrated to the United States in 1917 during the Great War. He used to mention that his family[9] decided to send him to MIT, instead of the University of Cambridge in England, to avoid the German attacks on ships.[10] He could travel safely from Mexico City to the US border in Laredo, Texas, and then to Cambridge, Massachusetts. He would take this journey by car multiple times during his summer vacations at MIT, driving along the road that connected Mexico City and the US frontier. Subsequently, that road would be part of the Pan-American Highway, which was planned to traverse the Americas from north to south.[11]

During the twenty-five years that he lived in the United States, Sandoval Vallarta became integrated into the scientific community of his adopted country. He even adapted his full name to US cultural uses, adopting a signature that omitted his first family name, Sandoval. From then on he would be Manuel S. Vallarta, known to his colleagues as Vallarta. He was part of the lucky generation of physicists—as his MIT colleague John Clarke Slater described this collective. As such, he had the privilege of traveling to the main centers of physics in Europe and benefited from Guggenheim Foundation fellowships commonly awarded to US scholars. Vallarta was one of those young professors who were strongly committed to strengthening the physics community and increasing the role of scientific research at MIT in the 1920s (even before the most celebrated period of this institution, which started when Karl Taylor Compton became its president).[12] This group of young scientists intended to reproduce at MIT their experience at academic institutions in Europe in order to encourage scientific research and training.[13] Their collective goal exposed Vallarta to the creation of a dynamic of scientific exchange between communities of physicists on both sides of the Atlantic.

Vallarta's personal commitment to the formation of scientific connections across borders was revealed especially when he participated in a scientific expedition organized by the Nobel Prize physicist Arthur Compton, who was starting his research program on cosmic radiation at the beginning of the 1930s. Vallarta was invited to help as a guide in Mexico during Compton's expedition.[14] However, he soon became involved in Compton's research program and even proposed a theory to explain the measurements that were obtained during the expedition. Furthermore, cosmic-ray research presented itself as a way to engage Mexican engineers and professionals who had an interest in physics research. Vallarta helped to establish this research program in Mexico by training Mexican scholars, acquiring scientific instruments and, in general, supporting the formation of Mexican scientific institutions.[15]

Vallarta also developed connections with Latin American colleagues. He was involved in various commissions related to Latin American cultural, scientific, and intellectual networks. Beginning when he was a student at MIT, he belonged to Latin American associations of professionals in the United States.[16] Moreover, he participated in (pan/inter-)American scientific congresses, which were organized as part of the structure of the inter-American system in the first half of the twentieth century. These meetings were diplomatically significant for the construction of the hemispheric union.[17] Vallarta's participation in these commissions, organizations, and conferences served to connect him with other Latin American intellectuals and influential individuals. Because of his expertise and academic status, he was also contacted by other Latin American physicists specializing in cosmic rays, for example, in Brazil and Argentina. In these ways, he expanded his association with and commitment to multiple scientific communities in Latin America.

Although he did not explicitly defend a particularly strong Latin Americanism, his interventions during World War II show that he embraced an integrationist vision of the Americas, as seen from his professional position in the United States. In this context, his transnationalism fitted with the goals of the US government with regard to its foreign relations with Latin America. It was in connection with the US war effort that Vallarta emerged as a mediator between Latin American and

US science through the Committee on Inter-American Scientific Publication (CIASP). This body was dedicated to the translation of scientific articles by Latin American scientists from Spanish and Portuguese into English for publication in US journals and to the promotion of standards of scientific writing among the Latin American scientific community.[18] This was chiefly a strategy to attract Latin American scientific production to US scientific journals (and to deter it from publication in European journals). It was a sustained effort that contributed to the expansion of the international influence of US science. As president of the CIASP, Vallarta built a network of scientists in the United States and Latin America. Building this network allowed him to meet with other scientists and to map the Latin American scientific context, contributing in this way to "the enterprise of knowledge" valuable to US interests.[19] Later, this knowledge about Latin American science was assembled in reference catalogs that included data from institutions, scientists, and scientific publications.

With the outbreak of World War II, the US government became concerned about the influence of Nazism and fascism in Latin America. That concern encouraged the creation of an organization dedicated to strengthening cultural relations with the Latin American countries, the Office of the Coordinator of Inter-American Affairs (OCIAA), directed by Nelson Rockefeller.[20] US president Franklin Roosevelt had already promoted his so-called Good Neighbor policy with Latin American governments to improve relations between them. The OCIAA extended the ties in the cultural and scientific fields (see chapter 10 in this volume) as part of a broader strategy for ensuring hemispheric unity. For this purpose, various US scientific institutions were enrolled. For example, the National Academy of Sciences called upon the Harvard physiologist Lawrence Joseph Henderson to head a commission focused on relations with Latin American scientists.[21] In the same vein, the director of the American Institute of Physics, Henry Barton, proposed encouraging Latin American physicists to subscribe to the journals published by the institute.[22]

Vallarta presented a proposal with an aim similar to Barton's, although his initiative was not restricted to physics publications but considered science in general. His main objectives were the following:

> First, to stimulate intellectual and scientific intercourse among the nations of the American continent; second, to promote the circulation of scientific jour-

nals published in any nation of the Western Hemisphere in other American nations, more specifically, the circulation of American scientific journals in the Latin American nations; third, to secure for publication in scientific journals of the United States as large a share as possible of papers written by scientists of Latin American nations. . . . Fourth, to print a reasonable number of representative scientific papers from the United States in existing scientific journals elsewhere in the New World.[23]

When Vallarta made his proposal, the Division of Cultural Relations of the OCIAA had received several other applications related to scientific publications. Division head and MIT history professor Robert Caldwell turned for advice to Henry Allen Moe, of the Guggenheim Foundation, who was also a consultant to the National Academy of Sciences and to the OCIAA and who had considerable experience and contacts with Latin American intellectuals, scientists, and artists. Moe replied that a consideration of science was important for the goals of US foreign policy in the Latin American countries and that he supported the initiatives presented in that sense. However, he found that Vallarta's proposal was more inclusive and had one main advantage over the other options: Vallarta was the only proponent who actually knew Latin America.[24]

Vallarta's associations and connections with Latin America were important for conducting this project and for strengthening scientific relations in the hemisphere. He also took advantage of preexisting currents of science exchange between Latin America and the United States and had the ability to align them with the common purpose of conducting knowledge across borders. This involved overcoming differences of language and cultures of scientific writing in the publication of articles in US journals. Thus, Vallarta created a space in which he could contribute to the war effort by taking advantage of his transnationalism, despite having been excluded from war research programs at MIT.

"Since You Are a Citizen of Mexico . . .": The Interrupted Journey to Latin America

In February 1942 Vallarta, accompanied by his wife,[25] traveled to Mexico as a US representative to the Inter-American Congress of Astrophysics. Afterward, he planned to travel through various Latin American countries and, in his capacity as head of the CIASP, to organize the Inter-American Academy of Sciences (IAAS). This project was part of the agreement between the OCIAA and MIT that supported and funded the activities of the Publications Committee.[26] Specifically, the IAAS was intended to bring together US and Latin American scientists who were influential in their respective national scientific communities.[27] They would thus promote a dynamic of knowledge exchange in which Latin American scientists would send their articles for translation and publication in US journals (in English), and US scientists would send summaries of their main articles to be published in Latin American journals (in Spanish or Portuguese). In addition, the IAAS would organize scientific meetings periodically on the continent. During his journey through Latin America, Vallarta would contact relevant scientists to persuade them to take part in the IAAS. In this manner, he intended to create new ways of connecting scientific communities in the United States with those in Latin American countries.

Vallarta planned to give a course at the Universidad Nacional Autónoma de México[28] later in March, whereupon he would travel to Lima, Peru, and continue to Santiago de Chile, staying ten days in each city. Subsequently, he would travel to Argentina for two and a half months and give lectures in Tucumán, La Plata, and Buenos Aires. Then, he would stay for one month in Montevideo, Uruguay, where he had been invited to give a talk by a local university. Soon after, he would travel to São Paulo and Rio de Janeiro, in Brazil, where lectures had also been arranged. Finally, he would stay in Caracas, Venezuela, and Bogotá, Colombia. He selected this itinerary according to invitations that he had received and because of the interest shown by scientists in these cities for the work of the CIASP and the organization of the IAAS.[29] However, as we will see, Vallarta had to confront several obstacles that hindered the fulfillment of his official trajectory.

Vallarta gave one of the opening speeches of the Inter-American Congress of Astrophysics, by invitation of the Mexican government. Other keynote speakers included the president of Mexico, Manuel Ávila Camacho, and Harlow Shapley, head of the Astronomical Observatory of Harvard, whose support was crucial for the provision of instruments to equip the Mexican Astronomical Observatory of Tonantzintla, whose inauguration motivated the organization of the aforementioned meeting.[30] The meeting was also significant for the US Good Neighbor policy: the State Department promoted the participation of American scientists as a way of demonstrating their desire to improve relations between the United

States and Mexico, as well as to promote inter-American relations.[31] Although the Mexican government, as organizer, claimed that the conference was intended to promote international scientific relations, its call was restricted to US and Mexican scientists.

George David Birkhoff, a mathematician from Harvard University, was another of the conference attendees with the US delegation. Like Vallarta, Birkhoff had plans to travel through different countries of Latin America with OCIAA support.[32] His aim was to establish a network of mathematicians from the United States and Latin America.[33] In contrast to Vallarta, Birkhoff was able to carry out his trip, circumventing difficulties with OCIAA officials. Around 1942 the OCIAA was beset by doubts about the real relevance of inter-American projects for the war effort and by an emerging change in geopolitical priorities, which displaced the focus on Latin America to other fronts in the world. In this context, projects such as those led by Birkhoff and Vallarta were scrutinized closely by OCIAA officials. Birkhoff had the support of Henry Allen Moe, who as head of the Guggenheim Foundation had negotiated a replacement grant when the OCIAA cut Birkhoff's funding.

In the case of Vallarta, the OCIAA decided to postpone the organization of the IAAS because of the sudden death of Lawrence Joseph Henderson, who was chair of Inter-American Affairs at the National Academy of Sciences.[34] OCIAA officials believed that the IAAS should be associated with the National Academy of Sciences, and since Henderson was its representative, it seemed fundamental to wait until Henderson's successor was appointed. Thus, Vallarta's trip through Latin America was temporarily suspended.

Like Birkhoff, Vallarta contacted Henry Allen Moe when he encountered difficulties in financing his trip.[35] Since the journey included courses and conferences in Argentina, Uruguay, and Brazil, he considered requesting a scholarship from the OCIAA Committee for Inter-American Artistic and Intellectual Relations. He discussed this possibility with Moe, who was on this committee, along with Frederick P. Keppel of the Carnegie Corporation and David H. Stevens of the Rockefeller Foundation. Vallarta was acquainted with Moe because he used to review Latin American applications for the Guggenheim Fellowship. According to Moe, although he tried to support Vallarta, his request was rejected since "the funds of the Keppel-Stevens-Moe Committee are, by the terms of our contract, limited to bringing citizens of the Latin American republics to the United States, or the sending of citizens of the United

States to one or more of the Latin American republics; and that being the case, since you are a citizen of Mexico we cannot make a grant of funds for you."[36] While Vallarta's transnationalism was a plus when it came to important inter-American scientific projects, at this point his nationality prevented him from receiving funds from some major US sources.

Meanwhile, Vallarta stayed in Mexico. Between March and April, he gave a series of lectures at the Universidad Nacional Autónoma de México's Faculty of Sciences.[37] He had a paid leave of absence from MIT that covered his activities as president of the CIASP, including his planned journey through Latin America. This allowed him to temporarily suspend his teaching duties, at least for the first term of the year. Although by March, there was uncertainty about the realization of his trip, this did not imply a definitive suspension. Both the OCIAA and MIT officials remained ambiguously open about the continuation of Vallarta's plans and expressed their support in case he finally got the funds and approval to proceed with the organization of the IAAS.[38]

The CIASP continued its program, with Christina Buechner working as the executive secretary in the office at MIT. The number of publications received and published throughout that year demonstrated its success, though this was not enough to renew the contract with the OCIAA for another year.[39] In June 1942 Vallarta was informed that the CIASP could continue in association with the Joint Committee on Latin American Studies, created for the promotion of Latin American studies with the support of the National Research Council, the Social Science Research Council, and the American Council of Learned Societies.[40] OCIAA officials discussed this proposal with Vallarta, who agreed but asked that the setting of policy and the internal procedures of the CIASP remain in his hands.[41] This condition was accepted, and he received a formal invitation to take part in this interinstitutional committee and to attend a meeting on September 11 of that year in New York.[42] However, MIT failed to guarantee the conditions he considered essential to his continued leadership of the CIASP.

War Effort Priorities at MIT

At the end of 1941 the United States declared war against the Axis nations after the attack on Pearl Harbor. This implied profound changes in many US institutions. MIT was particularly relevant to US mobilization efforts; just to mention one example, at the end of the war its Radiation Laboratory had a staff and budget comparable to those of the Manhattan Project.[43]

In addition to war-related scientific and technological projects, there was an urgent need to train science students, which demanded changes in teaching schedules.[44] To this end, in June 1942 K. T. Compton, as MIT president, pointed out, "The physicist who is training more physicists for the war effort is making a contribution which is just as essential as the man who is designing a new instrument."[45] Despite this statement, almost the entire MIT Department of Physics was involved in government commissions connected with the war effort. Because of this situation and considering the temporary cancellation of Vallarta's plans, John Clarke Slater, then chairman of the Department of Physics, asked him to return in September 1942 to cover regular courses during the fall term and replace professors who had joined war projects.[46]

Since the OCIAA had suspended the funding for his trip through Latin America, Vallarta began to receive letters informing him about the lack of physics professors at MIT. He agreed to come back as soon as possible, though both he and the MIT authorities kept open the possibility of finding a solution that would allow him to continue with his original plans. At first, he was informed that he would be in charge of some courses during the summer.[47] But later this request was withdrawn, considering that summer courses lasted only a few weeks and that at this stage Vallarta's OCIAA plans might still be possible. In any event, a short while later, MIT authorities put great emphasis on a more substantial request; they asked Vallarta explicitly and insistently to return in time for the start of the fall term, by mid-September.[48]

In addition, John Slater informed him by letter that because of the war Vallarta could not teach courses that were closely connected to his research; instead, "[i]t should be a course in partial differential equations and boundary values, and I do not doubt that you will do an excellent job teaching this. There does not seem to be anyone else on hand who could give this course and it seems to be more important than your usual course in cosmic rays and relativity under the present circumstances."[49] Vallarta would also be expected to teach theoretical physics courses, two introductory and one advanced, courses that in fact he had taught in previous years. Apparently, he was almost the only professor available for this task in the theoretical physics group at MIT, since the other members were involved in government commissions.[50]

Slater's proposal would allow Vallarta to resume his MIT academic position, in view of the complications and failure to continue the project of organizing the IAAS. In Slater's view, this assignment was a way in which Vallarta could help MIT to face the contingencies provoked by mobilization for war. But Vallarta did not see it that way.

Vallarta had wanted to participate in research activities connected with the war effort since 1940, when the radar project started at MIT. He had shown his enthusiasm for collaborating in these activities when MIT president K. T. Compton sent an urgent and confidential statement to the Department of Physics asking for suggestions about how the institution could contribute to the war effort.[51] Slater forwarded Compton's letter to a group of members of his department, including Vallarta, who replied:

> It goes without saying that I would be only too happy to collaborate, within my limitations, in any way that appears desirable in connection with the matter brought up in this memorandum, and to set aside or postpone indefinitely for this purpose our present research program on cosmic rays, the sun's magnetic field, the structure of the ionosphere, magnetic storms and other related matters. I think I would be qualified, as far as my knowledge of the subject goes, to devote all my effort to such matters as bomb and shell trajectories, problems of bomb sighting, problems of short wave propagation and reception, questions on mechanical vibration of airplane structures and the like.[52]

Although willing to participate, Vallarta was not called on to collaborate in the multiple research projects that were developed at MIT. Instead, as we have seen, he contributed to the war effort in another way, linked to his ability to build connections between scientific communities from the United States and Latin America. In this regard, he pointed out: "I have always hoped that our Committee on Inter-American Scientific Publication, which I have planned and organized from first to last, would be recognized as my own particular contribution to our joint war effort, bearing in mind that, in spite of the solidarity existing between our two nations [Mexico and the United States], my contribution as a research physicist was not made possible."[53] At first, Vallarta accepted Slater's offer for his teaching schedule, provided that he would be allowed to continue with his work leading the CIASP.[54] However, he did not receive a response concerning this condition. Slater considered that Vallarta's main contribution to the war effort should be his new teaching schedule at MIT, without conditions. Vallarta, however, believed that since he was not able to participate in war research, his main contribution to the war effort should be through his work as head of the CIASP. He did not return to MIT in September. In response, K. T. Compton informed him about a change in his special permission, which became a "leave of absence without salary."[55]

Mexico or the United States—"Those Who Can, Do; Those Who Can't, Teach"

Vallarta himself, his colleagues, and even professional historians have explained his return to Mexico from the United States in terms of his wish to promote the development of science in Mexico.[56] Historians of Mexican science have even contended that Vallarta decided to settle in Mexico because of his rejection of the military uses of science.[57] Vallarta himself contributed to spreading opacity and ambiguity about the circumstances that led him to remain permanently in Mexico, as the following quotation shows:

> In 1943 I started coming to Mexico. Previously, I came on vacation from MIT, but it was not for long. Already in 1942, I started coming more frequently; for a few years, between 1942 and 1946, I distributed my time between Cambridge and Mexico. But the time came when I realized that if I continued with that program I would not have a long life, as it was often necessary to travel. Our concern was then to see how we could raise the scientific level in Mexico and we came up with the idea of the old Commission for the Promotion and Coordination of Scientific Research.[58]

This version contrasts with the interpretation presented in this chapter. As noted a few pages back, Vallarta traveled to Mexico in 1942 with a specific agenda related to his work as head of the CIASP and the organization of the IAAS, both projects funded by the OCIAA. He changed his plans in the course of his stay in Mexico, after arriving in February 1942 and remaining there until the end of that year. He then accepted the proposal of the Mexican government to direct the Comisión Impulsora y Coordinadora de la Investigación Científica.[59] In this manner, he justified his settling in Mexico for the rest of the war and, later, because of multiple political, public, and diplomatic obligations.

Vallarta was confronted with a dilemma that implied a final choice regarding his placement in Mexico or the United States, constraining his hybrid identity spanning nations and cultures among which he transited and articulated connections. His situation highlights the circumstances in which the behavior of a truly transnational actor may start to appear incomprehensible and questioned. A war situation undoubtedly disrupts all aspects of life and brings about realignments and the resurgence of national identities that were previously subdued.

Vallarta illustrated his dilemma with a quotation from the play *Man and Superman* by George Bernard Shaw: "Those who can, do; those who can't, teach."[60] According to his interpretation, what Slater was offering as an option to return to MIT involved constraining him to be a regular professor of physics courses and to put aside his work as head of the CIASP. "Those who can, do" meant to him the possibility of doing both, because he did not refuse to give the courses Slater asked for, but he demanded to be allowed to continue simultaneously in the CIASP.

Meanwhile, the CIASP got funds for another year and the Joint Committee on Latin American Studies recommended the Harvard astronomer Harlow Shapley as a replacement for Vallarta. The latter would have no further relation to this project, notwithstanding his foundational work and the network of contacts in Latin America that he had generated.[61] The CIASP became an interinstitutional organization whose members were all US citizens.[62] During 1942 it placed in US journals forty-seven articles written by Latin American scientists, mainly from Argentina, Brazil, and Mexico, in several disciplines, such as physics, chemistry, biology, and medicine.[63] By the late 1940s, the committee was expanded to become an international program and was renamed the Committee on International Scientific Publication, funded by the State Department and the National Academy of Sciences. Its mission turned "on a world-wide basis, . . . following the activities which had originally been undertaken only in the inter-American field."[64] Thus, Vallarta's efforts had helped to establish a network and an infrastructure that were intended to extend the inter-American experience of scientific diplomacy to the rest of the world.[65]

With these developments, the most important factor in Vallarta's dispute with the MIT authorities was eliminated. They implied that if he returned to MIT, he would have no option but to accept the conditions set by Slater. Geopolitically, the sense of "Those who can't, teach" was located in the United States. In contrast, "Those who can, do" acquired in Vallarta's trajectory a new meaning associated with returning to Mexico.

"Too Much Divided in Your Loyalty and Interests"

By November 1942 tensions reached a high point and Vallarta was pressured to decide definitively between the United States and Mexico. Misunderstandings with MIT and OCIAA authorities continued, and personal reasons prolonged his stay in Mexico when his father suffered a heart attack. John Slater wrote to him to ask if he would come back for the next term by the beginning of 1943. This time, Slater accompanied his request with an extensive explanation of what he considered the institutional misdemeanors accumulated by Vallarta:

> Remember that all of Cambridge feels, whether rightly or not, that if you honestly wanted to get back on time in the fall, you would do it; that the reasons you have given for getting back late have been largely excuses, founded no doubt on fact, but mainly a result of your subconscious desire to find some argument for staying a little longer in Mexico. Remember that you were in Cambridge only nine weeks out of the last academic year; that you stayed in Mexico for the spring, while negotiations for a South American trip were going on in a way that was rather mysterious to all of us here. Remember that I had asked you to teach in summer school and you had not wanted to. Remember that I, and Dr. Compton, and Dean Caldwell, had repeatedly urged on you the necessity of coming back on time in the fall. Remember that there is a war on, that every other member of the department, almost without exception, worked this summer, and that all of us are carrying several different jobs this year as a matter of course. With this background perhaps you will realize that your presence on time this fall was more important than it had ever been before.[66]

Slater accused Vallarta of being inflexible and uncommitted to MIT when he refused to abandon the IAAS project and to resume a different schedule for his teaching duties in September. These actions were especially intolerable at a time when his cooperation with MIT was indispensable for the war effort.

In Slater's view, Vallarta had divided loyalties, and precisely because of the war, he had to point them out. Slater again offered him the option to come back to the United States, but this time he had to assume that if he wanted to be part of MIT, then he had to consider a substantial change in terms of his ties to Mexico:

> If you wish to come back, however, let me give you a word of warning. If I were you, I should . . . realize that if my job were here in Cambridge, my loyalty and interests should be here too. You are too much divided in your loyalty and interests, I believe. You have been tied so completely to Mexico that you have always wanted to get back there, rather than staying here and giving your best efforts to the Institute. This has been natural, but it has made a division in your interests which has made you always uncertain as to what you wanted to do next. The situation has now come to a head, and I believe you must, for your own peace of mind as much as for anything else, take a decisive step: either decide that your real interest is here and not in Mexico, and plan in the future to take your duties here much more seriously; or decide that fundamentally your interests are in your own country, and leave the Institute and take a position there. . . . whatever you decide, I hope you decide it definitely, for I think it is only in that way that you can take your place in the world that your ability should bring you.[67]

With these words, Slater collapsed Vallarta's hybrid identity into a primary allegiance to a single nation. Apparently, ambiguities in this respect were not tolerated in wartime. His words reflected a deep indignation because of Vallarta's actions, as well as misunderstandings between what each one considered essential or negligible. While for Vallarta his role in the construction of inter-American relations was his main contribution to the war effort, for Slater this was unclear, and instead, the only way that Vallarta could demonstrate his commitment to the war effort was through accomplishing his assignments at MIT without reserve. According to Slater, there was a contradiction of interests and loyalties in the background of Vallarta's dilemma. His national allegiance should be exclusively tied either to Mexico or to the United States; dual allegiance was not admissible in wartime.

Once this national dilemma was posed to him, Vallarta responded by referring to his war contribution from his own country, in contrast to what he could do from the United States: "I would only ask you to remember that we are fighting this war together. No matter at what cost I want to do something for our cause, something commensurate with my own ability. Had I accept[ed] the role which you had set aside for me last September, I might have done a bit for M.I.T., very little for our joint war effort and nothing at all for my own country."[68] Vallarta's decision about what side of the frontier he would be working in was then reduced to a choice between teaching regular physics courses at MIT or leading a Mexican national institution. Alluding to Shaw's phrase, at this point Vallarta could do nothing but teach in the United States, as Slater had offered to him.[69] Instead, in Mexico he could be a publicly and politically relevant scientist having the power to make decisions with national repercussions. As Slater pointed out, Vallarta preserved his interests in Mexico during the time he worked as a professor at MIT. Not only had he kept in touch with the Mexican scientific community, but also his family was important in Mexican politics and intellectual life. Thus, in principle it would not be difficult for him to settle in Mexico and enjoy a high level of political agency.

As mentioned before, Vallarta opted to organize a Mexican commission that was created to coordinate scientific research for the war's contingencies, the Comisión Impulsora y Coordinadora de la Investigación Científica.[70] Considering the fact that Mexico and the United States were allies during the war,[71] he justified his decision by stating that public service for the Mexican government was similar in nature and importance to the service provided by his MIT colleagues to the US government: "since this is a matter of public service intimately connected with Mexico's war effort, I must accept. Further, since Mexico is an ally of the United States, I don't see how MIT can refuse permission on the same basis as in other similar cases."[72]

Vallarta's appointment in Mexico did not imply his definitive resignation from MIT, but only the continuation of his unpaid leave of absence. The many positions he held in his native country until the end of the war in 1945 justified the continuation of his MIT license.[73] It was in 1946, when MIT started its reorganization after the war, that K. T. Compton informed him about a new internal policy that implied the formalization of his resignation.[74] Vallarta did not express disagreement with that resolution and even invited Compton to participate in an academic event in Mexico that he was organizing.[75] He also announced to Compton that he soon would travel to New York to participate in the meeting of the Atomic Energy Commission of the United Nations as a scientific representative of Mexico.[76] That commission would take him back to the United States, although in a significantly different situation in terms of his transnationalism. From then on, he developed an active role in multiple international forums as a scientist-diplomat representing the interests of the Mexican government. He thus reconfigured his professional trajectory by strengthening his identity as a Mexican scientist. He also recovered his signature with two surnames, as is usual in Mexico.

Concluding Remarks

In this chapter, I have explored a different interpretation of a turning point in the professional trajectory of Manuel Sandoval Vallarta, highlighting his role as a key promoter of inter-American scientific relations as part of the US war effort. We have seen that in connection with war and different forms of science mobilization (research, diplomacy, and teaching), he faced a challenge to the transnationalism that paradoxically had enabled his intervention as a mediator between US and Latin American science. This perspective brings to the fore aspects of Sandoval Vallarta's life and work, as well as aspects of science in general, that national historiographical traditions tend to disregard.

Sandoval Vallarta's professional trajectory was marked by his mobility across the United States and Mexico and his active role in establishing scientific connections beyond borders. This case study illustrates the importance of people like him whose trajectory has been defined by mobility, who establish ties in multiple national contexts, and display a capacity for shaping and producing knowledge by building connections among disparate cultural, disciplinary, and national worlds. Mobility by itself is not enough to become a transnational actor; this condition requires a deliberate interest in connecting across national borders, along with special factors that break up the homogeneous national identity that most actors have. What is more, once achieved, it must be maintained, and it can collapse under conditions that foster nationalism, such as war.

Focusing on Sandoval Vallarta's career from a transnational perspective allows us to recognize his role as a mediator. In addition, it raises questions regarding the asymmetrical historical depiction of this scientist in Mexico and the United States. In Mexico, Sandoval Vallarta is distinguished as one of the most prestigious scientists in modern national history. His historical relevance in Mexico was expressed when his remains were moved to the Mexican Rotunda of Distinguished Men a decade after he died in 1977. Even today, a statue representing Sandoval Vallarta guards the entrance of the National Council of Science and Technology in Mexico City. In contrast, the trace of Vallarta in the United States is notably weak. This bias in historical recognition might be because of the circumstances in which he abandoned MIT and the United States, but it is also because of the methodological nationalism prevailing in history of science. A consideration of why this kind of analysis has dominated in the case of Sandoval Vallarta reveals the need to balance the different historical narratives in those national contexts where migrant scientists transited. In addition, this leads us to assert transnational history of science as the appropriate analytical framework in which to reconsider the case of Manuel Sandoval Vallarta, and other similar actors, and to appreciate its value for our historical understanding of science.

We should ask ourselves how exemplary or particular this case study of Sandoval Vallarta is. Sandoval Vallarta's profile is useful to analyze or revisit other cases of scientists who migrated and worked in national contexts different from those in which they were born. It suggests a pattern regarding the impact of migration in the formation of hybrid identities. Obviously, when people migrate, something has to change in terms of their identity to allow adaptation to the new culture. In this manner, individuals can enrich and expand their understanding of different cultural codes. It is even possible for them to combine ties to different places and to keep a foot here and one there simultaneously. In this way, people develop their transnationalism. An additional effort could involve taking advantage of their situation to establish a dialogue and an encounter between the different cultures with which they are engaged, thus becoming transnational actors. While there are a number of well-known examples in history of science that follow this pattern, there is still a lack of perspective and focus, which impedes consideration of the analytical complexities involved.

As we have seen, there are circumstances which can operate to hinder, or conversely to nurture eventually, the making of truly transnational actors. The case analyzed in this chapter offers a privileged point of view from which to observe the complexities that migrant scientists with hybrid identities have to confront, as well as the particular political, cultural, and social conditions in which transnationalism is maintainable. Sandoval Vallarta's case illustrates how the stress on national identities in extreme situations exposes the uses, meanings, and limits of transnationalism. The context of war is probably the most obvious example of a situation in which transnationalism is confronted owing to national alignments, but this can emerge in other, less extreme situations that are worth investigating historically.

Last but not least, this case situates cultural diplomacy as a major field in which science was mobilized during World War II, with important implications as the Cold War took shape. The transnationalism of Sandoval Vallarta became practicable in this context and was fostered by the OCIAA, established for fear that Latin American countries would fall prey to Nazism and fascism. This type of institutional support enabled him to mobilize his prestige as a physicist at MIT to serve the US war effort and also to represent Mexico after the war at the UN Atomic Energy Commission, which was formed under the shadow of the emerging Cold War. The role of US-Latin America cultural diplomacy is a promising field for research on transnational phenomena in contemporary history. It also helps us situate the postwar history of science in a country like Mexico, for which developments in the 1930s are important for identifying intellectual, institutional, and political continuities that are often overlooked in Cold War historiography.

Notes

1. Amin Maalouf, *In the Name of Identity: Violence and the Need to Belong* (New York: Penguin Books, 2000), 5.

2. See, e.g., Brenda S. A. Yeoh, Katie D. Willis, and S. M. Abdul Khader Fakhri, "Introduction: Transnationalism and Its Edges," *Ethnic and Racial Studies* 26, no. 2 (2003): 207–217; Peggy Levitt and B. Nadya Jaworsky, "Transnational Migration Studies: Past Developments and Future Trends," *Annual Review of Sociology* 33 (2007): 129–156; David Bartram, Maritsa V. Poros, and Pierre Monforte, "Transnationalism," in *Key Concepts in Migration* (London: Sage, 2014), 140–144.

3. Andreas Wimmer and Nina Glick Shiller, "Methodological Nationalism and Beyond: Nation-State Building, Migration and the Social Sciences," *Global Networks* 2, no. 4 (2002): 301–334; Bernhard Struck, Kate Ferris, and Jacques Revel, "Introduction: Space and Scale in Transnational History," *International History Review* 33, no. 4 (2011): 573–584.

4. See, e.g., Dorotea Barnés and Alfonso Mondragón, eds., *Manuel Sandoval Vallarta: Homenaje* (Mexico City: Instituto Nacional de Estudios Históricos de la Revolución Mexicana, 1989); María de la Paz Ramos Lara, "La física en México: Homenaje a José Antonio Alzate y Manuel Sandoval Vallarta," *Boletín de*

la Sociedad Mexicana de Física 13, no. 4 (1999): 157–165; Luz Fernanda Azuela, "Manuel Sandoval Vallarta y la responsabilidad del hombre de ciencia," in *Humanismo mexicano del siglo XX*, vol. 1, ed. Alberto Saladino García (Toluca: Universidad Autónoma del Estado de México, 2004), 453–471.

5. S. S. Schweber, "The Empiricist Temper Regnant: Theoretical Physics in the United States, 1920–1950," *Historical Studies in the Physical and Biological Sciences* 17 (1986): 55–98; Katherine Russell Sopka, *Quantum Physics in America, 1920–1935* (New York: AIP Press, 1988); Larry Owens, "MIT and the Federal 'Angel': Academic R&D and Federal-Private Cooperation before World War II," *Isis* 81, no. 2 (1990): 188–213; Philip N. Alexander, *A Widening Sphere: Evolving Cultures at MIT* (Cambridge, MA: MIT Press, 2011). In reference works on the history of physics or MIT physics there is no mention of Sandoval Vallarta; see, e.g., Daniel J. Kevles, *The Physicists: The History of a Scientific Community in Modern America* (New York: Random House, 1979); David Kaiser, ed., *Becoming MIT: Moments of Decision* (Cambridge, MA: MIT Press, 2010); David C. Cassidy, *A Short History of Physics in the American Century* (Cambridge, MA: Harvard University Press, 2011).

6. This argument is stated in Struck, Ferris, and Revel, "Introduction: Space and Scale in Transnational History"; Emily S. Rosenberg, *Transnational Currents in a Shrinking World, 1870–1945* (Cambridge, MA: Belknap Press of Harvard University Press, 2014); AHR Forum, "Transnational Lives in the Twentieth Century," *American Historical Review* 118, no. 1 (2013): 45–139.

7. His non-US nationality was a distinctive characteristic that was referred to by his US colleagues and that also limited his access to certain political positions, such as, for example, when he was proposed to head the MIT Department of Physics in 1929 but received a negative evaluation owing to his Mexican nationality (Alexander, *Widening Sphere*, 338). It was also a practical concern, as, for example, with the adjustments of US migration law, which by the eve of World War II obliged him to request a visa.

8. Zuoyue Wang has put a similar emphasis on the circumstances that stressed transnationalism in science, particularly with regard to the Chinese scientists who were forced to decide whether to stay in the United States or return to China during the Cold War: Zuoyue Wang, "Transnational Science during the Cold War: The Case of Chinese/American Scientists," *Isis* 101, no. 2 (2010): 367–377.

9. Sandoval Vallarta belonged to an influential elite family that was closely tied to Mexican political and intellectual circles and whose members included relevant actors in the history of Mexico since colonial times. See Rodrigo-Alonso López-Portillo y Lancaster-Jones, "Los De Vallarta," *Club Social México*, Sept. 1991.

10. Manuel Sandoval Vallarta, "Reminiscencias," sección Personal, subsección Distinciones, Homenajes y Biografías, box 44, file 3, Archivo Histórico Científico—Manuel Sandoval Vallarta (hereafter AHC-MSV).

11. His records at MIT contain letters and telegrams he sent informing about the difficulties he faced during his drive to the United States on the Mexican portion of the Pan American Highway, mentioning floods and landslides that hindered his timely return. MIT Office of the President, AC4, box 228, file 4, Vallarta 1932–47, MIT Institute Archives and Special Collections (hereafter MIT Archives), Cambridge, MA.

12. In his autobiography, John Clarke Slater described Sandoval Vallarta as a member of the lucky generation; see John Clarke Slater, "A Physicist of the Lucky Generation," 442, John C. Slater Papers, MC189, box 1, MIT Archives. Sandoval Vallarta received his PhD at MIT in 1924 (he was the first Mexican with a degree in physics), and that same year he was appointed as research associate and started to teach in the Department of Physics. In 1926 he was promoted to assistant professor, becoming associate professor in 1931 and professor of physics in 1939 (Course Catalogues, 1924–40, MIT Archives).

13. Julius A. Stratton et al., "Is the European System Better?," *Bulletin of the American Association of University Professors* 15, no. 2 (1929): 150–154.

14. In my PhD dissertation, I expand on this point: Adriana Minor García, "Cruzar fronteras: Movilizaciones científicas y relaciones interamericanas en la trayectoria de Manuel Sandoval Vallarta (1917–1942)" (Universidad Nacional Autónoma de México, 2016), 85–118. See also Gisela Mateos and Adriana Minor, "La red internacional de rayos cósmicos, Manuel Sandoval Vallarta y la física en México," *Revista Mexicana de Física E* 59, no. 2 (2013): 148–155.

15. The first Mexican Institute of Physics was created in 1938. It was directed by one of Sandoval Vallarta's Mexican students at MIT (Alfredo Baños). For a general overview of the multiple ways in which a tradition in physics was developed in Mexico with strong connections to the United States and with Sandoval Vallarta's mediation, see Adriana Minor, "Shaping 'Good Neighbor' Practices in Science: Mobility of Physics Instruments between the United States and Mexico, 1932–1951," in *Scientific Instruments in the History of Science: Studies in Transfers, Use and Preservation*, ed. Marcus Granato and Marta C. Lourenço (Rio de Janeiro: Museu de Astronomia e Ciências Afins, 2014), 185–206.

16. During his years as a student, Sandoval Vallarta belonged to the Latin American Club at MIT. Also, he was one of the founders in Mexico of the Latin American Union of the Phi Iota Alpha fraternity.

17. Eckhardt Fuchs, "The Politics of the Republic of Learning: International Scientific Congresses in Europe, the Pacific Rim, and Latin America," in *Across Cultural Borders: Historiography in Global Perspective*, ed. Eckhardt Fuchs and Benedikt Stuchtey (Lanham, MD: Rowman and Littlefield, 2002), 205–244.

18. For a more detailed analysis of the articulation, operation, and meanings of the CIASP, see Adriana Minor García, "Traducción e intercambios científicos entre Estados Unidos y Latinoamérica: El Comité inter-Americano de publicación científica (1941–1949)," in *Aproximaciones a lo local y lo global: Amé-*

rica Latina en la historia de la ciencia contemporánea, ed. Gisela Mateos and Edna Suárez-Díaz (Mexico City: Centro de Estudios Filosóficos, Políticos y Sociales Vicente Lombardo Toledano, 2016), 183–214.

19. Ricardo D. Salvatore, "The Enterprise of Knowledge: Representational Machines of Informal Empire," in *Close Encounters of Empire: Writing the Cultural History of U.S.–Latin American Relations*, ed. Gilbert M. Joseph, Catherine LeGrand, and Ricardo D. Salvatore (Durham, NC: Duke University Press, 1998), 70–104.

20. Darlene J. Sadlier, *Americans All: Good Neighbor Cultural Diplomacy in World War II* (Austin: University of Texas Press, 2012); Gisela Cramer and Ursula Prutsch, eds., *¡Américas Unidas! Nelson A. Rockefeller's Office of Inter-American Affairs (1940–46)* (Frankfurt: Iberoamericana Editorial Vervuert, 2012).

21. Letter from Ross Harrison, president of the National Research Council, to Lawrence Joseph Henderson, June 12, 1941, Lawrence Joseph Henderson Papers, carton 4, file 2/2, Committee on Inter-American Relations, Baker Library, Harvard Business School Archives, Cambridge, MA.

22. Letter from Henry Barton, director of the American Institute of Physics, to Mr. Ross G. Harrison, president of the National Research Council (copy), Apr. 17, 1941, National Research Council Central Files, file "Foreign Relations, 1941, International Organizations: Committee on Inter-American Scientific Publication," National Academy of Sciences Archives (hereafter NAS Archives).

23. "Memorandum concerning a Proposal to Stimulate the Publication of Scientific Papers from Latin American Countries in Scientific Journals of the United States and Viceversa," presented by Manuel S. Vallarta, Mar. 23, 1941, MIT School of Humanities and Social Science, Office of the Dean records, AC20, box 4, file 201, MIT Archives.

24. Letter from Henry Allen Moe to Robert Caldwell, Feb. 25, 1941, National Research Council Central Files, file "Foreign Relations, 1941, International Organizations: Committee on Inter-American Scientific Publication," NAS Archives.

25. In 1933 he married María Luisa Margáin Gleason, who belonged to a Mexican elite family as he did. Their marriage expanded his kinship network ties, giving him even greater access to power elites in Mexico. They did not have children.

26. Letter from Lawrence H. Levy (OCIAA Legal Division) to Henry Allen Moe, Feb. 10, 1942, sección Personal, subsección Correspondencia, serie Científica, box 24, file 2, AHC-MSV.

27. "Plan for an Inter-American Academy of Sciences," presented by Robert G. Caldwell, Apr. 15, 1941, MIT Office of the President 1930–1958, Records 1930–1959, AC4, box 44, file 3, MIT Archives.

28. In English, National Autonomous University of Mexico, the largest and most prestigious university in Mexico then and now.

29. Letter from Manuel S. Vallarta to Robert G. Caldwell, MIT dean of humanities, Feb. 5, 1942, MIT School of Humanities and Social Science, Office of the Dean records, AC20, box 4, file 202, MIT Archives.

30. Program of the Inter-American Congress of Astrophysics, 1942, sección Personal, subsección Distinciones, Homenajes y Biografías, box 44, file 20, AHC-MSV.

31. For a detailed study of this meeting and its link with the Good Neighbor policy, see *Jorge Bartolucci, la modernización de la ciencia en México: El caso de los astrónomos* (Mexico City: UNAM and Plaza y Valdés, 2000).

32. Arthur Compton was another US scientist who also traveled to Latin America supported by the OCIAA (see chap. 10, this volume).

33. Eduardo L. Ortiz, "La política interamericana de Roosevelt: George D. Birkhoff y la inclusión de América Latina en las redes matemáticas internacionales (Primera parte)," *Saber y tiempo: Revista de historia de la ciencia* 4, no. 15 (2003): 53–112.

34. Letter from Robert G. Caldwell to Manuel S. Vallarta, Mar. 2, 1942, sección Personal, subsección Correspondencia, serie Científica, box 24, file 2, AHC-MSV.

35. Letter from Manuel S. Vallarta to Henry Allen Moe, Apr. 25, 1942, sección Personal, subsección Correspondencia, serie Científica, box 24, file 2, AHC-MSV.

36. Letter from Henry Allen Moe to Manuel S. Vallarta, Apr. 28, 1942, sección Personal, subsección Correspondencia, serie Científica, box 24, file 2, AHC-MSV.

37. Lecture given by Manuel Sandoval Vallarta, "Problemas escogidos de mecánica y teoría electromagnética," Mar. 1942, sección Personal, subsección Distinciones, Homenajes y Biografías, box 44, file 20, AHC-MSV.

38. Letter from George H. Harrison to Manuel S. Vallarta, May 22, 1942, MIT School of Humanities and Social Science, Office of the Dean records, AC20, box 4, file 203, MIT Archives.

39. Letter from Robert Caldwell to J. C. Beebe-Centre, June 15, 1942, MIT School of Humanities and Social Science, Office of the Dean records, AC20, box 4, file 203, MIT Archives.

40. Letter from Robert Caldwell to John M. Clark, Mar. 21, 1942, MIT School of Humanities and Social Science, Office of the Dean records, AC20, box 4, file 202, MIT Archives.

41. Letter from Manuel S. Vallarta to Robert Caldwell, June 25, 1942, MIT School of Humanities and Social Science, Office of the Dean records, AC20, box 4, file 203, MIT Archives.

42. Letter from Robert Redfield to Manuel S. Vallarta, Aug. 3, 1942, MIT

School of Humanities and Social Science, Office of the Dean records, AC20, box 4, file 203, MIT Archives.

43. David C. Cassidy, "The Physicist's War," in *A Short History of Physics in the American Century* (Cambridge, MA: Harvard University Press, 2011), 72–89.

44. Deborah Douglas, "MIT and War," in Kaiser, *Becoming MIT*, 81–102.

45. Notes of Karl Compton for his presentation "Research in Physics for the War Program," as part of a joint meeting of the American Physical Society, the American Association of Physics Teachers, and the Society for the Promotion of Engineering Education, June 25, 1942, Compton Papers 1906–1961, box 2, file 16, lectures and addresses, Jan. 1–Dec. 31, 1942, MIT Archives.

46. Letter from John C. Slater to Manuel S. Vallarta, June 29, 1942, MIT Office of the President, AC4, box 228, file 3, MIT Archives.

47. Letter from George R. Harrison to Manuel S. Vallarta, May 5, 1942, MIT Office of the President, AC4, box 228, file 3, MIT Archives.

48. Letter from George R. Harrison to Manuel S. Vallarta, May 22, 1942, MIT School of Humanities and Social Science, Office of the Dean records, AC20, box 4, file 203, MIT Archives.

49. Letter from John C. Slater to Manuel S. Vallarta, Sept. 1, 1942, sección Personal, subsección Correspondencia, box 21, file 17, AHC-MSV.

50. Letter from John C. Slater to Manuel S. Vallarta, June 29, 1942, MIT Office of the President, AC4, box 228, file 3, MIT Archives.

51. Memorandum from the Office of the MIT President, May 28, 1940, John Clarke Slater Papers, file "Compton, Karl T. #7," American Philosophical Society.

52. Letter from Manuel S. Vallarta to John C. Slater, June 1, 1940, John Clarke Slater Papers, file "Compton, Karl T. #7," American Philosophical Society.

53. Letter from Manuel S. Vallarta to Karl T. Compton, Aug. 24, 1942, MIT Office of the President, AC4, box 228, file 3, MIT Archives.

54. Letter from Manuel S. Vallarta to John C. Slater, Sept. 9, 1942, MIT Office of the President, AC4, box 228, file 3, MIT Archives.

55. Telegram from Karl Compton to Manuel S. Vallarta, Sept. 16, 1942, sección Personal, subsección Correspondencia, serie Científica, box 24, file 2, AHC-MSV.

56. For an analysis of the biographical sketches of Manuel Sandoval Vallarta, see Minor García, "Cruzar fronteras," 5–12.

57. Ramos Lara, "La física en México."

58. "En 1943 comencé a venir a México. Anteriormente venía en vacaciones de MIT, pero no lo hacía por mucho tiempo. Ya en 1942 empecé a venir más tiempo; durante unos años, entre 1942 y 1946, distribuí mi tiempo entre Cambridge y México. No obstante llegó el momento en que me di cuenta de que si seguía con ese programa no tendría yo muy larga vida, ya que era necesario viajar

a menudo. Nuestra preocupación entonces, fue ver de qué manera se podría levantar el nivel científico en México y entonces se nos ocurrió la idea de la antigua Comisión Impulsora y Coordinadora de la Investigación Científica." Manuel Sandoval Vallarta, "Reminiscencias," Nov. 17, 1972, sección Personal, subsección Distinciones, Homenajes y Biografías, folios 4–14, AHC-MSV.

59. In English, Commission for the Promotion and Coordination of Scientific Research. This commission was one of the first efforts of the Mexican government to articulate science at a national level and is considered a predecessor to the current Mexican Council of Science and Technology (Consejo Nacional de Ciencia y Tecnología).

60. Letter from Manuel S. Vallarta to Tenney Lombard Davis, Dec. 10, 1942, sección Personal, subsección Correspondencia, serie Científica, box 24, file 2, AHC-MSV.

61. Letter from Clarence Haring to Manuel S. Vallarta, Dec. 5, 1942, sección Personal, subsección Correspondencia, serie Científica, box 24, file 2, AHC-MSV.

62. Harlow Shapley, "The Committee on Inter-American Scientific Publication," *Science* 109, no. 2842 (1949): 603–605.

63. Letter from Christina Buechner to Manuel S. Vallarta, Dec. 14, 1942, sección Personal, subsección Correspondencia, serie Científica, box 24, file 2, AHC-MSV.

64. National Academy of Sciences, *Report of the National Academy of Sciences* (Washington, DC: National Academies, 1949), 62.

65. In a similar manner, Clark Miller has suggested that the Department of State configured its postwar scientific diplomacy in Latin America during World War II: Clark A. Miller, "'An Effective Instrument of Peace': Scientific Cooperation as an Instrument of U.S. Foreign Policy, 1938–1950," *Osiris* 21, no. 1 (2006): 133–160.

66. Letter from John C. Slater to Manuel S. Vallarta, Nov. 20, 1942, sección Personal, subsección Correspondencia, serie Científica, box 24, file 2, AHC-MSV.

67. Ibid.

68. Letter from Manuel S. Vallarta to John C. Slater, Dec. 8, 1942, sección Personal, subsección Correspondencia, serie Científica, box 24, file 2, AHC-MSV.

69. Letter from Manuel S. Vallarta to Tenney Lombard Davis, Dec. 10, 1942, sección Personal, subsección Correspondencia, serie Científica, box 24, file 2, AHC-MSV.

70. *Comisión Impulsora y Coordinadora de la Investigación Científica, anuario 1943* (Mexico City: La Prensa Médica Mexicana, 1944). It is important to mention that this was not the first time that Vallarta had a chance to return to Mexico. In 1931 a newly appointed secretary of education of the Mexican gov-

ernment invited him to head the Office of Technical and Industrial National Education, but he rejected this offer following the advice of a member of his family who considered that at that time government jobs in Mexico were not safe and well paid but were "exposed to the vagaries of politics." See letter from Ignacio Vallarta Bustos to Manuel Sandoval Vallarta, Nov. 23, 1931, sección Personal, subsección Correspondencia, serie Científica, box 21, file 5, AHC-MSV. In contrast, the possibility of directing the CICIC in 1942 was interpreted in a very different way by another of Vallarta's relatives: "This is your chance to work in Mexico and the opportunity for our country to take advantage of your ability." See note from Carlos Lazo to Manuel Sandoval Vallarta, Jan. 3, 1943, sección Personal, subsección Correspondencia, serie Científica, box 24, file 2, AHC-MSV.

71. Indeed, in 1941 both countries had signed a binational security agreement, and in 1942 the Mexican government had also declared war against the Axis powers. For a historical account of how the Mexican and US governments acted as allies during World War II, see Julio Moreno, *Yankee Don't Go Home! Mexican Nationalism, American Business Culture, and the Shaping of Modern Mexico, 1920–1950* (Chapel Hill: University of North Carolina Press, 2003); Josefina Zoraida Vázquez and Lorenzo Meyer, *México frente a Estados Unidos: Un ensayo histórico, 1776–2000* (Mexico City: Fondo de Cultura Económica, 2006).

72. Letter from Manuel S. Vallarta to Christina Buechner, Dec. 21, 1942, sección Personal, subsección Correspondencia, serie Científica, box 24, file 2, AHC-MSV.

73. In that period, Sandoval Vallarta was also director of the Institute of Physics of the Universidad Nacional Autónoma de México (1943–1945), founder of the Society of Physical and Mathematical Sciences (1943) and the National College (1943), and director of the National Polytechnic Institute (1944–1947) in Mexico. See his curriculum vitae, sección Personal, subsección Distinciones, Homenajes y Biografías, AHC-MSV.

74. Letter from Karl T. Compton to Manuel S. Vallarta, Mar. 13, 1946, MIT Office of the President, AC4, box 228, file 3, MIT Archives.

75. Letter from Manuel S. Vallarta to Karl T. Compton, Mar. 27, 1946, MIT Office of the President, AC4, box 228, file 3, MIT Archives.

76. For an analysis of how he became an "expert" in matters of nuclear energy diplomacy and the specific interests he negotiated as a scientist-diplomat in representing Mexico at this particular meeting, see Adriana Minor and Joel Vargas-Domínguez, "Mexican Scientists in the Making of Nutritional and Nuclear Diplomacy in the First Half of the Twentieth Century," *HoST—Journal of History of Science and Technology* 11 (2017): 34–56.

CHAPTER NINE

The Officer's Three Names

The Formal, Familiar, and Bureaucratic in the Transnational History of Scientific Fellowships

Michael J. Barany

Introduction: Fragment from a Cocktail Party

The archives of the iconic organizations of transnational science in the twentieth century are padded thick with desk work: forms, reports, accounts, letters, notes, and memoranda reposing for reference and posterity in neatly seriated files in row after orderly row of boxes and shelves. Most desk work is terribly dull, a litany of i's dotted and t's crossed, testament to the meticulous, multilayered patchwork of activity required to send money, people, and materials, along with high ideals and sweeping enterprises, across the globe. The officers and bureaucrats who made such desk work hum did not live for paper, but paper is, for the most part, what the archival historian (also a creature of desks and paper) gets to see.

Occasionally, however, these archives serve up a glimpse of those parts of organizational life that extend beyond paper and desk work. Such irruptions can, in turn, help one read the rest of the archive differently. Emissaries from another idiom—one ever present in these organizations but largely missing in their archives—records of non-desk work can emphasize different values, relationships, manners, and practices. This holds all the more if that non-desk work took place at a cocktail party.

In June 1957 the Rockefeller Foundation celebrated the sixty-second birthday and (more to the point) the twenty-fifth year of service of Dr. Harry Milton Miller Jr. with a gathering charged with cocktail-fueled reminiscences. The party's archival record takes the form of remarks offered by the foundation's high-profile science officer Warren Weaver, who attempted to indicate "a small fragment of that quarter century of [Miller's] remarkable service" and committed his remarks to writing for inclusion in the foundation's newsletter.[1] There was, indeed, much to cover. Born in Baltimore and educated in the American Midwest—completing a PhD in parasitology from the University of Illinois after a year and a half of service in the medical corps in France during the Great War—Miller was then nearing the end of a career with the Rockefeller Foundation that would span twenty-eight years (plus one month). Through wide-ranging travels and far-reaching networks of intelligencing and administration, he brokered the foundation's international fellowship programs in the natural sciences during a pivotal period in its history, marked by an expansive and epoch-defining intervention in transnational science.

"Dusty Miller (or Associate Director H. M. Miller, Jr., if you insist on being formal) joined the RF staff on June 1, 1932," Weaver began. His chummy overture suggests a basic insight into how Miller and the Rockefeller Foundation worked, which shall guide this chapter. Neither Weaver nor Miller put much stock in artificial formality, but they both knew the places and uses of being formal. Look at Miller's paper traces as he moved young men of science across the globe and you find a triple presence: the officer's three names.[2] There is Dusty Miller, the gregarious globe-trotter who quickly won affection wherever he went. There is the rigorous official Dr. Harry Miller, or Associate Director H. M. Miller Jr. (as Weaver put it), agent of a philanthropic titan with worldwide ambitions. These two sides of Miller dominated Weaver's remarks, for instance in his recollection that Miller "never made a visit without leaving behind him friendship and respect": the camaraderie of Dusty and the dignity of Dr. Miller.

Just as important was a third name, omitted by Weaver: the "HMM" who peppers notices and memoranda and who annotates reports as they travel through the Rockefeller Foundation's churning bureaucratic underbelly. What I shall call Miller's "bureaunym," a two- or three-letter name Rockefeller staff at all levels from typist to president deployed in internal communications, marks out the institutional connective tissue that underwrote and bound together Dusty's friendships and Miller's executive wherewithal alike. Each name represents a different way of moving through and intervening in the world of transnational science. Different, but not disjoint: each side of Miller's selectively splintered subjective posture interacts with the others, and the shifts and exchanges among Miller's names say as much about his work as do those names' contextual and operational distinctions. Miller was hardly unique in this regard, and his three names indicate a broadly applicable way of reading the personal and bureaucratic work behind science's transnational institutional architecture at midcentury.

Administrators in philanthropic, governmental, and academic institutions combined formal communications and inquiries with extensive informal travel and relationship building. They bound these channels of formal and informal communication together by intensively collecting, archiving, and cross-referencing reports, logs, diaries, and evaluations. One subjective posture rarely sufficed for long: "Dusty" received informal news, which he matched to reports filed and annotated by "HMM," prompting formal inquiries by "Dr. Miller," abetted by Dusty's back-channel maneuvers and tracked in HMM's official notes and observations.

Following Miller's names gives purchase to the multifarious forms of border crossing deployed and promoted by Rockefeller Foundation officers and their counterparts in related organizations. These officers sent both people and documents across borders, navigating social, political, and institutional constraints and relationships through personal contact in diverse venues and media. Paperwork helped officers to identify people and institutions for their interventions, to manage contingencies, and to open avenues for adaptation, at once documenting existing arrangements and transforming them to serve the officers' ends. Efforts to move bodies depended integrally on efforts to move papers, and vice versa. Miller's names index that movement, as the inscribed records of his personal travels and the means by which he projected his administrative presence across considerable distances through post, telegram, and other means.

I have come to know Miller's three names through my research into the midcentury emergence of an intercontinental mathematical discipline, defined by transits and exchanges of people and texts across multiple continents as an established feature of scholarly practice.[3] Mathematics was always a small part of Miller's fellowship portfolio, confined mostly to the decade and a half beginning in 1940 when the Rockefeller Foundation pivoted from a war-engulfed Europe to the comparatively tranquil frontier of Latin America for its interventions in science and medicine. Starting with the two leading mathematicians of Uruguay—Rafael Laguardia and José Luis Massera—and reaching to Argentina, Brazil, and Mexico, Miller shepherded a small Latin American mathematical elite through further training in the United States. Along the way, the peripatetic parasitologist helped establish the bureaucratic formations that joined Latin America to a newly integrated multicontinental community of mathematicians.

Miller and the Rockefeller Foundation worked amid a tangled network of private, governmental, and transnational bodies to assess and intervene in the developing world's scientific institutions. Fellowship programs like those of the Rockefeller and Guggenheim Foundations found common cause with, for example, the technical assistance programs of the new postwar United Nations Educational, Scientific and Cultural Organization (UNESCO). I have elsewhere elaborated this process in connection with the itineraries of two of Miller's South American charges, Uruguay's Massera and Brazil's Leopoldo Nachbin, showing how mathematical institution building in the early Cold War rested integrally on the assessment and circulation of people—a pair of activities at which Miller was especially adept.[4] Circulation, in the context of Rockefeller fellowships, has a specific and important operational meaning: officers were centrally preoccupied with ensuring that fellows returned home—completing a circuit—under conditions that allowed them to become institutional leaders for nations and a continent in transition. Opening and closing these circles was hard work, requiring the hard paperwork of circulating documents to and from the foundation's offices and consolidating many kinds of hard-won relationships into orderly and mobilizable files.

Revisiting these fellowship files—and selected others—from the perspective of Miller's multiple names shows how the Rockefeller officer mixed subjective postures through paper and personal relations to establish cross-border connections in the face of a variety of obstacles. This perspective underscores the personal, institutional, and infrastructural values that took precedence over a wide range of other considerations, from the political, to the intellectual, to the ideological. In Miller's three names, we glimpse the plural foundation of transnational science, the layered informal, formal, and bureaucratic voices and selves that put certain forms of transnational exchange at the center of a vast reconfiguration of science in the mid-twentieth century.

Hearts and Ping-Pong

Miller earned his stripes in fellowship administration in the Rockefeller Foundation's interwar Paris office, which Weaver portrayed as a site of frenetic network building punctuated by lunchtime card games and evening table sports. "We were young then," he explained, "and very full of beans." Weaver insisted that the Paris office "worked even harder than we played" and was not "all hearts and ping pong."[5] But hearts and ping-pong is not so bad a description of the interwar practice that eventually underwrote the Rockefeller Foundation's postwar fellowship programs, especially in less resource-intensive fields like mathematics. Foundation officers bounced from site to site, collecting information on who held what cards (people, resources, needs, and priorities), winning useful friendships as they went.

The Paris office became the lynchpin of Rockefeller's transatlantic efforts through a range of interventions under the International Education Board (IEB) in the 1920s, which gave Rockefeller philanthropy a foothold in Europe. IEB officers aimed to rebuild war-devastated European institutions on an American model featuring interinstitution competition, mobility for junior scholars, and entrepreneurial elitism.[6] The Rockefeller Foundation would attempt the same, under somewhat different conditions and with new political and ideological motivations, after World War II.[7] Between the wars, the IEB and the RF focused on establishing or reinforcing centralized institutions—for mathematics, most notably in Paris and Göttingen—and on supporting programs of travel and exchange that allowed junior scholars to circulate within Europe to benefit from such concentrations of resources and expertise.

IEB officers established a pattern of intelligencing that Miller learned in the Paris office of the Rockefeller Foundation's Division of Natural Sciences, which took over many of the IEB's activities in the sciences in the 1930s.[8] Officers traveled extensively, and they intensively documented their travels for future reference. They supplemented this firsthand information by systematically supporting travel by trusted experts, who then supplied reports of their observations and assessments, and by systematically collecting information from experts whose travels brought them into the vicinity of Rockefeller officers. At the same time, they sustained close advisory relationships with selected scientific elites, whose regular professional activity gave them a broader view of their disciplines that could inform Rockefeller programs.

Grand tours helped officers build personal relationships and develop an overarching sense of the systems, needs, and obstacles relevant to their projects. In 1923–1924, for example, the IEB's first president, Wickliffe Rose, undertook a nineteen-country trip through Europe. Rose generated a steady stream of reports on local institutional conditions and on authority figures who could help broker exchanges and provide their own expert assessments in the future. At the same time, with an eye toward developing scientific personnel, he noted promising younger scholars, along with their research and training conditions, employment prospects, and related considerations.

While Rockefeller officers generally had advanced scientific training, the most relevant parts of that training for their Rockefeller work concerned the culture, sociability, and institutional organization of science rather than any specific topical knowledge. Miller, for his part, was well aware of how little his scientific background prepared him to evaluate mathematicians, remarking of José Luis Massera's candidacy that he was "certainly in no position to judge [the proposal] on the technical side."[9] Officers consulted scientific experts in place of cultivating specialist knowledge of their own. This vastly expanded the number of research areas to which the foundation could direct resources, while also helping officers champion new or emerging fields. The approach tended to concentrate intellectual authority in those well-placed individuals who possessed the most institutional authority and who were best positioned to guide the foundation's evaluations of unproven researchers and topics.

Rockefeller Foundation and IEB officers extended trusted experts' potential as informants by amplifying those experts' own professional intelligencing, using an influx of resources to turn modest ventures into more ambitious ones. When the dean of American mathematics, Harvard's George David Birkhoff, planned to make his very first trip to Europe in 1926, he requested a thousand dollars of supplemental support from the IEB in exchange for a detailed report on conditions in the countries on his itinerary.[10] Birkhoff shared his expert opinion on who the leaders were in each locale, where investments might be profitably directed, and other topics of interest to the IEB. Despite his robust correspondence with European mathematicians (by the standards of his period), he was surprised by many aspects of the situation on the ground, and his expert observations covered topics like sabbatical leave policies that had not occurred to IEB officials in their own inquiries.[11]

Where bureaucratic or expert grand tours were not forthcoming, officers made do with the assembled expertise of those nearer to hand. Any mathematician passing through New York and known to the foundation could count on an invitation to stop by the Rockefeller Foundation offices and to have notes from any conversations typed and filed. Correspondence with those the foundation identified as leaders of the field gave rise to thick dossiers of tables, lists, maps, and reports.[12] Rose's IEB practice turned on using his formal position to establish informal advisory relationships, transforming those informal relationships selectively into formal ones for producing administrative data, and transmuting those data through bureaucratic practice into formats that could guide projects. This multifarious practice, which Miller would learn in Rockefeller's Paris office and would come to reflect in his three names, drew on the social and institutional norms of academic research and of philanthropy. The outcome was not entirely of either world, but as philanthropic involvement in transnational science grew in scale and significance, an important segment of academic practice came to conform to this hybrid approach to intelligencing and intervening.

A Friendly Invasion

When Europe descended once more into war, the Rockefeller Foundation withdrew from what had hitherto been their primary base of operation abroad. Miller and Weaver's Division of Natural Sciences, as well as the International Health Division (which Rose had launched before moving on to spearhead the IEB), redeployed personnel and resources from Europe across the American hemisphere, in part toward the American war effort (including preparations for war before the United States' formal entry into the conflict) and in part toward what Weaver termed "a friendly invasion of Latin America."[13] Rockefeller officials latched on to the latter theater as an outgrowth of a burgeoning turn to hemispherism among American scholars, businesspeople, and policy-makers in the twentieth-century.[14] Most immediately identified with President Franklin Roosevelt's Good Neighbor policy of inter-American hegemony, the foundation's interest in Latin American science and mathematics derived more from an ideal of cultural exchange and solidarity than from previously dominant motives to understand the region as such. There did not need to be a place-specific reason to do mathematics in South America in order to justify its inclusion in Rockefeller programs—that mathematics was an important part of American scholarship was reason enough. Just as American interventions in Europe presumed a fundamental commonality as the basis for navigating differences, later Latin American programs took a basic homology between respective societies and scientific institutions for granted.[15]

That presumed homology shaped everything from grand programmatic ideals to routine aspects of administration. Each form of intelligencing just identified with the IEB in Europe thus found a parallel in Latin America. The translation extended across formal and informal aspects of officers' work, even to the drinks that lubricated long evenings on the road. As Weaver recalled at the 1957 cocktail party, he and Miller were certainly acquainted with "the marvelous tranquilizing power of triple Pisco Sours," a potent Andean cocktail.

Scientific experts, too, proved portable. George Birkhoff, for one, followed his interwar European tour with a 1942 passage through Mexico, Peru, Chile, Argentina, and Uruguay—sponsored by the US Office of Inter-American Affairs (OIAA) rather than the Rockefeller Foundation.[16] Birkhoff's former student Marshall Stone made an OIAA tour of his own the next year to Peru, Bolivia, Argentina, Uruguay, Paraguay, and Brazil.[17] The Rockefeller Foundation obtained Stone's OIAA report, in turn, through its role as a principal sponsor of the American Mathematical Society's War Policy Committee, which Stone chaired. Miller's annotations, tagged with his bureaunym "HMM," show that Stone's report provided important background for the first mathematical fellowships under Miller's watch.[18]

The very first such fellowship went to Rafael Laguardia of Montevideo, Uruguay, who moved among a number of institutions in the US Northeast. His fellowship was successful but left little direct paper trail. Its success is attested indirectly, through the many and lasting correspondence relationships he was able to establish with US mathematicians, which helped him to become a significant participant in a variety of postwar international mathematical formations. Though he was a meticulous self-archiver, the only records he retained of the fellowship itself were a generic pamphlet the foundation gave to each fellow and a packing list showing that he shipped a trunk with one hundred pounds of personal effects and books.[19]

The Rockefeller Foundation's own dossier for Laguardia is comparably slim. Judged to have nothing of lasting administrative consequence, the file's correspondence and reports were purged sometime between 1957 and 1964, according to a stamp near the bottom of a set of six index cards filled front and back with more than fifty entries in a chronological catalog of Laguardia's records.[20] His fellowship file itself now contains only two application forms, dated respectively October 27, 1941, and November 20, 1943. The first explained that Laguardia had studied in Paris in 1928, a time when the Rockefeller Foundation was deeply involved in Paris mathematics, and proposed a visit to the Institute for Advanced Study in Princeton, New Jersey. That proposal went unrealized, but his second application—marked as read by "HMM" in February 1944—notes that the foundation was then funding Laguardia's international study closer to home, in Rosario, Argentina. There, he worked with Italian mathematician Beppo Levi and Spanish mathematician Luis Santaló, both of whom had migrated to Argentina to escape European fascism, as well as Argentine statistician Carlos Dieulefait. Once in the Rockefeller orbit, one tended to remain there, and so it is not surprising that the second application led to a fellowship in the United States, where Laguardia was expected to visit a variety of institutions to prepare to direct research and instruction in mathematics and statistics in Montevideo's Faculty of Engineering.[21]

Another early Rockefeller mathematics fellow, Guillermo Torres Diaz of Mexico, left a more complete paper trail in the Rockefeller archives that shows the kinds of correspondence required of straightforward fellowships.[22] Torres began pursuing his PhD at Princeton, advised by Solomon Lefschetz, under a fellowship from the US State Department between the fall term of 1947 and the spring term of 1949. Nearing the end of his State Department fellowship, he applied to the Rockefeller Foundation in order to be able to complete his degree, a task that, according to an "HMM" annotation on his application form, would be feasible with a further nine months of Rockefeller funding. A handful of letters from Lefschetz furnished "Dr. Miller" with a glowing assessment of the candidate and several pertinent points of background information, especially pertaining to the candidate's circumstances at Princeton. "HMM" annotated these letters with further details and evaluations, and the file also includes reports labeled "HMM" of Miller's personal meetings with Lefschetz in 1949 and Torres in 1950. The latter note records Miller's discussion with Torres about possible modifications to his fellowship that might allow him to attend the 1950 International Congress of Mathematicians, which was to take place in Cambridge, Massachusetts, a few months past the planned end of his Rockefeller support. As a general rule, formal information flowed on paper to "Dr. Miller," and "HMM" integrated it into his fellowship apparatus along with gleanings from personal conversations.

Deseos de Perfeccionar

Laguardia's Montevideo colleague José Luis Massera had a much more difficult time than Laguardia did and so left a paper trail that shows Miller's bureaucratic process in much greater detail. I have elsewhere elaborated on these difficulties, which ranged from transiting in wartime to finding a suitable supervisor to navigating diplomatic and FBI objections to his Communist politics, with their numerous mathematical and political contexts and consequences.[23] Despite many complications, Massera's fellowship was ultimately quite successful by the Rockefeller Foundation's standards. It connected him to prominent researchers in the United States, leading to lasting ties through correspondence, publication, and other professional activities. Here, I shall revisit Massera's Rockefeller dossier to catalog the many different kinds of personal and administrative relationships Miller cultivated and called upon to make the fellowship succeed.

Massera's Rockefeller paper trail begins with Stone's April 1944 OIAA report, filed with other documents from Stone's War Policy Committee. Early in the letter, Stone complained that South America's "political and geographical divisions" made it difficult for intellectual elites to cross borders in the continent for training and work—a circumstance Stone hoped the Rockefeller Foundation could remedy by promoting intracontinental exchanges in addition to bilateral exchanges with the United States. Miller here noted in red (as "HMM") that Laguardia had gone from Uruguay to Argentina under the sort of arrangement Stone proposed. For a variety of institutional reasons, Stone recommended initially focusing on sending Latin American mathematicians to train in the United States, rather than sending US mathematicians south to influence faculties there. Stone assessed the state of mathematics instruction and research, noting, for instance, that Argentina's Julio Rey Pastor "has a very wide knowledge of mathematics and great gifts as an expositor" but "does not seem to have a gift for organizing and promoting the group interest of mathematicians."

Of Montevideo's mathematicians, Stone noted that Laguardia was then at Harvard (with Rockefeller support) and "[a]ssociated with him is Professor Massera." He observed, "Both men are under forty, have very pronounced mathematical interests, excellent training, and good judgment about their problems," and added some further remarks about their potential as researchers. Later in the letter, Stone included Massera on a list of six mathematicians whom he considered priorities for fellowships in the United States.

Stone's letter helped establish Massera's credibility as a fellowship candidate, but it was not sufficient to initiate his candidacy. However, Stone did not just visit and report on South American institutions. He also spoke with mathematicians there about the United States and opportunities (including those supported by the Rockefeller Foundation) for American training. Such conversations gave ambitious young scholars ample opportunity to learn about potential fellowships, both directly and indirectly. They had comparatively less information about how to solicit such fellowships, and so the personal contacts Miller and his colleagues established in the region played a secondary role in connecting prospective fellows to fellowship programs. Walter S. Hill, an Uruguayan professor of engineering and physics in Montevideo described by one visitor from the United States as "a local big shot,"[24] played that role for Massera, contacting Miller directly and also furnishing Massera with Miller's contact information.

Hill's letter reached Miller first and was initially filed with other correspondence involving Montevideo physicists.[25] Massera himself wrote a few days later, and an English translation of Massera's letter along with a typed transcript of the portion of Hill's letter concerning Massera formed the seed for the latter's Rockefeller fellowship dossier.[26] (Though Miller was, according to Weaver's recollections, fluent in Spanish, Portuguese, and French, his annotations show that he worked with Spanish correspondence primarily in English translation arranged by Rockefeller staff.) The recommendation from Hill was brief. He described Massera as "a young man of exceptional talent and background" who was "a brilliant student" with "great aptitude for teaching." Massera himself wrote at greater length, describing his research and expressing "mis deseos de perfeccionar mis conocimientos." The translator rendered this as "my desire to broaden my knowledge," though the Spanish more precisely indicates what Massera sought: *perfeccionar* typically connotes training and development and *conocimientos* indicates familiarity and understanding beyond just factual knowledge. Significantly, the translation and its reception show that Massera vastly overestimated Rockefeller Foundation officials' interest in the specifics of his research and their capacity to understand it. He offered to send reprints and referred to publications and terminology which the translator's skewed renderings show were unfamiliar and which did not elicit comment from Miller.

Instead, Miller noted several points of missing information that for him were much more relevant. After receiving Massera's translated letter, Miller wrote to Hill to ask for information about Massera's marital and employment status, and specifically whether he had a wife who would expect to join him in the United States and whether he could expect to return to a full-time position at the conclusion of a fellowship.[27] The latter was a crucial point for prospective fellows, emphasized as well in Laguardia's paperwork, as the foundation aimed to support those who already had a firm footing in their local academic institutions and could thus use their fellowship experience and training immediately to improve those institutions. Rockefeller officers wanted to encourage circulation, not migration. Miller also referred to his own upcoming visit to Uruguay and his intention to meet with Massera (as well as Hill and other local contacts) in person. As he had in Europe, Miller continued to use such trips to make and reinforce personal connections that could be carried on by post from Rockefeller headquarters in New York.

Miller did not, however, wait for Hill (or, for that matter, Massera) to supply the missing information. The same day he wrote to Hill he also wrote to Laguardia, referring to "our friend Walter Hill" and requesting the same details in the hope of being able to offer a preliminary assessment of Massera's fellowship prospects before traveling to Uruguay.[28] The precaution of writing to Laguardia proved unnecessary, as Hill shared Miller's inquiries directly with Massera, who himself wrote in short order to Miller with detailed answers.[29] On the question of full-time employment, Massera explained that "in this country we have no clear idea of the exact meaning of this expression" and described his teaching and academic schedule as well as his regular political activities.

He explained as well that he was married and wondered if his wife (an artist) might accompany him "not as a companion but as a fellow of the Rockefeller Foundation." Massera also arranged for the dean of the Faculty of Engineering to write Miller a letter of support.[30] Laguardia, too, replied to Miller's request for information, commenting on Massera's family situation, employment status, and political inclinations.[31]

For Massera's formal application, medical examination, and associated screening and paperwork, Miller turned to the nearest Rockefeller officials based in the area: L. W. Hackett and C. W. Wells of the International Health Division, in Buenos Aires. Wells, on their behalf, conveyed Massera's health records and commented on his itinerary and visa application, which appeared routine.[32] Massera followed up with Miller shortly thereafter to elaborate on his logistical arrangements and academic plans, enclosing a copy of a letter from his prospective supervisor at Stanford.[33] Miller replied to Wells to make sure Massera's itinerary included stopovers in the appropriate mathematical centers of South America along his intended route and to emphasize the importance of pursuing Massera's visa with the consulate in Montevideo.[34] In April 1945, as Massera's visa case soured, Miller returned to his archival copy of this letter and annotated it with references to Wells's official diary and its November entries regarding assurances he had obtained regarding the visa.

But that is getting ahead of ourselves. At this stage, when Massera's fellowship appeared routine, Miller alternated principally between "HMM"—the assembler and annotator of background information from the Rockefeller paper mill—and "Dr. Miller," "Mr. Miller," or "H. M. Miller, Jr."—the comparatively formal officer corresponding with foundation contacts in an official capacity. Though the salutations and signatures from his Hill and Laguardia correspondence do not appear in the excerpts in Massera's file, Miller likely took an approach closer to the informal "Dusty" in his informal information gathering with these informants. The formal mechanics of the fellowship coordinated by "Dr. Miller" here rested upon work as "HMM" and "Dusty" to establish Massera's viability as a candidate and anticipate upcoming needs. Massera's paper trail is so long and rich in part because "HMM" and "Dusty" fell somewhat short in this latter task, leaving Miller with a series of conundrums whose resolution would require resorting to all three subjective postures.

A Visa Difficulty

During November and December 1944, Miller continued his correspondence with Massera over the academic details of his fellowship, with the dean of the Faculty of Engineering over Massera's institutional situation at home, and with Wells over the visa and other required hurdles. A December 5 letter from Wells sparked a flurry of discussion at Rockefeller headquarters by noting that Massera's hitherto-unspecified political activities included a leading position in the local communist party.[35] As the letter was passed around and its contents discussed, officials in New York added their initials to the document and annotated it with new developments.

Here, the significance of a relatively anodyne letter to "Mr. Miller" emerged only after the letter had churned through Rockefeller back rooms under the sign of "HMM" and other bureaunyms. A note that appears to be in Miller's hand asks (apparently to Division of Natural Sciences associate director Frank Blair Hanson), "what is your reaction?" and elicited a penciled response that the foundation did not bother about a fellow's political affiliation "except as they affect the fellow's future in his own country." (Massera's communism would, indeed, famously land him in prison in his own country, though that would have been hard to predict late in 1944.) Another note, also apparently in Miller's hand, records that the "Communist aspect [was] cleared with TBA," the bureaunym of Rockefeller Foundation vice president Thomas B. Appleget. As the letter and annotations made clear, Massera's communism was recognized as an annoyance but not considered a barrier to his fellowship.

This kind of assessment was relatively unremarkable for the Rockefeller Foundation in the mid-1940s. Notes about São Paulo's Omar Catunda, whom Miller (as "HMM") would describe in a 1950 note as a "rabid Communist," focused more in 1946–1947 on his relatively advanced age, his weak command of English, and his (rather more favorable) qualities as a mathematician.[36] Indeed, his age so dominated Rockefeller discussions that his politics are hard to discern in his initial batch of documentation. One would expect Catunda's politics to become more notable for American officials between 1947 and 1950, a period of rapid intensification in official American anticommunism. But the seemingly complete lack of Rockefeller concern for his politics at the start of this period nonetheless signals how far those like Miller were able to remove themselves from the period's anxieties and suspicions over communist ideology.

At Wells's urging, a counselor from the US Embassy in Montevideo wrote to Miller on January 4, 1945, to advise him that the Embassy was prepared to grant Massera a visa and that he should follow up with the State Department in Washington, DC.[37] Lewis Hackett of the International Health Division returned to the Buenos Aires office in mid-January and wrote a personal note to "Dusty" to advise of his plans to visit Montevideo with Wells and to joke that Miller had better warn the people at Stanford of Massera's communism, lest he "try to convert" them.[38] Addressing his letter to "Dusty" allowed Hackett to switch registers, to share a joke, and to commiserate outside their formal exchanges regarding administrative details of the fellowship. After the Montevideo trip, Wells updated "Mr. Miller" on Massera's itinerary and medical examinations.[39] All appeared to be progressing easily enough, under the circumstances.

Then, in March, a State Department official in Washington wrote a confidential letter to "Dusty" to explain that Massera's communism had indeed prevented a visa from being issued.[40] Note, here, both that Miller was on sufficiently familiar terms with someone in the State Department to receive such a notice with his nickname in the salutation, and that this familiarity did not help Miller to avert an embarrassing misreading of Massera's prospects for a visa. In accordance with the State Department's advice, the Rockefeller Foundation canceled the fellowship offer in April. An internal foundation memorandum that month, signed "HMM," showed a series of discussions, with Rockefeller officials (including Miller) expressing umbrage to their State Department counterparts and pressing for a resolution.[41] One possibility they considered was admitting Massera with partial FBI surveillance, and this became the US government's course of action after protracted negotiations. Wells continued as the foundation's point person with the Montevideo Embassy and corresponded with "Mr." or "Dr." Miller about new developments. In the meantime, the foundation resolved to wait for an affirmative assurance from the State Department before resuming fellowship plans on Massera's behalf.

Their patience paid off at the start of January 1946, when the cultural attaché at the Montevideo Embassy telephoned Wells to inform him that the US Attorney General had granted Massera permission to study mathematics at an institution selected by the Rockefeller Foundation.[42] Here, the foundation's established reputation among US government agents as a reputable and apolitical broker of international scientific exchanges—a reputation forged through extensive interwar and (especially) wartime contacts in both formal and informal contexts—left room for officials at the very top of the American immigration bureaucracy to delegate to the foundation the prerogative of guaranteeing an outspoken communist's legitimate scientific intentions. Wells informed Miller and set the gears in motion for Massera's prospective departure. The State Department confirmed the development to "Dr. Miller" in March, and Miller, Massera, and Wells then resumed contact in a formal register with academics in Montevideo and Stanford to revise Massera's itinerary and make logistical arrangements.[43] Miller returned to Montevideo as part of a South American tour in October 1946 and reported, to his frustration, that Massera's political activity—which now included running for the Uruguayan Congress as a Communist—continued to make his travel precarious: "Embassy officials . . . feel that they will be damned if they do issue the visa and damned if they do not." To flag the observation for discussion within the Rockefeller bureaucracy, addressing the matter to Warren Weaver, Miller annotated it bureaunymically "To discuss with WW—HMM."[44]

The same Embassy visit that furnished Miller with that bit of intelligence, however, also gave him a means to act. He telegraphed his Embassy contact on December 2 and was able to secure a letter dated December 4 assuring him that Massera would be granted a visa in the end, though the letter's transmission required a further telephone inquiry to a State Department contact in Washington.[45] Massera was on his way to Stanford by March.

Contacts and Reverberations

Miller's work did not end with Massera's departure, however. Wherever possible, he attempted to meet with Massera and record a note for his Rockefeller file when they crossed paths—whether in transit through Central America or once situated in California.[46] The Federal Bureau of Investigation made regular contact with Miller, in his capacity as "Dr. Harry M. Miller," after taking an interest in Massera's case early in 1947.[47] Interpolating from attributed comments, it is likely that Miller was the bureau's anonymized "Source K" of inside information from the

Rockefeller Foundation regarding Massera, including a note "that the foundation was aware of MASSERA's Communist background but that the fellowship had been granted solely because of his intellectual attainments and without regard for his political beliefs."[48] If his FBI file is to be believed, Massera ranged widely over the United States during his fellowship for a combination of academic, political, and touristic purposes. While Massera's mathematical contacts were unreserved in praise of his seriousness and scholarly dedication, there was not much Miller could do to verify or respond to academic matters. The lone exception came when Massera and his Stanford supervisor mutually determined that Massera would be better off studying at New York University and Princeton, at which point Miller turned to his familiar routine of logistical brokering to enable Massera's relocation.[49] After the fellowship ended, Miller again met Massera in Montevideo and recorded his latest academic and political activities, with the latter including a brief imprisonment.[50]

Massera's politics had a number of ramifications for others in and beyond the Rockefeller Foundation. Both he and Laguardia hoped to return to the United States in the summer of 1950 for the first postwar International Congress of Mathematicians, hosted principally at Harvard.[51] They worried initially about finances and were able to secure support for their travel from funds dedicated to bringing foreign mathematicians to both the congress and an organizing meeting for a revived International Mathematical Union held in conjunction with it. With less than a month remaining before the start of the congress, however, neither Laguardia nor Massera had been approved for visas to travel to the United States. They considered Massera a lost cause at that late stage, and Laguardia quickly found that his professional association with Massera was enough to raise doubts about his own ideology for the American consul in Montevideo. Indeed, Miller had learned as early as 1948 that the US ambassador in Montevideo considered the two Uruguayan mathematicians to be politically linked. Despite Miller's assurances, the ambassador vowed to denounce any future Rockefeller support for Laguardia, and Miller quipped that his "dossier in the Embassy may now indicate that he [Miller], too, is a Communist."[52]

With time running short, Laguardia protested to the consul that "I have no political activity whatsoever, [and] I do not belong to any political party or cultural organization with political implications."[53] Meanwhile, American mathematicians (including Stone), together with Guggenheim Foundation officers, lobbied the Montevideo consul, the US

State Department, and others to seek a swift resolution. This multi-pronged effort resulted in a visa by the middle of the month, just in time for Laguardia to depart and participate in both the International Mathematical Union meeting and the congress. While in New York for the former gathering, Laguardia met with Miller, who made a note of the affair in the Rockefeller Foundation's files.[54] Even after this series of events, officials in the US government retained suspicions of Laguardia. Assessing another mathematician, Paul Halmos, visiting Uruguay from the United States in 1951–1952, the FBI recorded that "a usually reliable source" advised their informant in Montevideo that Laguardia—who had helped to arrange the visit—was "a reported Communist sympathizer."[55]

Halmos was well aware during his visit of American officials' ongoing preoccupation with suspected communism in the region. The US cultural attaché in Montevideo, reported Halmos, tried to enlist him in "a little spy work, junior grade," which involved infiltrating the Federación de Estudiantes Universitarios del Uruguay. Halmos awkwardly declined that assignment but was later asked by that same attaché to comment on a claim "that Laguardia is a communist," to which Halmos "said that I thought it was the damndest nonsense I ever heard."[56] During his visit, Halmos reported to Marshall Stone, "From personal observation I wouldn[']t know that Massera is a communist." He and Massera would "either discuss the weather over a cup of tea, or else he tells me what were contents of his course on Hilbert space and what part of differential equation theory he is currently working on." Halmos found Massera "remarkably pleasant" and "placid," not prone to "spreading propaganda" but rather "the only one at the institute who really works."[57]

At the same time, Halmos met and traded tales with "a certain Mr. Miller of the Rockefeller foundation" and learned from Laguardia that the institute's ambition "of being able to obtain some Rockefeller money . . . exploded one day violently when Miller found out that Massera is a big wheel in the institute and Massera is a communist," though "[a] vague friendship between Laguardia and Miller continues, but with no indication of money ever changing hands."[58] To another correspondent, Halmos observed that Massera was "a real nice, warmly [*sic*], friendly guy, and quite a good mathematician," who "makes no secret of" being "a very active member of the local communist party" but "manages to keep his professional life and his political life carefully separated." All the same, Halmos relayed that Laguardia found Massera's communist activities "somewhat embarrassing." Indeed, "[w]hile the commys [*sic*] are not feared and hated so much here as in the States, they are nevertheless far from popular, particularly in . . . the circles where money comes from." Halmos alleged that the Rockefeller Foundation learned of Massera's communism "on a certain Monday and broke off negotiations" over funding that very Tuesday.[59] Thus, while Rockefeller officials could afford not to be demonstrative over politics, they nonetheless acted in ways discernible to outsiders to insulate themselves from the ideological baggage of funding Cold War rivals too freely or openly.

Miller's South American wrangling became a key case in point for Rockefeller Foundation officers' internal assessments of programs and policy with regard to communist and fellow traveler grantees. A November 1949 note from WW (Warren Weaver) to CIB (the bureaunym for Rockefeller Foundation president Chester I. Barnard) filed under "Program and Policy—National Security" explained that "HMM has had informal conversations with some Cultural Attaches" on the continent so that he might "know in advance if it seems quite clear that a certain man would be refused a visa."[60] As Weaver elaborated in a later discussion, Miller's approach to maintaining such contacts informally and with as light a touch as possible exemplified a general principle in Rockefeller diplomacy, crafted to preserve their "exceedingly valuable reputation as an apolitical agency."[61] Where Barnard appeared to disclaim awareness of past communist grantees, Weaver was obliged to correct him by pointing out that Massera "has been very active in the Communist Party" since concluding his Rockefeller fellowship. "It remains true that he is an excellent mathematician," Weaver explained, "but it is also true that our assistance to him has been criticized."[62]

In a subsequent note, Weaver acknowledged the special delicacy from a national security perspective of supporting theoretical physics, especially "atomic or nuclear physics," which the foundation did not generally underwrite. The exception to this rule was the foundation's support for physicists in São Paulo, as it was "the only place south of the Rio Grande where any really modern physics is being done," and equally pertinent, it was in a nation then on good diplomatic terms with the United States.[63] By the early 1950s Weaver could cast the Rockefeller Foundation's broader move away from the physical sciences in favor of biology and agriculture as a simplification "from the point of view of security content," as the latter "are largely innocuous."[64] The FBI did, in fact, express concern over Massera's potential to learn nuclear physics in California, and a side effect of officers' coarse and superficial familiarity with fellows' scholarship was an inability to say whether such concerns had merit.[65]

Weaver's evaluation of Massera's communism—that it was a deeply held commitment but did not interfere with his excellent scholarship—would seem directly at odds with the position the Rockefeller Foundation would take just a couple years later in response to allegations from US officials and lawmakers that they had knowingly supported communist individuals and institutions. Replying to the Cox Committee of the US House of Representatives, Rockefeller officers couched their approach to communist scientists in terms of their commitment to "sound, scholarly scientific procedure" and "objectivity," which, they asserted, communists, as such, "cannot be trusted" to maintain. They noted their "informal and confidential" practice of inquiring into foreign candidates, which they tied to their "status as a private nonpolitical organization." Any further ideological test, they asserted, risked denying "the importance of the non-conformist in the advancement of human thought," which "is the antithesis of Communism." Objectivity alone would be a sufficient safeguard against "totalitarian ideology."[66]

How can one square this with the foundation's own assessment of Massera? A case could be made that, while clearly a communist, Massera did not fit the totalitarian model on which the foundation's Cox Committee defense was premised. Nor, for that matter, would many Soviet scientists. The foundation assembled digest files on everyone who could possibly be construed as suspicious by the committee, including Massera, Laguardia, and Catunda, and found each funding decision defensible one way or another.[67] Adopting the committee's sweeping caricature of communism gave room to excuse or explain away the foundation's embrace of individual communist and fellow traveler grantees.

A more complete account of the foundation's apparent inconsistency, however, can be rooted in the plural subject positions we have come to recognize in Miller's fellowship administration, reflected and recapitulated here in broader Rockefeller Foundation policy discussions. Politics and ideology, here, are safely confined to informal and bureaunymic exchanges. These back-room acknowledgments of contradiction and complexity enabled officials to adopt formal claims on the foundation's behalf that advanced the foundation's interests without compromising its ability to maneuver behind the scenes. Here, the continuity of the foundation's interwar bureaucratic practice and sociability strongly tempered discontinuities in the political and ideological environments in which officers operated. Each formal assertion that communism cannot enter into Rockefeller programs is coupled, in the Cox responses, with an assertion of the foundation's prerogatives to operate informally and without rigid ideological guideposts—to entertain the participation of potential communists so long as their communism stayed out of the foreground.

Conclusions

The foregoing tripartite analysis of Harry Miller's Rockefeller Foundation work hinges not on the possibility of a rigorous division of roles and selves but on a recognition of the constant mutual imbrication of figures that are nonetheless distinguishable. Executing Massera's fellowship, Miller operated at different points as the formal Dr. Miller, the informal Dusty, and the behind-the-scenes bureaucrat HMM, adopting each posture according to the relationships and dictates of the situation at hand. Consider all that these plural voices bound together: Miller undertook both formal and informal contacts at institutions in North and South America, advocated across government and consular offices in multiple countries, and coordinated interventions by officials from multiple Rockefeller offices and operations. The work of fellowship administration required intercontinental travel by Miller and international travel by Wells and Hackett, and built as well on the travels of Laguardia, Stone, and others. Political questions blended into logistical questions, which merged with intellectual and institutional questions. Even when nearly everyone supported the fellowship, one recalcitrant entity (in this case, the Montevideo consulate, which reneged on Massera's visa) made trouble that rippled across every layer of negotiation and organization.

Massera's fellowship arrangements were complicated, but one should not lose sight of the fact that they had a particular form of complication, one that speaks as well to the many fellowships that ran without major incident. The complexities of his fellowship reverberated through a diplomatic-administrative network that was mostly well adapted to accommodate political and economic obstacles and ambiguities. But that network had its own less-visible weaknesses and blind spots. Rockefeller officers had little control over or ability to adjudicate Massera's scholarly activities as a fellow. Their reliance on informal assessments of potential barriers to his fellowship meant that they were late to recognize the consequences of Massera's political commitments and activities. Other particularities had to do with who Massera was: his combination of youth, enterprise, financial and family stability, and other affordances allowed him to gain notice and weather delays and complications in ways that other prospective fellows could not.

Though Rockefeller programs in the physical sciences and mathematics waned in the 1950s, they helped set a model and laid a personal and institutional infrastructure that other organizations—including UNESCO, the Guggenheim Foundation, and the Fulbright Commission—would continue well into the latter half of the twentieth century, with resources and objectives that combined earlier philanthropic models with new structures and prerogatives of the Cold War. The Rockefeller Foundation, moreover, played a pivotal early part in identifying and reinforcing a scientific elite in and beyond South America whose foreign training and connections and institutional wherewithal assured them ongoing authority in their respective locales. The lasting effects of Rockefeller institutional arrangements make their administrative idiosyncrasies all the more pertinent. Officers' means of gathering information for programs and fellowships created a feedback loop in which those who were better connected to the organization had more opportunities to reinforce and build upon those connections.[68] Institutional elites who, by chance or design, had the kinds of international connections that placed them in the foundation's orbit could turn their Rockefeller ties into substantial material and institutional resources.

Harry Miller used his ability to win friends and maintain informal scholarly networks to lay a broad groundwork for Cold War transnational science. Informal relationships let him broker travel and exchanges that reached far beyond what he managed through formal channels alone. Although the Cold War changed many formal diplomatic practices and considerations, it left these vital informal means comparatively unaltered. It therefore mattered tremendously who got to be Miller's friend, or whose sociabilities were compatible or incompatible with his. He wove far-reaching bureaucratic infrastructures with memoranda and reports, and the interlocutors of HMM shaped programs and policies at great distances. All the while, he spoke for and established the place of a towering foundation: his formal address marshaled resources, made careers, and guided institutions.

The lesson of this analysis for the transnational history of science is, at its most basic, that borders are always crossed multifariously. For

Massera and other fellows to travel as they did, a large volume of papers and a large number of other people had to travel between countries and institutions many times over. Those like Miller crossed borders by both formal and informal routes; they amassed and deployed other border crossings through bureaucratic amalgamations that were never simply records of crossings past. As they moved between nations, they never moved as just one subjectively singular and coherent person: crossing was always also a matter of splitting, of dividing and recombining positions and agencies. Like the transnational movements characterized elsewhere in this volume, the people and things of transnational mathematics could never simply go unaltered from place to place. Even bilateral exchanges built on extensive multilateral transits of people, paper, and resources. Teasing apart the plural subjective faces of these interactions can illuminate the occluded operation of power, the improvisation and contingency, and the reverberating connections that underwrote transnational endeavors.

Acknowledgments

I thank Margaret Hogan, Emily Merchant, the participants in the workshop that gave rise to this edited volume, and most especially John Krige for their contributions to the evidence and analysis represented here. Portions of this research were funded by a National Science Foundation Graduate Research Fellowship, Grant No. DGE-0646086.

Notes

1. Weaver remarks for newsletter, 1957 (hereafter Weaver Miller remarks), Rockefeller Foundation Archives (hereafter RF), Biographical File (FA485), box 5, "Miller, Harry M., Jr." folder, Rockefeller Archive Center, Sleepy Hollow, NY. I thank Margaret Hogan of the Rockefeller Archive Center for locating and sending me this file. I have reconstructed Miller's employment history from this file and from his listed positions in the Rockefeller Foundation annual reports, online at rockefellerfoundation.org.

2. See the polypartite analysis of Stephen Hawking's subjectivity in Hélène Mialet, *Hawking Incorporated: Stephen Hawking and the Anthropology of the Knowing Subject* (Chicago: University of Chicago Press, 2012), which in turn in-

vokes the classic formulation of Ernst Kantorowicz, *The King's Two Bodies: A Study in Mediaeval Political Theology* (Princeton: Princeton University Press, 1957). For a more detailed summary of the multiplication of selves in Mialet's *Hawking Incorporated*, see my review of the book, *British Journal for the History of Science* 46, no. 3 (2013): 544–546.

3. Michael J. Barany, "Distributions in Postwar Mathematics" (PhD diss., Princeton University, 2016).

4. Michael J. Barany, "Fellow Travelers and Traveling Fellows: The Intercontinental Shaping of Modern Mathematics in Mid-twentieth Century Latin America," *Historical Studies in the Natural Sciences* 46, no. 5 (2016): 669–709.

5. Weaver Miller remarks.

6. Reinhard Siegmund-Schultze, *Rockefeller and the Internationalization of Mathematics between the Two World Wars: Documents and Studies for the Social History of Mathematics in the 20th Century* (Basel: Birkhäuser, 2001); Reinhard Siegmund-Schultze, "The Institute [*sic*] Henri Poincaré and Mathematics in France between the Wars," in "Regards sur les mathématiques en France entre les deux guerres," ed. Liliane Beaulieu, special issue, *Revue d'histoire des sciences* 62, no. 1 (2009): 247–283.

7. John Krige, *American Hegemony and the Postwar Reconstruction of Science in Europe* (Cambridge: MIT Press, 2006), chap. 4.

8. See Siegmund-Schultze, *Rockefeller*, chap. 2.

9. Miller to Massera, Apr. 16, 1947, Archivo privado José Luis Massera, folder 5A, Archivo General de la Universidad de la República, Montevideo.

10. Birkhoff to Rose, May 12, 1925, International Education Board Archives (hereafter IEB), ser. 1, box 12, folder 171, Rockefeller Archive Center. Further documentation for the trip is in this same folder. Rockefeller also underwrote bilateral exchanges—for example the reciprocal visits of G. H. Hardy (Oxford) and Oswald Veblen (Princeton) documented in IEB, ser. 1, box 17, folder 247, Rockefeller Archive Center.

11. E.g. Trowbridge to Rose, Oct. 1, 1926 (discussing Birkhoff's report), IEB, ser. 1, box 12, folder 171.

12. E.g., IEB, ser. 1, box 8, folder 110.

13. Weaver Miller remarks. Some Rockefeller engagements in the region, particularly in connection with the International Health Division, predated Weaver's "friendly invasion." A key example was the foundation's role in the 1934 establishment of the Universidade de São Paulo. See Maria Gabriela S. M. C. Marinho, *Norte-americanos no Brasil: Uma história da Fundação Rockefeller na Universidade de São Paulo, 1934–1952* (Bragança Paulista: Editora Autores Associados, 2001). On Miller's later ties to this university, leading to an honorary degree in 1951, see Marinho, *Norte-americanos no Brasil*, chap. 4. The Guggenheim Foundation's Latin America program also began before the outbreak of World War II, but it expanded significantly in 1940.

14. See Ricardo D. Salvatore, *Disciplinary Conquest: U.S. Scholars in South America, 1900–1945* (Durham, NC: Duke University Press, 2016).

15. This presumed homology was fundamental to asymmetric, hegemonic aspects of the respective interventions, not just those aspects considered placeless or universal. I discuss this phenomenon under the rubric of mathematical and scientific colonialism in Barany, "Fellow Travelers and Traveling Fellows," 674–681.

16. Eduardo L. Ortiz, "La política interamericana de Roosevelt: George D. Birkhoff y la inclusión de América Latina en las redes matemáticas internacionales," *Saber y tiempo: Revista de historia de la ciencia* 4, no. 15 (2003): 53–111, and 4, no. 16 (2003): 21–70.

17. Karen H. Parshall, "Marshall Stone and the Internationalization of the American Mathematical Research Community," *Bulletin of the American Mathematical Society* 46, no. 3 (2009): 459–482.

18. Stone to Moe (copy), Apr. 13, 1944, RF, Record Group 1.1, ser. 200D, box 127, folder 1561. On the lasting ramifications of the War Policy Committee for mathematicians' ties to policy and government, see Michael J. Barany, "The World War II Origins of Mathematics Awareness," *Notices of the American Mathematical Society* 64, no. 4 (2017): 363–367.

19. Archivo privado Rafael Laguardia (hereafter Laguardia Papers), box 5, folder 10, Archivo general de la Universidad de la República.

20. RF, Record Group 10.2, fellowship recorder cards, box 18, "Uruguay: Laguardia (Carle), Mr. Rafael."

21. RF, Record Group 10.1, ser. 337E, box 219, folder "Laguardia Carle, Rafael." Laguardia's "prospective position" in the Faculty of Engineering is listed near the top of his fellowship recorder card.

22. RF, Record Group 10.1, ser. 323E, box 88, folder "Torres Diaz, Guillermo."

23. Barany, "Fellow Travelers and Traveling Fellows"; Barany, "Distributions," chap. 4. See also Vania Markarian, "José Luis Massera, matemático uruguayo: Un intelectual comunista en tiempos de Guerra Fría," *Políticas de la memoria* 15 (2014–2015): 215–224.

24. Halmos to Stone, Nov. 15, 1951, Paul R. Halmos Papers, box 4La74, "Uruguay" folder (hereafter Halmos Uruguay file), Archives of American Mathematics, Dolph Briscoe Center for American History, University of Texas at Austin.

25. Hill to Miller, May 15, 1944, RF, Record Group 10.1, ser. 337E, box 56, "Massera, Jose Luis" folder (hereafter RF Massera).

26. Massera to Miller, May 19, 1944, with translation completed by "BTR" on June 1, 1944, RF Massera.

27. Miller to Hill, June 2, 1944, RF Massera.

28. Miller to Laguardia, June 2, 1944, RF Massera.

29. Massera to Miller, June 15, 1944, with translation completed by "BTR" on June 28, 1944, RF Massera.

30. Magi to Miller, June 22, 1944, with English translation (likely July 11, 1944), RF Massera.

31. Laguardia to Miller, July 2, 1944, RF Massera.

32. Wells to Miller, Oct. 20, 1944, RF Massera.

33. Massera to Miller, Oct. 30, 1944, trans. "MLS" Nov. 17, 1944, RF Massera.

34. Miller to Wells, Nov. 3, 1944, RF Massera.

35. Wells to Miller, Dec. 5, 1944, RF Massera.

36. RF, Record Group 10.1, ser. 305E, box 20, "Catunda, Omar" folder.

37. Sparks to Miller, Jan. 4, 1945, RF Massera.

38. Hackett to Miller, Jan. 23, 1945, RF Massera.

39. Wells to Miller, Feb. 8, 1945, RF Massera.

40. Pierson to Miller, Mar. 7, 1945, RF Massera.

41. Inter-Office Correspondence, Apr. 27, 1945, RF Massera; see also the handwritten note by "HMM" dated Nov. 30, 1945, beneath Inter-Office Correspondence of Nov. 29, 1945. As Miller was the first recipient of the Inter-Office document, his amendment became part of the record for its other readers.

42. Wells to Miller, Jan. 3, 1946, RF Massera.

43. Pierson to Miller, Mar. 4, 1946; Miller to Wells, Sept. 13, 1946; and intervening letters; all in RF Massera.

44. HMM Diary, Oct. 23–26, RF Massera.

45. Massera Sparks to Miller, Dec. 4, 1946, enclosed in Caldwell to Miller, Jan. 8, 1947, RF Massera.

46. Miller to Massera, Jan. 10, 1946; HMM Diary, June 13–14, 1947; both in RF Massera.

47. E.g., Federal Bureau of Investigation, Headquarters File 100-HQ-341838, sec. 01, Office Memorandum, Apr. 8, 1947; sec. 02, Scheidt to Director, Mar. 6, 1948; and Lemaitre report Feb. 7, 1949.

48. Federal Bureau of Investigation, Headquarters File 100-HQ-341838, sec. 01, SF (San Francisco bureau) file number 100-27215, report of Jan. 2, 1948, p. 9. Cf. Headquarters File 100-HQ-341838, sec. 01, Office Memorandum, Apr. 8, 1947.

49. Miller to Massera, Sept. 10, 1947; HMM Diary, Sept. 15, 1947; GRP (Gerard Pomerat) Diary, Oct. 6, 1947; LWH Diary, Mar. 12, 1948 (meeting with Laguardia in Montevideo); HMM Diary, Apr. 27, 1948, with several annotations; HMM notes, June 1, 1948; all in RF Massera.

50. HMM Diary, Oct. 13–15, 1948, RF Massera.

51. See Barany, "Distributions," 223–228, for further details and background on Massera's and Laguardia's attempts to attend the congress.

52. HMM Diary, Oct. 13–15, 1948, RF Massera.

53. Laguardia to Kline, Aug. 8, 1950, Laguardia Papers, box 17, folder 9.

54. Record Group 1.1, ser. 200D, box 125, folder 1546, RF.

55. Excerpt from Federal Bureau of Investigation Bufile 100-387157, Chicago, IL, report of Jan. 16, 1953, p. 15, released pursuant to the author's FOI/PA request no. 1305216-0.

56. "A Mathematician in Uruguay" (typewritten account), pp. 42–43, Halmos Uruguay file.

57. Halmos to Stone, Oct. 15, 1951, Halmos Uruguay file.

58. Ibid.

59. Halmos to Ambrose, Oct. 23, 1951, Halmos Uruguay file.

60. WW to CIM, Nov. 1, 1949, Record Group 3.1, ser. 900, box 25, folder 199, RF. I thank John Krige for alerting me to these records and sharing his copy.

61. WW to CIB and LFK [Vice President Lindsley F. Kimball], Feb. 6, 1952, RF, Record Group 3.1, ser. 900, box 25, folder 200.

62. WW to CIM, Nov. 1, 1949, RF, Record Group 3.1, ser. 900, box 25, folder 199.

63. WW to CIB, Jan. 4, 1950, RF, Record Group 3.1, ser. 900, box 25, folder 199.

64. WW to CIB and LFK, Feb. 6, 1952, RF, Record Group 3.1, ser. 900, box 25, folder 200.

65. Federal Bureau of Investigation, Headquarters File 100-HQ-341838, sec. 01, memo, Mar. 4, 1947, and report of Charles G. Campbell, Aug. 21, 1947.

66. RF, Record Group 3.2, ser. 900, box 14, folder 89, pp. 46–48, 55.

67. RF, Cox and Reece Investigations (FA418), ser. 1.1.

68. Cf. Robert K. Merton, "The Matthew Effect in Science," *Science* 159, no. 3810 (1968): 56–63.

CHAPTER TEN

Scientific Exchanges between the United States and Brazil in the Twentieth Century

Cultural Diplomacy and Transnational Movements

Olival Freire Jr. and Indianara Silva

Ten years ago, while discussing the influence of American science on Europe after World War II, historian John Krige remarked on the gulf between works on diplomatic and economic history, on the one hand, and works on history of science, on the other.[1] This observation may easily be extended to the historiography of Latin American science and its relationship to the United States.

Let us take the Brazilian case. It is widely acknowledged that around World War II foreign influence on scientists in Brazil shifted abruptly from European to American. It is also known that during the Cold War this influence was reinforced, while diplomatic relations between Brazil and the United States were characterized by periods of conflict and periods of rapprochement. However, few works on the history of science in Brazil deal simultaneously with Brazil's diplomatic, economic, and Cold War histories of the same period.[2] In this chapter we seek to remedy this by focusing on the role played by physicists as social actors who circulated to and fro across national borders. All these individuals were well connected with their disciplinary communities and enrolled in international scientific networks. Some Americans and Brazilians among them played roles which were fully aligned with the foreign policies of their respective countries, while others acted in stark contrast to prevailing diplomatic relations. By describing the circumstances of their *travels*, a feature of the transnational approach emphasized in this volume, we open important perspectives otherwise obscured in stricter, more traditional approaches.[3]

The transnational approach and world or global history became prominent among American historians following the Cold War and the coalescence of a new phase of globalization around 1990. As remarked by Michael Barany and John Krige in their afterword to this volume, "it was born in the United States in the 1990s in the wake of the Cold War, in deliberate methodological and ideological opposition to American exceptionalism." Traces of such an approach may also be found in new approaches to history that emerged in the late twentieth century.[4] Historians of science were well placed to take into account such conceptual novelties. If, as remarked by Simone Turchetti and colleagues, "transnational history is a loosely defined term indicating the effort to produce novel historical accounts by focusing on the flows of people, goods, ideas or processes that stretched over borders," this resonated with the study of the diffusion and circulation of science as well as scientific networking not only inside national borders but also across them. It is also true, as Turchetti et al. put it, that "historians of science have practised transnational history without, however, paying sufficient attention to the theoretical foundations of this approach."[5]

The argument we develop here places the deep engagement of physicists with the norm of scientific internationalism in dialogue with the opportunities for—and constraints on—transnational circulation engendered by the foreign policies of national governments. Scientific internationalism—epitomized in the persons of Albert Einstein and Niels Bohr—helped build an international community whose loyalty to sharing knowledge was often at odds with their attitude to the policies of their governments. Bringing scientific internationalism to the fore in this chapter is not without problems, as the use of the term "scientific internationalism" by historical actors as well as by historians bestows upon it a polysemic, historically contingent meaning. Paul Forman has pointed out that, during World War I, scientists who had supported scientific internationalism were also engaged with their own countries in war, concluding that internationalism in personal relations was compatible with nationalism at large. Patrick D. Slaney has emphasized the difficult coexistence in the United States of internationalism with the anticommunism typical of Cold War times, when a person's political allegiance could exclude him or her from the international community of scholars. John Krige, for his part, has stressed that it could serve the broader interests of the American government. In Eisenhower's Atoms for Peace program, for example, scientific internationalism meant both "openness and security, sharing knowledge or technology and implementing regimes of surveillance," so that "rather than being an unnecessary complication, transnational engagements were a strategy that was essential to [the United States'] securing and maintaining" its hegemony. Nor was anticommunism the only counter to international cooperation, as Alexis De Greiff has shown in his study of the boycott of Abdus Salam's International Centre for Theoretical Physics in Trieste, Italy, by American and Israeli scientists.[6] In addition to this multifaceted meaning of scientific internationalism, in our case study physicists in the United States and Brazil, our historical protagonists, did not claim a general ideal of scientific internationalism to justify their interactions. Thus, we use—as an analyst's category—the term "scientific internationalism" in a broader sense, meaning by it mainly scientific exchanges between physicists in the United States and Brazil, including solidarity with scientists whose political views created difficulty for them with their governments.

We are all aware of the conflicts between the scientific community and the American government in the early 1950s, when a virulent anticommunism challenged core scientific values.[7] As Mario Daniels emphasizes in this volume (chapter 1) the US national security state clamped down on the free circulation of eminent scholars in and out of the United States, using visas and passports as tools to regulate their movement. The present chapter places this well-known period in a broader historical context, from the first enthusiastic engagements between the two countries to the tensions that emerged in the 1950s—but it does not stop there. Brazil as a haven for left-leaning scientists in the early Cold War was transformed into Brazil as a place of persecution under the dictatorship. Now Brazilian scientists fled abroad, including to the United States—if they could. The taint of communist affiliations pursued some of them both at home and abroad long after McCarthyism had lost its potency. Indeed, as Daniels emphasizes, the national security state was ever vigilant regarding the transnational circulation of "knowledgeable bodies," especially physicists. Those of left-wing or communist affiliation were always under suspicion. They could not escape the punitive gaze of either the American State Department or their local authorities in Brazil, who sometimes worked together to constrain their movement.

The first part of this chapter deals with World War II. We turn our attention to the American and Brazilian physicists who crossed borders during the war, including Arthur Compton, Gleb Wataghin, Mário Schönberg, and Paulus Pompeia, emphasizing the facilitating role of American foundations as agents of cultural diplomacy in Latin America.[8] In the second part, as we move into the Cold War period, the transnational approach reveals its full power as a tool for historians. Cold War border crossings of certain Brazilian and American physicists conflicted with the official policies of both regimes at various times. David Bohm, the talented American physicist who fled the glare of McCarthyism in the early 1950s, was supported by Brazilians during his exile. The Brazilian dictatorship (1964–1985) precipitated an exodus in the opposite direction. The American physics community offered widespread support to its Brazilian colleagues who sought to leave the country either out of fear of losing their jobs, even their lives, or to look for better job opportunities. These expressions of solidarity went against the (mostly covert) support in Washington for the coup in Brazil, but they were tolerated up to a certain extent by both governments.

This chapter illustrates how transnational movements between the United States and Brazil changed over the course of the twentieth century, fluctuating with changing historical and political contexts. During and soon after World War II, the two countries were fully aligned as allies and their borders were mostly open to each other, resulting in a moment of high cultural and scientific exchange. Conversely, during the Cold War and the Brazilian dictatorship, this scenario changed significantly. Now boundaries were not always open. The possibility of transnational knowledge flows varies according to time and space, actors' beliefs and political flags, and the foreign policies of governments. At one moment national boundaries can be crossed; at another, invisible walls appear between countries, which may constrain cultural and scientific exchange. By following individuals across borders, we can see the changing weight of scientific internationalism as it dovetails with but also comes up against political norms regulating transnational movement. The national security state accepts the commitment to the open circulation of knowledge bearers only up to a point, a threshold that depends on the kind of knowledge that is traveling and the "loyalty" of the individual in whom it is embedded.

Scientific Exchanges and Diplomatic Policies during World War II

In the early twentieth century exchanges between Brazilian and American scientists were limited to a certain number of fields, while Brazilian and European exchanges dominated. As for trade, during the 1930s the Brazilian government looked for better deals with both Germany and the United States. Historian Maria Ligia Prado argues that a "strategy of ambiguity" best describes these times.[9] When modern universities were set up in Brazil, particularly in São Paulo, their founders did not turn to the United States to look for teaching staff. For the opening of the University of São Paulo, state officials sought French scholars in the domain of social sciences and humanities, Germans for chemistry, and Italians for mathematics and physics. In physics a school working on cosmic rays flourished under the leadership of Gleb Wataghin and of Giuseppe Occhialini, both of whom came from Italy with the explicit support of Mussolini's regime to help found the Physics Department in 1934. Their first Brazilian students were sent for short stays in Europe—Mário Schönberg to Italy and Marcelo Damy to the United Kingdom. World War II dramatically changed this landscape of intellectual trajectories.[10]

In the early 1930s President Roosevelt reoriented American foreign policy by establishing the Good Neighbor approach in an attempt to build tighter bonds between the United States and Latin America. The fear that Latin American countries would align themselves with the Axis powers in World War II led to a further step in this direction with the creation in August 1940 of a new federal body: the Office of the Coordinator of Inter-American Affairs (OCIAA). To head this office, Roosevelt called on Nelson Rockefeller, an eminent businessman with political aspirations and well known for his corporate relationship with, and cultural interest in, Latin American countries. Its program was intended to supplement an expanding presence of military missions and attachés in a determined effort to counteract the "growing Axis propaganda menace" in the Americas. By 1942 Rockefeller and his assistants were allotted $38 million for a program of cultural diplomacy, mobilizing soft power to "get our message across." The United States came in with "a program of truth in answer to enemy lies," an initiative that would "persuade the Latin American nations to follow the lead of the United States in its opposition to the Axis powers, and to integrate the economy and politics with those of the United States in a Pan American system."[11]

Historians are familiar with the effects of such changes in US foreign policy. The Brazilian government aligned itself with the United States and broke diplomatic relations with the Axis powers, thus putting an end to a strategy of ambiguity toward Washington. The government of Brazil allowed the United States to set up a military airport in Natal in order to supply the battlefront in Northern Africa, and eventually Brazil sent troops to fight the Germans on Italian soil.[12] Similarly, the role of the OCIAA as an instrument of cultural diplomacy is widely known.[13] Shortwave radio, newspapers, translations of major literary works, and motion pictures were used to promote the American way of life. Goodwill tours to Brazil were arranged for Hollywood personalities such as Walt Disney and Orson Welles. The exchange was in both directions. If you visit the Hispanic Reading Room at the Library of Congress, you will see Candido Portinari's works painted in the early 1940s, a time when the singer Carmen Miranda toured the United States as an iconic figure of Brazil, a tropical country.[14]

Less known are the OCIAA cultural initiatives in science, technology, and education as part of its efforts to strengthen links among people in the Americas. In the Brazilian case, in certain fields at least, this period heralded a change from scholarly and professional exchanges with European countries that were tarnished by their relationships with the Axis powers to exchanges with the United States. Let us illustrate this with a few examples.

With Europe at war, Wataghin, the director of the Physics Department in São Paulo, and his team stepped up their interaction with American physicists.[15] In early 1940 he went to the United States on vacation, taking the opportunity to garner support for his research in cosmic-ray physics. He made contact with Arthur Compton, who had been working in the field since the 1930s and was now the American leader in cosmic rays and the hub of an international network dedicated to the search for mesotrons.[16] Wataghin was a member of this network.[17] The interaction was mutually beneficial. Wataghin and his team's early results on penetrating showers (i.e., the production of diverse particles resulting from the interaction of cosmic rays in the upper levels of the atmosphere) were first sent to Arthur Compton and then published in the *Physical Review*.[18]

Compton wrote back to Wataghin, saying, "Thank you for sending on to me the manuscript of your 'Letter' to the Editor of the PHYSICAL REVIEW on 'Showers of Penetrating Particles.'"[19]

Compton's work was funded by American foundations such as the Rockefeller Foundation and the Carnegie Corporation. During his 1940 vacation Wataghin, who was tuned into new trends in US foreign policy, approached the Rockefeller Foundation to sound out the possibilities of funding. An officer at the foundation reported his conversations with him, noting, "Following the idea that the United States is more interested in South America than heretofore, W[ataghin] wants to give us some information about his laboratory."[20] These personal contacts bore fruit.

In 1941 Compton's hopes to do more cosmic-ray measurements in the south were met with an invitation from Wataghin to make these measurements in São Paulo using balloons and to attend a conference in Rio de Janeiro. Brazilian diplomacy and the Brazilian Academy of Science were equally engaged in the project. As a matter of fact not just Compton but also some colleagues and their spouses made the trip to Brazil. Among Compton's party was Harry Milton Miller Jr., an officer from the Rockefeller Foundation's Division of Natural Sciences (see chapter 9, this volume).[21]

Compton's 1941 visit to Brazil figures in the lore among Brazilian physicists of the São Paulo team in cosmic rays. Its achievements eventually led to the participation of César Lattes, another student of Wataghin's, in the discovery of the pi-meson after the war.[22]

Compton's visit was part science, part diplomacy—perhaps even more diplomacy than science. The trip was supported and funded by the OCIAA as soon as Nelson Rockefeller learned of the invitation. Compton later reported that "the most successful aspect of our South American work with regard to its influence on international relationships seems to have been the Symposium on Cosmic Rays, held under the auspices of the Brazilian Academy of Science at Rio de Janeiro." Rockefeller himself testified to the success of Compton's diplomatic mission. Compton reported him as saying that this "expedition of ours, according to his reports, was the most successful one that had been undertaken thus far with regard to its stimulation of good relations between the Americas." Diplomatic procedures also explain the presence of spouses in Compton's party. As Compton put it, "in view of the cultural relations objectives of the expedition, I have encouraged the married members of the party to take their wives with them. . . . From our experience in Mexico and from our contacts with Latin-Americans in this country, we believe that this will be of real advantage."[23] In brief, Compton's role in his visit to Brazil was not dissimilar to those of Walt Disney and Orson Welles. His was part science, part diplomacy; theirs were part art, part diplomacy.

Mário Schönberg, a former student of Wataghin's, spent two years in the United States, beginning in 1940. His trip was supported by the Guggenheim Foundation. While abroad he made important contributions to the quantitative study of stellar evolution, collaborating with both Subrahmanyan Chandrasekhar and George Gamow. The exchange was favored by new trends among American philanthropic foundations that were marching in step with their government's foreign policy. As Donald Osterbrok has pointed out, Schönberg was one of several Latin American astronomers welcomed at Yerkes Observatory in the implementation of the government's Good Neighbor policy.[24]

Compton's visit triggered another important process for the development of physics in Brazil: funding from the Rockefeller Foundation for the Physics Department in São Paulo, which ran from 1941 to 1949.[25] Some figures will illustrate how significant this funding was. In 1942 the annual budget for research at the university department was about $1,200. The funding from the Rockefeller Foundation was double this. After the war, the Rockefeller funding increased to $75,000 between 1946 and 1949. The total funding from the foundation from 1942 to 1949 amounted to $82,500, which today would be in the region of $2 million.[26]

Compton was the scientific guarantor of the expertise of Wataghin's team. In February 1942 the Rockefeller Foundation's Harry Milton Miller Jr. (whose activities in Latin America are explored at length by Michael Barany in this volume) described him as "definitely the most promising center of physics activity in South America." "In spite of considerable difficulties with regard to remoteness and lack of necessary equipment," Miller went on, "Wataghin and his colleague, Occhialini, have not only carr[i]ed on experimental and theoretical studies of real value, but have also trained a group of young Brazilian students, including Desouza Santos and Paulus Pompeia, who when they come to this country are found quite capable of holding their own among our own graduate students."[27] In fact, Pompeia stayed in Chicago to work with Compton on cosmic rays.

The Rockefeller Foundation also played a diplomatic role in line with the aims of the OCIAA. In September 1941, before returning to the United States, Harry Miller put pressure on Warren Weaver, director of the Division of Natural Sciences, to approve and deliver the initial funds because of the "psychological effect" that it would have on physicists in Brazil—this "psychological effect" being related to the Good Neighbor policy.[28] Furthermore, the Rockefeller Foundation, as well as other foundations, was fully attuned to the OCIAA's policy. Indeed, one of the first initiatives undertaken by Nelson Rockefeller was to bring together some of the major American foundations to support Latin American scientists and artists, leading to the creation of the Committee for Inter-American Artistic and Intellectual Relations in February 1941. This committee was composed of Henry Allen Moe from the Guggenheim Foundation, David H. Stevens from the Rockefeller Foundation, and Frederick P. Keppel from the Carnegie Corporation.[29]

These activities were not interrupted during the war years. Supported either by the consortium of the three foundations—Rockefeller, Guggenheim, and Carnegie—or directly through the OCIAA, a number of American and Brazilian researchers and students in higher education and the sciences (not only physics) went back and forth between their respective countries in the early 1940s. A full list of names is not appropriate here but a few may illustrate the extent of these exchanges. The philosopher Willard V. Quine, the anthropologist Charles Wagley, and the sociologist Donald Pierson worked in Brazil for extended periods during the war.[30] Three American engineers, Allan Bates (Westinghouse), Robert Mehl (Carnegie), and Arthur Phillips (Yale), all spent time in São Paulo between 1943 and 1944, each giving a three-month course to consolidate the recently created course of metallurgical engineering at São Paulo.[31] On the Brazilian side, the list includes names such as Manoel Frota Moreira, Euryclides de Jesus Zerbini, Aristides Leão, André Dreyfus, and Miguel Ozório de Almeida, all of them researchers in the medical sciences. These names are a sample of outstanding researchers, some with already impressive academic achievements and others with their achievements still ahead of them. Indeed, just taking the case of the medical doctors, Moreira was later to become the science director of the Brazilian federal agency for funding research (Conselho Nacional de Pesquisas); Zerbini would perform the first heart transplant in Brazil; Leão would make breakthroughs in electrophysiology and become president of the Academia Brasileira de Ciências; Dreyfus would lead Brazilian research in genetics; while Almeida was already well known in Brazil and Europe as a medical researcher.[32]

To summarize, in the early 1940s the tilt toward Latin America led by Roosevelt's efforts to align the region with the United States encouraged the construction of transnational networks with outstanding Brazilian scientists to mutual advantage. American foundations, and the Rockefeller Foundation in particular, served as arms of the administration, working closely with the government to actively foster cultural diplomacy through transnational scientific exchanges beginning in World War II. These transnational exchanges not only fostered scientific excellence but also built social capital for Brazilians at home, enhancing careers and creating opportunities for scientific and institutional leadership for those who helped construct cross-border networks.

Cold War Times

In 1945 World War II was over, the Axis powers were defeated, and Europe was devastated. The Cold War would soon begin. Against this backdrop it is no surprise that American hegemony in Latin America would be dramatically reinforced. In the years after the end of the war a number of Brazilian scientists moved to the United States, this time looking for full doctoral training. These included the scientists José Leite Lopes, Jayme Tiomno, Sergio Porto, Hervásio de Carvalho, and Walter Schützer, in addition to mathematicians Mauricio Peixoto, Leopoldo Nachbin, and Elon Lages Lima. Princeton was considered a hub for interactions among Brazilian and foreign scientists. Figure 10.1 shows the meeting among Brazilian doctoral students, including César Lattes, who was in the United States working with Eugene Gardner at Berkeley on the detection of the pi-meson, and Hideki Yukawa, the discoverer of the nuclear force and mesons, who was then at Columbia University. American physicists also came to Brazil for long and short stays. Richard Feynman spent one year in Rio de Janeiro at the new Centro Brasileiro de Pesquisas Físicas. David Bohm stayed in São Paulo for three years. Among the short stays, we should mention the illustrious cases of Isidore Rabi and Eugene Wigner, who came to Rio de Janeiro for a conference in 1952. Indeed, Rio de Janeiro had become a center that also attracted

FIGURE 10.1. César Lattes, Hideki Yukawa, Walter Schützer, Hervásio de Carvalho, José Leite Lopes, and Jayme Tiomno at Princeton in 1949.
Source: Archives of Centro Brasileiro de Pesquisas Físicas.

visitors from Europe, such as the physicists Léon Rosenfeld and Cécile DeWitt-Morette.

The history of these exchanges after World War II may still be framed in the field of the cultural diplomacy between the United States and Brazil but should not be fully identified with it. The first reason for this is that, while Brazil was part of the Western camp under the leadership of the United States, their foreign policies were not always fully aligned during the Cold War. For Maria Ligia Prado the "strategy of ambiguity" that had characterized the Brazilian attitude toward the United States in the interwar years now became a "strategy of confrontation," expressed through Brazilian nationalistic projects that conflicted with

Washington's expectations.[33] This strategy of confrontation clearly appeared in the two most important Brazilian political crises after World War II, the first leading to the suicide of President Vargas in 1954 and the second in the putsch against President Goulart in 1964. The second reason is that in many situations scientists were not totally in tune with the policies adopted by their own countries. This was unlike the previous period, during World War II, when not only were Brazil and the United States strongly aligned, but Brazilian and American physicists were also strongly committed to the Allied war effort. This was particularly the case with Compton, who was engaged in the atomic bomb project, but also Wataghin and Pompeia, whose relationship with Compton testified to their commitment to the Allied cause.[34] The same can be said of Schönberg and Occhialini, who were well known for their antifascist standpoint.

The example of David Bohm is illustrative of the more complex situation in the 1950s. Bohm, who left the United States after being hounded for his communist sympathies, was welcomed in Brazil. The host government was less concerned about the political allegiance of "knowledgeable bodies" than was the American national security state. The capacity to develop new weapons was deemed essential to American superiority in Cold War rivalry with the Soviet Union, but the US government was determined to weed out left-wing and communist scientists, notably physicists.

Bohm's Exile in Brazil

In the late 1940s David Bohm was one of the most promising American theoretical physicists, having been one of J. Robert Oppenheimer's students and later hired by Princeton University. After the war he was caught up in the anticommunist hysteria in the United States, later called McCarthyism. Subpoenaed to testify before the House Un-American Activities Committee about his links with the Communist Party, he declined to answer questions (fig. 10.2). Accused of contempt of Congress, he was jailed, then released on bail and eventually cleared. In the meantime, Princeton University suspended his contract. As it seemed unlikely that he would get a job in American academia in the context of rising McCarthyism, Bohm looked abroad. He became an illustrious American scientific exile in the twentieth century.[35]

At this juncture, some random events played a role in Bohm's tra-

FIGURE 10.2. David Bohm after refusing to testify before the House Un-American Activities Committee.
Source: Library of Congress, New York World-Telegram and Sun Collection, courtesy of AIP Emilio Segrè Visual Archives.

jectory. Jayme Tiomno, who was finishing his PhD at Princeton, invited him to come to the Universidade de São Paulo, an invitation strongly supported by the full Physics Department and the university. Bohm also had a recommendation letter from Einstein. After some delays Bohm got his US passport and departed for Brazil. He spent three years in Brazil, a crucial period for his scientific work, as he had just finished a new and alternative interpretation of the quantum theory. Nevertheless, his troubles with the American authorities were not over. In São Paulo the US consul confiscated his passport, telling him it would be given back to him on his return to the United States. Apparently, Bohm's exile in Brazil was not an issue for the United States, as Brazil was geopolitically aligned with them and was not seeking nuclear capabilities.[36] Fortunately, other physicists came to work with Bohm in Brazil, namely, Jean-Pierre Vigier, Mario Bunge, and Ralph Schiller. He also worked with Tiomno and Schützer. But these interactions were not enough for him. Bohm hoped to travel abroad to discuss his new interpretation of quantum theory in the main world centers of physics.[37]

In mid-1954 Bohm got a job offer from Nathan Rosen of Technion, in Israel, supported again by Einstein. To take it up, however, he needed a passport. Helped by his Brazilian friends he was able to get Brazilian citizenship in record time, and in early 1955 he left for Israel, from where he went to England in 1957. There he stayed until the end of his days. The United States canceled his American citizenship when he got his Brazilian passport. This bizarre situation lasted for thirty years until Bohm eventually regained his American citizenship in court in 1986. Bohm and his Brazilian and American friends had refused to bow before the restrictions on scientists imposed by the US administration in the early Cold War, finding a way around the sweeping powers of the national security state, notably its control over the issuing of visas and passports (see chapter 1, this volume). Bohms's travels first to Brazil, then to Israel, and finally to the United Kingdom epitomized the solidarity that was engendered by scientific internationalism and its capacity to thwart the reach of government power.[38]

The Brazilian Dictatorship, 1964–1985

Let us move now to the context of the Brazilian military dictatorship, which prevailed between 1964 and 1985. Some background information is necessary, however, before considering the fortunes of some Brazilian physicists. Widely documented in the historiographical literature is the deep engagement of the United States in preparations and support for the coup d'état that overthrew President João Goulart in March 1964. Although it took more than ten years for documentary evidence to begin to surface, all the main actors of the 1964 events, including supporters and opponents of the coup, knew, at least to some extent, of the American involvement.

The first documentary evidence of US efforts in support of the overthrow of the Brazilian government was brought to light in 1977 thanks to the American Phyllis Parker, who was studying recently declassified documents at the Lyndon Johnson Presidential Library. Operation Brother Sam, as it was code-named, was supported by both Presidents Kennedy and Johnson and included military and logistic support to the rebels in case of resistance. Since 1977 there has been a stream of revelations from the US archives, the last by the Brazilian Carlos Fico, who found evidence that preparations were under way well before the 1964 coup.[39] Indeed, from all the documentary evidence, it appears that American support for the Brazilian dictatorial government abated only when Jimmy Carter took office in 1977.

In the first years of the Brazilian dictatorship the American and Brazilian governments were very close, as close as they had been during World War II. While during the war Brazil had distanced itself from Germany for closer relations with the United States, now Brazil abandoned a foreign policy which contemplated relations with nonaligned countries and the Soviet Union and aligned itself completely with the United States. In the fields of science and education, for instance, Brazil created a science attaché position at its embassy in Washington, and the United States and Brazil signed an agreement for American consultancy on the reform of Brazilian universities. These shifts in foreign policy were not without casualties. Before the dictatorship a number of Brazilians were trained in engineering and science in the Soviet Union. Many of these students ran into trouble because of the origin of their diplomas once the dictatorship was entrenched. This happened to Paulo Miranda, for example—the physics teacher of one of us (OFJ) in Bahia. He graduated from the University Patrice Lumumba in Moscow and later lost his job at the university in Bahia because of the difficulty he had in validating his degree.[40]

The Brazilian dictatorship brought conflicting changes to the universities. Its policy may be described as one of authoritarian modernization. This is not the occasion to present and analyze these changes.[41] Instead, we will touch on just one aspect of this complex and varied process, which concerned the restrictions on civil rights, focusing on the case of physicists and physics students. Some senior physicists were forced to take early retirement and kept away from Brazilian state universities. This was the case for some of the physicists already cited in this chapter, namely, Jayme Tiomno, José Leite Lopes, and Mário Schönberg. Elisa Frota Pessoa, Sarah de Castro Barbosa, Plinio Sussekind da Rocha, and Celso Diniz Pereira were also expelled from Brazilian universities.[42] In addition some physics students and junior researchers were targeted by the dictatorship and were either expelled from universities or sent to jail. Beginning in 1969 the authoritarian regime evolved into a barbaric system, jailing, torturing, and killing activists and even people who were not engaged in political activities against the government.[43]

Given this scenario, a number of people, physicists included, looked for safe havens abroad, leaving the country in a kind of voluntary exile. Some went to the United States, either to work or to study. Leite Lopes spent one year at Carnegie Mellon University in Pittsburgh, invited by Lincoln Wolfenstein and Sergio De Benedetti. Tiomno went to Princeton, invited by John Archibald Wheeler, who had been his former supervisor in the late 1940s. In 1969 Luiz Davidovich, the current president of the Academia Brasileira de Ciências, was expelled from the Catholic University in Rio de Janeiro, where he was working on his master's degree, and moved to Rochester, where he finished his PhD at the University of Rochester.[44] Silvio Salinas took a leave of absence from the Universidade de São Paulo, where he was an untenured teacher, to do his doctorate at Carnegie Mellon.[45]

Some physicists left the country in the early 1960s, before the 1964 coup, because of the poor salaries of academics in Brazil. This was the Brazilian version of the brain drain. Among them were Sergio Porto, who went to Bell Laboratories, and Herch Moysés Nussenzveig, who went to New York University. After the coup d'état Nussenzveig decided it was not safe to return to the country, and he obtained a permanent position at the University of Rochester, where he stayed till the mid-1970s, when he returned to Brazil. Many others went to other countries, such as Argentina, France, Chile, Italy. Leite Lopes, for instance, after spending one year in Pittsburgh, went to the University of Strasbourg in France, invited by Michel Paty.[46]

In all these movements across borders, Brazilian physicists were supported by colleagues in the countries where they sought exile and with whom they had formed personal and professional relationships during earlier periods of exchange. The history of this solidarity with Brazilian academics remains to be written. Fortunately, in the case of solidarity in the United States, this lacuna has already been filled by the superb book *We Cannot Remain Silent: Opposition to the Brazilian Dictatorship in the United States, 1964–1985*, written by the historian James Naylor Green from Brown University. The names of many American physicists who were active in this movement should be mentioned: we will illustrate with just one telling example.[47]

In early December 1970 a couple of physicists from the University of São Paulo, Ernesto and Amelia Hamburger, were detained by the military and kept incommunicado. The immediate cause of their detention was related to their sheltering of a young guest who was a political activist. The Hamburgers themselves were democratic and left-inclined people but not engaged in political activism. At that time, however, detainees in these circumstances would be tortured and risked losing their lives. The Brazilian press was then under severe censorship, and it was very important for the Hamburgers' survival and safety to let the military authorities know that the public was following the events. This was a strategy that saved lives. Warned by Herch Nussenzveig of the case, on December 12, a Saturday morning, John Archibald Wheeler cabled the president of Brazil, General Emilio Medici: "We would greatly appreciate your efforts to secure legal and human rights for our nuclear physics colleague and fellow member of the American Physical Society, Professor Ernesto Hamburger and his wife Emilia [*sic*] of Sao Paulo who we understand from two independent sources are being held incommunicado." Wheeler signed the cable as professor of physics at Princeton, past president of the American Physical Society, and member of the US National Academy of Sciences. In addition to Wheeler's cable, colleagues of the Hamburgers in Pittsburgh—where they had studied in the previous decades—approached the US Department of State. Indeed, Nathan Melamed approached a Pennsylvania senator who was well connected with the executive. This in turn led the American embassy in Brazil to request information from the military government regarding the Hamburgers. Although we are unable to follow the chain of events that eventually led to their release, Wheeler and the Hamburgers' colleagues in Pittsburgh were part of a wide network that proved instrumental in saving lives and that, years later, contributed to the democratization of the country.[48]

In contrast to the support among American physicists for their Brazilian colleagues persecuted by the military dictatorship, the US government acted with political selectivity. In general the rule to be followed, with a few exceptions, was to support those scientists who had a liberal political stance and to refuse support for those with leftist or communist inclinations. Thus, when Mário Schönberg was imprisoned after the coup, there were several demonstrations against the action around the world, including among American physicists. However, as reported by the historian Rodrigo Motta, the US Department of State refused to intercede. Indeed, Washington had been advised in a cable from the US consul in São Paulo that "Schönberg is confirmed, publicly known and dangerous communist, whose arrest congen [US general consul in São Paulo] considers is in US interest."[49] Later, Schönberg had difficulty getting a visa to attend an astrophysics conference in the United States.[50]

Schönberg's and Bohm's cases nicely illustrate that the transnational movement of scientists cannot be taken for granted and that "knowledge does not move by itself" in a global scientific community. Changes in foreign policy and the wider geopolitical situation affect knowledge circulation. In the early 1940s Brazil, the United States, and the Soviet Union were in the same geopolitical camp, and Schönberg was received in the United States without any restrictions.[51] In the 1960s the harshest Cold War times in Latin America, the United States considered Schönberg's unlawful arrest in accordance with its interests. As for Schönberg, he always remained a committed Marxist.[52] Bohm, for his part, had his movements around the world constrained for more than thirty years by the conflict between his political commitments and US domestic and foreign policies, McCarthyism and the Cold War.

Conclusion

Brazilian scholars showed solidarity with David Bohm at the apex of McCarthyism. American scholars reciprocated during the Brazilian dictatorship. In both cases scholars acted against the overt or covert foreign policies of their respective countries. Such episodes cannot be reduced simply to stories concerning either cultural diplomacy or foreign affairs: these individuals were inspired by the ethos of scientific internationalism that had been so actively promoted by their most eminent leaders and that transcended loyalty to government policies restricting the freedom of movement that they found objectionable. Furthermore, taking into consideration both the history of science and history in general can shed light on the role played by individuals and their communities and institutions. During World War II Brazilian and American physicists acted in line with their governments. However, during the Cold War period and in the cases we have considered, some scientists acted against the diplomatic positions taken by the authorities. Brazilians welcomed an eminent physicist like Bohm even though he was accused in the United States of communist sympathies. Eminent American physicists such as Wheeler skillfully exploited a public space in the United States that advocated freedom and democracy to welcome colleagues fleeing a dictatorship whose anticommunist and anti-Marxist credentials had at least the covert support of senior American leaders such as Presidents Kennedy and Johnson. There was a limit to this, however: scientific solidarity had no answer to the collusion between the Brazilian authorities and the US State Department to restrict the travel of a "dangerous communist." Wheeler would probably not have objected anyway.

In this chapter we have used a transnational approach to the history of science, as opposed to purely national histories or histories of international relations. Indeed, we have focused on people who crossed boundaries either contrary to, or in agreement with, their government's policies. Using a transnational approach does not efface the role played by states, a fact well exemplified in Schönberg's and Bohm's cases. On the contrary, as noted by Barany and Krige, "the state emerges as an actor *precisely because* its borders are being crossed." More specifically, through highlighting policies on visas, citizenship, and passports, transnational studies highlights continuities and discontinuities in national identities under different circumstances. Thus, still according to Barany and Krige, "when knowledgeable bodies move across borders, the potential contradictions between these selves can be the source of considerable anxiety to them and their colleagues and of suspicion to immigration officials."[53]

We would like to conclude by suggesting that a transnational approach, which explicitly examines the movement of people, ideas, and things across borders, is able to bring to the foreground actors and organizations whose diverse motives may provide greater intelligibility and richness to our understanding of the global circulation of knowledge. It is also valuable for writing a history of science that seeks to study the intersection of the local with the global—in this case, the personal trajectories of Brazilian and American scientists and the general regulatory regime of the American national security state.

Acknowledgments

Earlier versions of this chapter were presented at the "Workshop on Writing the Transnational History of Science and Technology," Atlanta, GA, November 2–3, 2016, at the 25th International Congress of History of Science and Technology, Rio de Janeiro, July 23–29, 2017, and the 132nd Annual Meeting of the American Historical Association, Washington, DC, January 4–7, 2018. We are thankful for comments and editorial suggestions from John Krige, Michael Barany, Josep Simon, Adriana García, Rodrigo Patto Sá da Motta, Mirella Longo, Seth Garfield, and anonymous referees. We are thankful to the Conselho Nacional de Desenvolvimento Científico e Tecnológico (CNPq) for their support for this research.

Notes

1. John Krige, *American Hegemony and the Postwar Reconstruction of Science in Europe* (Cambridge, MA: MIT Press, 2006), 1.

2. While this overlapping of history of science and diplomatic and economic history has not been widely exploited, this landscape has begun to change. On the history of physicists and mathematicians, in both Brazil and Mexico, see, e.g., Adriana Minor García, "Manuel Sandoval Vallarta en la encrucijada entre Estados Unidos y México," *Ludus Vitalis* 23, no. 43 (2015): 125–149; Adriana Minor and Joel Vargas-Domínguez, "Mexican Scientists in the Making of Nutritional and Nuclear Diplomacy in the First Half of the Twentieth Century," *HoST—Journal of History of Science and Technology* 11 (2017): 34–56; Simone P. Kropf and Joel D. Howell, "War, Medicine and Cultural Diplomacy in the Americas: Frank Wilson and Brazilian Cardiology," *Journal of the History of Medicine and Allied Sciences* 72, no. 4 (2017): 422–447; Ana Maria Ribeiro de Andrade, "Os raios cósmicos entre a ciência e as relações internacionais," in *Ciência, política e relações internacionais—Ensaios sobre Paulo Carneiro*, ed. Marcos Chor Maio (Rio de Janeiro: Editora Fiocruz, 2004), 215–242; Olival Freire Jr. and Indianara Silva, "Diplomacy and Science in the Context of World War II: Arthur Compton's 1941 Trip to Brazil," *Revista brasileira de história* 34, no. 67 (2014), http://www.scielo.br/rbh; Eduardo Ortiz, "La politica interamericana de Roosevelt: George D. Birkhoff y la inclusion de America Latina en las redes matemáticas internationales," *Saber y tiempo: Revista de historia de la ciencia* 15 (2003): 53–111, 16 (2003): 21–70.

3. On the role of individuals as a scale for transnational movements, see Bern-

hard Struck, Kate Ferris, and Jacques Revel, "Introduction: Space and Scale in Transnational History," *International History Review* 33 (2011): 573–584, at 577.

4. For an extended view on the transnational approach in history, see Akira Iriye and Pierre-Yves Saunier, eds., *The Palgrave Dictionary of Transnational History: From the Mid-19th Century to the Present Day* (New York: Palgrave Macmillan, 2009). See also the call made by Gilbert M. Joseph, twenty years ago, while dealing with the cultural history of US-Latin American relations, for "new questions about the nature and outcomes of foreign-local encounters." Gilbert M. Joseph, "Close Encounters—Toward a New Cultural History of U.S.-Latin American Relations," in *Close Encounters of Empire—Writing the Cultural History of U.S.-Latin American Relations*, ed. Gilbert M. Joseph, Catherine C. Legrand, and Ricardo D. Salvatore (Durham, NC: Duke University Press, 1998), 3–46.

5. Simone Turchetti, Néstor Herran, and Soraya Boudia, "Introduction: Have We Ever Been 'Transnational'? Towards a History of Science across and beyond Borders," *British Journal for the History of Science* 45, no. 3 (2012): 319–336. Works and discussions on transnational, global, and world history may be found in the special issue of the *British Journal for the History of Science* just cited as well as in the Isis Focus sections "Global Histories of Science," *Isis* 101, no. 1 (2010): 95–158; and "Global Currents in National Histories of Science: The 'Global Turn' and the History of Science in Latin America," *Isis* 104, no. 4 (2013): 773–817.

6. Paul Forman, "Scientific Internationalism and the Weimar Physicists: The Ideology and Its Manipulation in Germany after World War I," *Isis* 64, no. 2 (1973): 150–180; Patrick D. Slaney, "Eugene Rabinowitch, the *Bulletin of the Atomic Scientists*, and the Nature of Scientific Internationalism in the Early Cold War," *Historical Studies in the Natural Sciences* 42, no. 2 (2012): 114–142; John Krige, "Atoms for Peace, Scientific Internationalism, and Scientific Intelligence," *Osiris* 21 (2006): 161–181; John Krige, "Building the Arsenal of Knowledge," *Centaurus* 52 (2010): 280–296; Alexis De Greiff, "The Politics of Noncooperation: The Boycott of the International Centre for Theoretical Physics," *Osiris* 21 (2006): 86–109. For a review of the subject, see Brigitte Schroeder-Gudehus, "Nationalism and Internationalism," in *Companion to the History of Modern Science*, ed. R. C. Olby, G. N. Cantor, J. R. R. Christie, and M. J. S. Hodge (London: Routledge, 1990), 909–919. For an essay review on other works dealing with scientific internationalism, see Paul K. Hoch, "Whose Scientific Internationalism?," *British Journal for the History of Science* 27 (1994): 345–349.

7. See, e.g., Jessica Wang, *American Science in an Age of Anxiety: Scientists, Anticommunism, and the Cold War* (Chapel Hill: University of North Carolina Press, 1999).

8. Mário Schönberg mostly used the spelling Schönberg when signing scien-

tific papers. In Brazil and in public relations, he used Schenberg, which is why both forms will appear in this chapter according to the sources.

9. Maria Ligia Coelho Prado, "Davi e Golias: As relações entre Brasil e Estados Unidos no século XX," in *Viagem incompleta—a experiência brasileira (1500–2000)—a grande transação*, 3rd ed., ed. Carlos Guilherme Mota (São Paulo: SENAC, 2013), 319–347.

10. On the creation of the Universidade de São Paulo, see Simon Schwartzman, *A Space for Science: The Development of the Scientific Community in Brazil* (University Park: Pennsylvania State University Press, 1991). On the French influence, see Patrick Petitjean, "Autour de la mission française pour la création de l'Université de São Paulo (1934)," in *Sciences and Empires: Historical Studies about Scientific Development and European Expansion*, ed. Patrick Petitjean, Catherine Jami, and Anne M. Moulin (Dordrecht: Kluwer, 1992), 339–362.

11. Gerald K. Haines, "Under the Eagle's Wing: The Franklin Roosevelt Administration Forges an American Hemisphere," *Diplomatic History* 1, no. 4 (1977): 373–388, at 380.

12. Boris Fausto, *História do Brasil* (São Paulo: EDUSP, 2012); Thomas Skidmore, *Politics in Brazil, 1930–1964: An Experiment in Democracy* (New York: Oxford University Press, 1967); Luiz Alberto Moniz Bandeira, *Presença dos Estados Unidos no Brasil* (Rio de Janeiro: Civilização Brasileira, 2007); Gerson Moura, *Autonomia na dependência: A política externa brasileira de 1935 a 1942* (Rio de Janeiro: Nova Fronteira, 1980).

13. However, it is only recently that the role played by the OCIAA and Rockefeller in establishing the American hegemony in the region has been exploited in works such as Haines, "Eagle's Wing"; Moura, *Autonomia na dependência*; Gerson Moura, *Tio Sam chega ao Brasil: A penetração cultural Americana* (São Paulo: Brasiliense, 1986); Antonio Pedro Tota, *The Seduction of Brazil: The Americanization of Brazil during World War II* (Austin: University of Texas Press, 2009); Neill Lochery, *Brazil—the Fortunes of War: World War II and the Making of Modern Brazil* (New York: Basic Books, 2014); Lira Neto, *Getúlio 1930–1945—do governo provisório à ditadura do Estado Novo* (São Paulo: Companhia das Letras, 2013); Coelho Prado, "Davi e Golias"; Freire and Silva, "Diplomacy and Science."

14. Tota, *Seduction of Brazil.* Orson Welles's connection to Brazil is familiar to movie lovers through the documentary *It's All True*, a film edited and released in the late 1980s. Though the original footage was produced in the early 1940s in Brazil, when Welles visited the country, the film was abandoned after the tragic drowning of one of the actors (a raft man), an event that was revisited when *It's All True* was finally released. See Catherine L. Benamou, *It's All True—Orson Welles's Pan-American Odyssey* (Berkeley: University of California Press, 2007). As for Portinari, his works at the Library of Congress would

become part of the Brazilian painter's credentials in the United States. He was later invited to paint the *War* and *Peace* murals at the UN headquarters.

15. The information given here on Compton's trip is widely based on Freire and Silva, "Diplomacy and Science."

16. We should recall that cosmic-ray physics was a hot topic at the time, as it could shed light on Yukawa's hypothesis of the existence of a new particle. This was called the mesotron, later the meson, with an intermediate mass between the electron and the proton; it was hypothesized to be the mediator of a new force, the nuclear force.

17. A. H. Compton to F. A. Keppel, July 2, 1940, letter plus a six-page report titled "Cosmic Rays at the University of Chicago," Record Group 1.1, ser. 216 D, University of Chicago—Cosmic Ray Study, box 9, folder 127, Rockefeller Foundation Archives, Rockefeller Archive Center, Sleepy Hollow, NY.

18. Wataghin began to publish in *Physical Review* in 1935. The first of a series of papers reporting the discovery of the penetrating showers was Gleb Wataghin, Marcelo D. de Souza Santos, and Paulus A. Pompeia, "Simultaneous Penetrating Particles in the Cosmic Radiation," *Physical Review* 57 (1940): 61. For a discussion of these discoveries, see Marcelo A. L. de Oliveira and Nelson Studart, "Cosmic-Ray Air Showers and the Emergence of Experimental Research in Brazil," paper presented at the 25th International Congress of History of Science and Technology, Rio de Janeiro, July 2017.

19. Compton to Wataghin, Jan. 4, 1941. Previously, Wataghin had thanked Compton for a letter to the journal's editors: "I want to thank you also for sending a letter to the Editor of Physical Review, and for the very interesting information on recent cosmic ray results obtained in your laboratory." Wataghin to Compton, Nov. 12, 1940, ser. 02, box 017, folder South America Expedition, 1941, Arthur Holly Compton Personal Papers, University Archives, Department of Special Collections, Washington University Libraries, St. Louis, MO.

20. "Thursday, March 7 1940—Professor Gleb Wataghin, University of São Paulo, Brazil," Record Group 1.1, ser. 305 D, Projects Brazil, box 13, folder 116, Rockefeller Foundation Archives.

21. "Excerpt from HMM's int. with Dr. ERNESTO DE SOUZA CAMPOS at Rio de Janeiro, Brazil, August 13, 1941," Record Group 1.1, ser. 305 D, Projects Brazil, box 13, folder 116, Rockefeller Foundation Archives.

22. On Lattes, see Cássio L. Vieira and Antonio A. P. Videira, "Carried by History: Cesar Lattes, Nuclear Emulsions, and the Discovery of the Pi-Meson," *Physics in Perspective* 16 (2014): 3–36.

23. Compton to W. P. Jesse, May 11, 1942, series 03, box 02, folder: South America, 1941–1942; Compton to W. P. Jesse, May 11, 1942, series 03, box 02, folder: South America, 1941–1942; Compton to R. G. Caldwell, April 26, 1941, series 02, box 017, folder: South America Expedition, 1941. All these documents

are at Arthur Holly Compton Personal Papers, University Archives, Department of Special Collections, Washington University Libraries.

24. Donald E. Osterbrok, *Yerkes Observatory, 1892–1950: The Birth, Near Death, and Resurrection of a Scientific Research Institution* (Chicago: University of Chicago Press, 1997), 261. On Schönberg's contributions to the study of stellar evolution, see Davide Cenadelli, "Solving the Giant Stars Problem: Theories of Stellar Evolution from the 1930s to the 1950s," *Archive for History of Exact Sciences* 64 (2010): 203–267.

25. The role played by the Rockefeller Foundation in Latin America in the domain of health sciences has been well studied, but there remains a lacuna to be filled in studies on its role in physical sciences. On health sciences, see, e.g., Marcos Cueto, ed., *Missionaries of Science—the Rockefeller Foundation and Latin America* (Bloomington: Indiana University Press, 1994); Maria Gabriela Marinho, *Norte-americanos no Brasil: Uma história da Fundação Rockefeller na Universidade de São Paulo (1934–1952)* (Campinas, SP: Autores Associados, 2001); Ana Paula Korndöfer, "Para além do combate à ancilostomíase: O diário do médico norte-americano Alan Gregg," *História, ciências, saúde—Manguinhos* 21, no. 4 (2014): 1457–1466; Thomas F. Glick, "O programa brasileiro de genética evolucionária de populações, de Theodosius Dobzhansky," *Revista brasileira de história* 28, no. 56 (2008): 315–325; Lúcia Grando Bulcão, Almir Chaiban El-Kareh, and Jane Dutra Sayd, "Ciência e ensino médico no Brasil (1930–1950)," *História, ciências, saúde—Manguinhos* 14, no. 2 (2007): 469–487; Carlos Henrique Assunção Paiva, "Imperialismo & filantropia: A experiência da Fundação Rockefeller e o sanitarismo no Brasil na Primeira República," *História, ciências, saúde—Manguinhos* 12, no. 1 (2005): 205–214. On mathematics, see Michael J. Barany, "Fellow Travelers and Traveling Fellows: The Intercontinental Shaping of Modern Mathematics in Mid-twentieth Century Latin America," *Historical Studies in the Natural Sciences* 46, no. 5 (2016): 669–709.

26. Fernando de Azevedo to H. M. Miller Jr., Sept. 1, 1942, Record Group 1.1, ser. 305 D, Projects Brazil, box 13, folder 117, University of São Paulo—Physics, 1942–1943; H. M. Miller Jr. to RBF, Jan. 31, 1946, Record Group 1.1, ser. 305 D, Projects Brazil, box 14, folder 129; both in Rockefeller Foundation Archives.

27. A. Compton to H. M. Miller Jr., Feb. 16, 1942, Record Group 1.1, ser. 305 D, Projects Brazil, box 13, folder 117, Rockefeller Foundation Archives.

28. "I hope this can go through promptly, for the psychological effect, something tangible following Compton's cooperative research with Wataghin." H. M. Miller to Warren Weaver, Sept. 12, 1941, Record Group 1.1, series 305 D, box 13, folder 116, Rockefeller Foundation Archives.

29. "Renewal of contract with the Committee for Inter-American Artistic and Intellectual Relations," Dec. 19, 1941, box 1170, Record Group 229, Records of the Office of Inter-American Affairs, 1918–1951 (hereafter ROIAA), series:

Central Files, 1940–1945, National Archives and Records Administration, College Park, MD.

30. On Quine, see letter from H. Moe to Nelson Rockefeller, June 10, 1942, boxes 1149–1158, ROIAA. On Pierson, see letter from Charles A. Thomson to William Schurz and Richard F. Pattee, Nov. 23, 1943, box 418, ROIAA. On Wagley, see Tota, *Seduction of Brazil*; and on his collaboration with the Museu Nacional, boxes 1149–1158, ROIAA.

31. Kenneth Holland to Gregory Comstock, June 2, 1943, box 418, ROIAA. From the São Paulo Office [Arnold Tschudy] to the Coordinator [Nelson Rockefeller], Aug. 2, 1944, Subject: Recommendations for the continuation of the project to send American professors to the Escola Politécnica, box 418, ROIAA. See also Olival Freire Jr., "Diplomacia cultural no contexto da Segunda Guerra: O caso da Engenharia Metalúrgica na USP," *Revista brasileira de história da ciência*, forthcoming.

32. For Frota Moreira, see letter from Nelson Rockefeller to Hugh Cumming [Pan American Union], July 14, 1941; letter from Hugh Cumming to Nelson Rockefeller, July 15, 1941, boxes 1159–1162. For Zerbini, see letter from A. Tschudy [São Paulo Office] to the Coordinator [Nelson Rockefeller], July 17, 1944, box 62. For Leão, see letter from Ross McFarland [Harvard University] to John Clark [OCIAA], Dec. 27, 1941, box 82. For Dreyfus, see letter from Cecil M. Cross to Secretary of State, Mar. 21, 1944, boxes 1149–1158. For Ozório de Almeida, see letter from H. Moe and D. Stevens to Nelson Rockefeller, June 10, 1942, boxes 1149–1158. All these documents are at ROIAA.

33. Coelho Prado, "Davi e Golias."

34. Freire and Silva, "Diplomacy and Science," 12–14.

35. On Bohm and McCarthyism there is an extensive literature. See R. Olwell, "Physical Isolation and Marginalization in Physics—David Bohm's Cold War Exile," *Isis* 90 (1999): 738–756; Shawn K. Mullet, "Little Man: Four Junior Physicists and the Red Scare Experience" (PhD diss., Harvard University, 2008); F. D. Peat, *Infinite Potential: The Life and Times of David Bohm* (Reading, MA: Addison Wesley, 1997); Ellen Schrecker, *No Ivory Tower: McCarthyism and the Universities* (New York: Oxford University Press, 1986); Wang, *American Science in an Age of Anxiety*; Olival Freire Jr., *The Quantum Dissidents—Rebuilding the Foundations of Quantum Mechanics (1950–1990)* (Heidelberg: Springer, 2015); Christian Forstner, "The Early History of David Bohm's Quantum Mechanics through the Perspective of Ludwik Fleck's Thought-Collectives," *Minerva* 46, no. 2 (2008): 215–229.

36. The release of Bohm's files at the FBI and the CIA could shed light on the exact US position with respect to Bohm.

37. On Bohm's exile in Brazil, see Freire, *Quantum Dissidents*, 17–74.

38. On Bohm's saga concerning his passport and citizenship, see ibid., 55–57.

39. Phyllis R. Parker, *Brazil and the Quiet Intervention, 1964* (Austin: Uni-

versity of Texas Press, 1979), 72–87; Carlos Fico, *O Grande Irmão—da Operação Brother Sam aos anos de chumbo—o governo dos Estados Unidos e a ditadura militar Brasileira* (Rio de Janeiro: Civilização Brasileira, 2008). On Fico's book, see on 295–309 the transcriptions of the document "Proposed Contingency Plan for Brazil, January 1964," and fragments of "Brazil—Country Internal Defense Plan." Parker's book was first published in Brazil as *1964 O papel dos Estados Unidos no golpe de estado de 31 de março* (Rio de Janeiro: Civilização Brasileira, 1977). On the American engagement in Goulart's overthrow and the ongoing support for the Brazilian dictatorship, see also James N. Green, *Apesar de vocês—Oposição à ditadura brasileira nos Estados Unidos, 1964–1985* (São Paulo: Companhia das Letras, 2009), 46–86, translated as *We Cannot Remain Silent: Opposition to the Brazilian Military Dictatorship in the United States* (Durham, NC: Duke University Press, 2010).

40. Rodrigo Patto Sá Motta, *As universidades e o regime militar—Cultura política brasileira e modernização autoritária* (Rio de Janeiro: Zahar, 2014), 110–147. The first Brazilian science attaché was Paulo de Góes, a scientist from Rio de Janeiro politically identified with the dictatorship. On Paulo Miranda's case, see ibid., 215.

41. On the modernization of Brazilian universities under the 1964–1985 dictatorship, see ibid.; Schwartzman, *A Space for Science.*

42. Motta, *As universidades e o regime militar*, 164–175.

43. On the Brazilian dictatorship and the actions taken against the universities, see ibid.; Green, *We Cannot Remain Silent*; Elio Gaspari, *A ditadura escancarada* (São Paulo: Companhia das Letras, 2002).

44. Ildeu C. Moreira, "A ciência, a ditadura e os físicos," *Ciência e cultura* 66, no. 4 (2014), http://dx.doi.org/10.21800/S0009-67252014000400015.

45. On Salinas's case, see Motta, *As universidades e o regime militar*, 232–233.

46. On the brain drain in Brazil in the 1960s, see Herch M. Nussenzveig, "Migration of Scientists from Latin America," *Science* 165, no. 3900 (Sept. 26, 1969): 1328–1332. On Nussenzveig, see Climério P. Silva Neto and Olival Freire Jr., "Herch Moysés Nussenzveig e a ótica quântica: Consolidando disciplinas através de escolas de verão e livros-texto," *Revista brasileira de ensino de física* 35 (2013): 2601. On Leite Lopes in France, his invitation by Michel Paty, and Paty's own tribulations in Brazil during the military dictatorship, see Olival Freire Jr., "Michel Paty e o Brasil: Uma história de amizade e de parceria," in *Filosofia, Ciência e História—Michel Paty e o Brasil, uma homenagem aos 40 anos de colaboração*, ed. Mauricio Pietrocola and Olival Freire Junior (São Paulo: EDUSP, 2005), 473–487.

47. Green, "Apesar de vocês."

48. On the Hamburgers' case, see Olival Freire Jr., "Amélia Império Hamburger (1932–2011): Ciência, educação e cultura," in *Mulheres na física—Casos históricos, panorama e perspectivas*, ed. E. M. B. Saitovitch, R. Z. Funchal,

M. C. B. Barbosa, S. T. R. Pinho, and A. E. Santana (São Paulo: Livraria da Física, 2015), 171–183. Motta (*As universidades e o regime militar*, 137) cites the Pittsburgh professors calling the US government. The reference to Melamed is from Fernando de Souza Barros to Olival Freire, e-mail, Sept. 20, 2017.

49. Rodrigo Patto Sá Motta, *As universidades e o regime militar*, 123; cable from US consul to the Department of State, Sept. 24, 1965, RG 59, box 1944, folder 1, National Archives and Records Administration II. Correspondence from the US consul in São Paulo concerning Schönberg fully approves his persecution by the Brazilian military. I am thankful to Rodrigo Motta for sharing these documents with me.

50. On Schönberg's visa, see Malcolm P. Hallam, Confidential—Memorandum of Conversation [with Mário Schönberg], Nov. 28, 1968, RG 59, box 1944, National Archives and Records Administration II.

51. Similarly, the communist mathematician José Luis Massera was allowed to enter the United States immediately after the war, but not without difficulties. See Barany's chapter (9) in the present volume and his "Fellow Travelers and Traveling Fellows."

52. On Schönberg's political commitments, see Dina L. Kinoshita, *Mario Schenberg—o cientista e o politico* (Brasília: Fundação Astrojildo Pereira, 2014).

53. Barany and Krige's afterword to the present volume.

CHAPTER ELEVEN

The Transnational Physical Science Study Committee

The Evolving Nation in the World of Science and Education (1945–1975)

Josep Simon

In 1964 Robert I. Hulsizer became director of the Science Teaching Center of the Massachusetts Institute of Technology, after more than a decade at the University of Illinois and after playing a key role in the development of the Physical Science Study Committee (PSSC). He had graduated as a physicist during World War II, worked at the Radiation Laboratory, and received his doctorate at MIT. On coming back to MIT he advocated the preparation of a new freshman physics course, which he characterized as "like trying to describe an evolving nation. It exists and therefore can be characterized at its present state. Yet one's view of the course is a mixture of past tradition, past and present hopes and partial realization of these hopes."[1]

Another major actor in the making of that MIT course, Anthony P. French,[2] emphasized in his proposal MIT's "well-developed tradition of strong and respected courses in introductory physics." This tradition had started in the 1930s with Nathaniel Frank's courses, followed by Francis Sears's textbooks and Jerrold Zacharias and Francis Friedman's PSSC. The new course would be another step in that tradition, for the service of MIT and the nation. French's narrative was an essentially local and national one radiating out from MIT and the United States to the rest of the world.[3]

The PSSC, the major project to which Hulsizer's and French's experiences and imaginaries refer, developed a new physics course for American high schools between 1956 and 1960. The PSSC was supported by grants from the National Science Foundation (NSF), the Sloan Foundation, and the Ford Foundation, with headquarters at MIT and the University of Illinois. It was a large and complex project that used a military-industrial management model, a major novelty in the production of teaching materials. In 1960 the first edition of the PSSC's *Physics* was published. A few years later, almost half of the nation's schools were using PSSC materials, and its textbook had been translated into a dozen languages.[4]

French and Hulsizer's testimonies mark the PSSC as a major and recent milestone in an "evolving nation," which was further evolving through the production of new courses at MIT. In some of the pedagogical landmarks, mentioned by French, the goals and approaches localized in a particular institution (MIT) could be projected onto the whole nation. In other cases, such as the PSSC, there was a larger interaction and integration of aims and methods from different institutions and communities of practitioners in order to produce a pedagogical package able to build the nation and eventually to go beyond it. Indeed, this project, which sought to collapse US science education into one single national course able to boost the production of scientists, managed to spread its impulse of educational reform to Europe, Latin America, Asia, and Oceania. Today, the PSSC is seen by science educators as a common heritage of science education worldwide.[5]

The enthusiasm for physics teaching and for developing an exportable physics course that could spread throughout the world was in tune with the vision of MIT being advocated at the time by Gordon Brown, MIT's dean of engineering. For Brown MIT was at the intersection between the past, present, and future. It had matured into a major center for engineering education, and its educational programs were ripe for exporting nationally and internationally. They were of a piece with the programs being advocated by MIT's Ford Foundation–funded Center for International Studies directed by Max Millikan and Walt Rostow, who would promote modernization theory and nation building as tools of US foreign policy around the world.[6] Hulsizer and French were imbued with that same MIT "idea." Their characterization of the MIT freshman physics course coupled the development of pedagogy and of the nation, and it located its force in both the continuation of tradition and the imagination of new configurations. Theirs is a powerful illustration of the strong connections between science education and the nation, and the national imaginaries of Cold War physicists, which fit perfectly with classic historical research on the nation and nationalism.[7]

Nationalism has not always characterized the writing of US national history. Historians of nineteenth-century American science studied the making of national institutions, but they also paid attention to the American observation and appropriation of the German, British, and French experiences.[8] In contrast, analogous work for the Cold War period favors the view that after World War II, US science and education had grown to become autonomous.[9] Despite the global importance of American science and science education in this period, it is worth asking if these facts constitute in themselves a sine qua non condition for the congruence of our historiographical perspectives with the national unit.[10] Undeniably, a nation exists with regard to other nations, and as a state, with regard to other states.[11] Educational reform is particularly illustrative of this, thanks to its cross-national tradition, since all types of nation-states (nonhegemonic and hegemonic) have commonly looked abroad for relevant experiences before undertaking substantial changes in education practices at home.[12] The PSSC illustrates well this dynamic in national and international affairs.

Decades ago, one of the prime movers in the development of transnational history was the urge to open American history to the world by overcoming its traditional exceptionalism.[13] Transnational perspectives could contribute to shaking the foundations of historical characterizations that were insufficient but had long held sway. This chapter takes the PSSC as an object allowing us to unravel the historical and historiographical elements that may justify narrating the history of US science and education from a local, national, international, or transnational perspective. Central to my approach is the discussion of the advantages and shortcomings of each of these perspectives.

Accordingly, I provide a big picture of the PSSC and its contemporary history in the United States and abroad. The first section of the chapter examines the making of the PSSC from a local and national perspective. The second section deals with the committee's internationalization. In the final section I discuss the findings presented in the previous two sections and the potential of applying a transnational approach to this case study.

The American PSSC

The PSSC was born first as a local and personal initiative, but it rapidly expanded into a collective effort that assembled a large number of universities, colleges, and schools, physics professors and teachers, and educational researchers, technicians, managers, and consultants. It befitted the "signs of the times" and found major support in institutions such as the NSF, the American Institute of Physics (AIP), and the American Association of Physics Teachers (AAPT), which were developing programs to tackle the same problems. In this section I want to stress that although the PSSC was developed on US soil, its national (American) character was not a given natural quality but the result of a particular geopolitical infrastructure and the agency of certain project members.

Since the early decades of the twentieth century, some academic physicists in the United States had expressed dissatisfaction with the school curriculum, textbooks, and teacher training, emphasized the need to control and homogenize college requirements, and remarked on the tension between training physicists and training citizens and on the excessive reliance on European physics professors.[14] The decentralized nature of the American school system and the lack of interest among most university physicists had prevented any large-scale reform.[15]

By the mid-1950s, however, a series of events had merged to promote a widely generalized opinion in the US Congress that the country required a large investment in school science reform. The large number of specialized conferences held since the early 1950s and the National Defense Education Act of 1958 were key elements in this movement.[16] The newly created NSF (1950) managed to become the favored institution to tackle the problem of scientific manpower.[17]

Several initiatives were launched to improve physics teaching. In March 1956 Jerrold Zacharias, an MIT physicist, sent a memo titled "Movie Aids for Teaching Physics in High Schools" to the president of his institution. His project was couched in the language of atomic physics, it focused on experiment and film, and its pretension was mainly local and personal.[18] In parallel, and beginning in 1955, a series of conferences supported by the AAPT, the AIP, and other agencies had also begun to tackle the main problems of physics teaching.[19] A central actor in these conferences was Walter C. Michels of Bryn Mawr College,

AAPT's president-elect. He chaired a Joint Committee on Teaching Materials for High School Physics established by the AIP, the AAPT, and the National Science Teachers Association.[20] From 1960 Michels also chaired the AIP-AAPT Commission of College Physics, which developed a national survey of physics teaching programs with the aim of improving them. As national federations of physicists' societies and national organizations composed of regional sections in each US state, the AIP and AAPT had both the legitimacy and the capacity to conduct such an endeavor. MIT physicists were manifestly absent from these conferences,[21] a fact which limits the role that the historiography has traditionally given to MIT and its professors in the making of national science education and suggests instead the agency of other institutional and individual actors in this process.

A week after the 1956 AIP-AAPT Conference on Physics in Education, Zacharias sent a proposal to the NSF.[22] He was able to receive the support of MIT's president and chancellor and to use their connections and his war acquaintances to ensure the success of the application.[23] By September 1956 the project had taken its final name: Physical Science Study Committee. The Committee expressed interest in working with the Educational Testing Service (ETS)[24] in the development of its materials and the design of examination tools.[25]

The PSSC rapidly expanded through the incorporation of members from Cornell, Caltech, Illinois, and Bell Laboratories.[26] The team eventually included several hundred physicists, high school teachers, instrument makers, filmmakers, photographers, editors, typists, and educational test designers. To deal with the daily requirements of the project, Educational Services Incorporated (ESI), a nonprofit company, was created with its own staff.

By the late 1950s the PSSC preliminary materials had been tried in schools in Pennsylvania (three), Massachusetts (two), New Hampshire (one), New York (one), and Illinois (one).[27] Without undermining the importance of the official leaders of the project, Jerrold Zacharias and Francis Friedman at MIT, I should mention two other actors who, from a national perspective, had a major role in the making of the PSSC. Their relevance has often been downplayed in official accounts of the PSSC project, which has been characterized by a national pretension but a local (MIT) narrative.[28]

One of them was Walter Michels, whom we met a moment ago. Michels played a major role in the coordination and supervision of the production of PSSC materials, especially with regard to their testing in pilot schools. Pennsylvania contributed three of the eight pilot schools and Michels's work was fundamental in this context. Furthermore, he had a knowledge of, and contact with, the nation's physics-teaching community, which other members of the PSSC did not possess. The physicists leading the project at MIT were not part of that community, nor did they demonstrate any particular interest in getting to know it.

The other major actor was the group at the University of Illinois, which led the production of the PSSC Teacher's Guide and the supervision of evaluation in the pilot schools. The University of Illinois had had a laboratory high school since the 1920s, developing close collaboration between teachers and university professors in science and education (which was atypical). It hosted the earliest project of science curriculum reform in the United States after World War II: the University of Illinois Committee on School Mathematics (1951–1961), funded by the US Office of Education, the NSF, and the Carnegie Corporation.[29] Some of its members subsequently joined the PSSC. As we saw at the outset of this chapter, before joining MIT, Hulsizer was a professor at the University of Illinois and a member of the PSSC group there. Not only his previous MIT training but especially his experience at Illinois with the day-to-day operations of the PSSC arguably played a major role in his subsequent hiring as director of MIT's Science Teaching Center.

Michels's and Hulsizer's actions at their institutions, working together and networking with schools, colleges, and teachers in Pennsylvania and Illinois, and extending beyond to a large number of institutions and practitioners in other US states (where Michels was very effective), really did contribute to shaping the nation through science education reform. If the PSSC became an "evolving nation" able to map a large amount of US political and educational territory, it was not exclusively because of the political power and scientific prestige of MIT physicists but especially because of the agency of other actors such as the aforementioned.

By 1958 eight teachers had been using preliminary versions of the PSSC course with around three hundred students. Summer training programs were offered by five universities for around three hundred teachers. By 1959 there were more than 10,000 students using the trial materials. The ESI bulletin that year included a US map displaying this expansive distribution.[30] In 1965 there were around 5,000 teachers and 200,000 students using the PSSC program of study, accounting for almost 50 percent of the secondary school students enrolled in high school physics courses in the United States.[31]

In parallel, regional groups met to study the PSSC materials. They emerged in all the states except Alabama, Arkansas, Hawaii, Kentucky, Mississippi, Montana, Nevada, New Mexico, North and South Dakota, Tennessee, Utah, Vermont, and Wyoming. Some of these meetings were organized around further state divisions or large metropolitan areas such as New York, San Francisco, Los Angeles, Chicago, San Diego, Philadelphia, and Boston.[32]

The largest increase in the number of schools using PSSC materials occurred along the East and West Coasts, in the Midwest (then called the North Central region), and especially around urban areas. The rest of the country—with the exception of Florida—did not use PSSC materials and accounted for roughly half of the nation's physics high school population. Some of these schools expressed their reluctance to adopt PSSC courses and their preference for other projects of curriculum development such as Harvard's Project Physics.[33]

During the early implementation of the PSSC course in American schools, the Committee had to negotiate with the College Entrance Examination Board (CEEB) to create a special achievement test for those who had followed the course. The CEEB was founded in 1899 by twelve eastern colleges as a way to regulate and rationalize the variety of examinations applied by colleges to select their students.[34] During the first half of the twentieth century, the CEEB expanded to win a national coverage.[35] In 1926 it began administering the Scholastic Aptitude Test (SAT). After World War II, the CEEB tests were published by ETS. The development of tests with regional or national aspirations in the United States had been especially boosted by the two world wars, to ensure that army recruits met a minimum educational standard. The tests were soon adapted to school and college management. The implementation of these test programs generated heated debates because they could lead to standardization of curricula and interference with the states' administration, and there was no consensus as to their purpose and value.[36] All the same, by 1966 around 800 colleges and 250 scholarship programs used CEEB tests in their admissions processes. This did not deter the PSSC team, who argued that the standard CEEB tests were designed to assess a traditional physics course and so were inappropriate for their students. To solve the problem for future school years, they worked with the CEEB and ETS to produce a unified physics test suitable for all students.[37]

We see then that to produce an American physics course, the PSSC had to rely on infrastructure and collaboration provided by other initiatives which were also aiming at building the US nation by standardizing evaluation in its schools and universities. ETS had been involved in the PSSC since its inception.[38] ETS tests were not only an end product but also a fundamental technique to shape the PSSC course, since the tests were seen as an objective technology to measure the course's excellence. The tests were also an advertising tool to promote the idea of the pedagogical superiority of the PSSC option, favoring its adoption over other courses in the nation's schools.[39]

The leaders of the PSSC did not all agree on what its national role should be. Although it aimed at targeting the largest number of American schools, some of the PSSC founding members had more restrictive and elitist views: "The course should be directed to the top 25 per cent of high school students with the aims of inducing more of them to move into advanced work and of creating in the others a cultural climate favorable to scientific activity."[40] Its pursuit of national supremacy was also challenged by other scientists and educators, who had different views on what physics teaching and American education should be. Many suggested the worth of more humility or denounced the presumptuousness of the PSSC endeavor and its leaders.[41] Moreover, there were other competing projects in physics teaching, and they all wanted to be adopted nationally.

The quotation from Hulsizer at the beginning of this chapter was from a special issue of *Physics Today* dealing with "introductory physics education." Among the wide range of perspectives presented,[42] a revolt against the PSSC national discourse was clear in some of the contributions. Some authors considered that what society required from American schools was "educating philosopher-scientists" instead of an army of professional scientists and engineers.[43] Others were against the PSSC pretensions of national sovereignty, calling for course diversity as a desired reflection of the pedagogical and national virtues of American culture.[44] Among these, Harvard's Project Physics would become one of the PSSC's main competitors at both a national and an international level. In the early 1960s, while the PSSC was implementing its strategy of national expansion, Jerrold Zacharias chaired a panel on education, as part of the President's Science Advisory Committee. In addition to eulogizing the PSSC program, he revealed some of his views on national schooling: "The school 'system' is a natural unit for reform. The system is an organic, semi-autonomous unit of education, with pension plans and supervisors, principals, promotion and hiring procedures, specification of jobs, adoption committees. It has electoral responsibilities, public relations problems, budgetary experience. World War II measured armies by divisions because the division was the smallest military unit that included all services—infantry, artillery, tanks, and air. The school system is the 'division' of education."[45] These analogies were not rare in the 1960s.[46] They were grounded in a cultural context shaped by wartime experiences, which had brought the nation together as an integrated system to fight foreign enemies. After World War II these alignments survived in the minds of many people who had played a major role in the war effort, such as some of the PSSC leaders.

Victory in World War II and the start of a space race reinforced a nationalistic perspective in the United States that enhanced political, economic, and institutional support for endeavors such as the PSSC and contributed to shape the ethos of many of the PSSC team members.[47] Thus, in his President's Science Advisory Committee report, Zacharias felt entitled to omit two major aspects. First, he ignored any educational research produced before the 1960s. Second, he dismissed any contemporary study produced in Europe, arguing that they would be useful only if they could demonstrate their relevance to the American context, thus stressing US autonomy.[48] The declining influence of Europe was being replaced by the emerging rivalry with the Soviet Union that configured the Cold War and its historical narratives. The launch of *Sputnik* by the Soviet Union did not start projects such as the PSSC, but it did benefit them, at least in providing further impulse and support to the resolution of accumulated concerns about science education.

CIA and NSF reports on the efficiency of the Soviet Union's centralized system of high school education and university training made the comparative assessment of the failures of American science education even more dramatic. According to these reports, unlike the Soviet Union, the United States, with the political autonomy of its state governments and the stratification of its school system, could hardly aspire to produce a significant number of scientists in a short period of time, as required by national interest. Zacharias would surely have agreed with that. Opposition between American democracy and Soviet totalitarism was a frequent argument in NSF reports regretting sourly the US lag revealed by the *Sputnik* affair. However, other US experts considered that while Soviet education was shaped by ideological indoctrination, this had little effect on the training of students in subjects such as mathematics, physics, and chemistry. Scientific laws and technological problems were in fact the same whether presented in a communist or in a capitalist guise.[49] These reports obviously simplified the key features of American versus Soviet (or Russian) science education by reducing them to democracy versus authoritarism: both countries had national cultures of science, education, and politics that were more diverse and complex than captured by these two adjectives. What matters for our purposes here is that this line of reasoning represents another way of making the nation (by reference to an external enemy). In this framework (international) comparison was relevant but was instrumentalized to serve a predominantly ideological, rather than educational, agenda.

Cold War historiography has greatly emphasized US-Soviet confrontation, in a narrative loaded with exceptionalism and a basic bipolarity.[50] If we look beyond the timeline imposed by the post-1945 emergence of superpower rivalry, however, we encounter longer-term narratives that can give a richer account of the historical phenomena relevant to understanding science education during the Cold War.[51] Comparative studies made with a view to learning more, rather than to establish superiority, appear to be a fundamental tool in the development of all national networks of education since the nineteenth century. Observers circulated officially or secretly across nations to compare the unknown with the known and to draw conclusions able to improve teaching and research back home. The United States was no exception.[52] Comparison involves a type of observation that is never symmetric (an observer is always subjective and politically biased) but it can at least be productive of new insights rather than simply used to dismiss the other. In this context, there was an international context for science education that was rapidly expanding, in which the United States would come to play a major role, but nonetheless, in which there could be reciprocal learnings, as the imperfect geometry of the "international" suggests.

The International PSSC

In 1959, alongside a map of the PSSC's distribution in America, ESI's report included a picture of Prime Minister Nehru examining PSSC materials accompanied by US officers at an exhibition organized in India.[53] Two years earlier a translation into Thai of the PSSC textbook's first volume had been made by a recent Harvard physics PhD who would subsequently occupy important government positions in Thailand.[54]

The goal of the PSSC had been to develop American curriculum reform. During the late 1950s, however, in the course of developing the PSSC materials, the project started to receive expressions of interest from foreign individuals and governments. ESI responded to these demands. As they grew in number the Committee was obliged to develop a plan for its international projection. It conceived of the international zone as divided into three types of countries: (1) "Advanced Nations—where there is something for both sides to learn": Sweden, Norway, Denmark, New Zealand, Yugoslavia, Spain, Israel; (2) "Intermediate Nations—The problem is primarily one of adapting the PSSC course": Japan, India, Latin America, or "Countries with relatively well established systems of education"; (3) "Emerging Nations—Where considerable aid work has to be done before PSSC can be of benefit: African Nations" or "Underdeveloped countries." Projects of the first type could be funded with the help of the NSF. Those of the second and third types would require funding from other agencies.

Just before the publication of the PSSC course materials, ESI reported having received requests for information from 350 individuals in foreign countries (plus 200 from Canada). That year, ten foreign visitors participated in PSSC summer institutes and publicized the project in Denmark, Germany, Finland, and England. In 1960 the number of visitors was expected to multiply by six. Three Spanish-speaking countries, Japan, and Sweden requested permission for the production of literal translations of the PSSC textbook into their national languages. Other countries, such as England, Canada, Germany, and Brazil, asked permission to adapt the course. The US Information Agency wanted to have PSSC materials (including films) for distribution in their network.[55]

By 1966 more than fifty foreign teachers had attended PSSC teacher-training programs in the United States.[56] This mode of operation produced results. Thus, for instance, a summer institute visit by a Swedish representative had a major role in the development of a trial program in Sweden aiming to adapt the PSSC materials. Moreover, Norwegian teachers joined the project to form a Scandinavian team cooperating to produce new teaching materials.[57] A similar experience occurred in New Zealand.

ESI contended that some of the pilot countries could adapt the materials to their educational needs by including additional topics,[58] and that "direct translation . . . will rarely be the optimal solution."[59] Some of the foreign editions adopted this view: the Norwegian edition incorporated an additional chapter, extracted from the PSSC Advanced Topics program,[60] and the Spanish edition was published in two volumes in order to be used in a two-year course (instead of the original one-year PSSC course). By 1964, only the Italian translation included all the course materials.[61]

After 1960 the bulk of the PSSC internationalization program was devised through the development of courses abroad.[62] Between 1960 and 1964 there were summer institutes in Israel, England, New Zealand (three), Brazil (two), Sweden, Italy (three), Nigeria, Uruguay, Costa Rica, and Chile, and conferences on the PSSC (or partially dealing with it) in India, Austria, France, Israel, Italy, Japan, and Southern Rhodesia.[63]

Foreign editions of the PSSC text were published in Denmark, Italy, Israel, Japan, Brazil, India, Sweden, Colombia, Canada, Spain, Norway, Turkey, (French) Canada, and France. Some of the films were in the course of being translated into Italian, they were purchased in India, and one of the films was translated into Spanish and shown at the 1963 Interamerican Conference on Physics Education (Rio de Janeiro) and at major universities in Mexico, Uruguay, Costa Rica, and Puerto Rico.[64]

In 1960 the PSSC began to develop a relevant on-site involvement in Europe at the request of the Organization for European Economic Cooperation (OEEC) and the NSF.[65] That year, the OEEC had organized at its headquarters in Paris a conference on physics education with the support of the International Union of Pure and Applied Physics. A report and a plan to develop pilot projects on science education in Europe were produced. Established in 1948 as a permanent institution to manage the Marshall Plan aid, the OEEC was now convinced that an economic recovery plan should involve the reform of school science education in its European member states.[66]

The PSSC team was approached by OEEC officials, and after a visit to England by Friedman, plans started to take shape for the organization of a PSSC summer institute in Cambridge (United Kingdom). It was held in August 1961 with the participation of teachers from France, Spain, Portugal, Ireland, Italy, Austria, Germany, Turkey, Greece, Iceland, the Netherlands, Switzerland, Norway, Denmark, the United Kingdom, Yugoslavia, Belgium, and Sweden.

The aim of the PSSC delegation, led by Uri Haber-Schaim, was to have truly intensive sessions, allowing participants to leave the institute with a range of written documents leading to the development of pilot projects in the different countries. These meetings were also conceived as places where the rights for translation of the PSSC materials would be negotiated. The Cambridge meeting indeed produced some of these drafts, developed not from a national perspective but through multinational teacher teams (except for a report on Yugoslavia).[67] Subsequently, Haber-Schaim considered that an international organization like OEEC would not have the capacity to develop such a project. It was preferable to leave the initiative to national groups as exemplified by the model experience of the (American) PSSC.[68]

The circulation of PSSC staff members across the world also played a major role in the internationalization of its products. Friedman and Haber-Schaim were arguably the members of the project who had a greater input in the development of the project abroad. Haber-Schaim led summer institutes in Europe, Latin America, Africa, and Japan. He would subsequently lead the preparation of the second and third editions of the PSSC course in the United States. Friedman traveled to the United Kingdom, India, and Pakistan and prepared the implementation of the PSSC there.

In addition, the project benefited from the international impact of US physics research and the worldwide circulation of the physicists connected to it. Thus, Philip Morrison, a member of the PSSC since its inception and a physics professor at Cornell, was in Europe, Israel, India, and Japan in 1960 for research purposes. During his trip he distributed PSSC materials and publicized the project.[69] MIT physicists not directly connected to the PSSC program did the same.[70] During his trip to India, Morrison expressed his surprise about Friedman having arrived in that country earlier than him and thus overtaking him in introducing the PSSC there—he used a metaphor which illustrated precisely the political and commercial substance of the PSSC international mission: "Had Columbus met the Admiral of Cadiz in Havana harbor he would have a little greater surprise."[71]

The earliest foreign editions of the PSSC course were translations into Spanish and Portuguese. In Latin America, there were three PSSC translations used in physics teaching a few years after the release of the PSSC materials in the United States. The first translation of the PSSC textbook was produced in 1962 in Spain and marketed in Spain and Latin America by the publisher Reverté.[72] It was used, for instance, in

Mexico, where knowledge of the PSSC was surely introduced early on by Luis Estrada, a Mexican PhD student who was a visiting student at MIT between 1958 and 1960.[73] During the 1960s Mexican physicists such as Estrada and Francisco Medina Nicolau conducted workshops on the PSSC at the Universidad Nacional Autónoma de México, and after the reform of its physics degree in 1966, a new general physics course was introduced which included PSSC course experiments and the replication of some of its instrument kits.[74]

The second translation of the PSSC course into Spanish was published in 1964 in Colombia by a team of ten Colombian MIT alumni and a group of physics and engineering professors from the major universities in Bogotá, with the support of the Organization of American States (OAS), MIT-Club Colombia, and the Colombian Association of Universities.[75] It was led by Alberto Ospina, a military naval engineer trained in electronics at MIT, who had witnessed the early development of the PSSC before returning home in 1958.

A few years earlier, the PSSC had been published (between 1962 and 1964) in Portuguese in Brazil. It was the result of a long-standing effort among Brazilian scientists and educators to improve science education, which was helped by the support of UNESCO in the creation of the Instituto Brasileiro de Educação, Ciência e Cultura (IBECC),[76] the development of ambitious plans to produce and distribute science kits in schools, and the support of US funding (Rockefeller Foundation, Ford Foundation) and inter-American organizations based in Washington, DC (OAS). The IBECC had a major role in the development of science education programs in Brazil and across Latin America during the 1960s and 1970s.

The IBECC was created in 1946 in Rio de Janeiro to administer UNESCO's projects in Brazil. Its involvement in science education came through the subsequent establishment of a São Paulo branch and the initiatives, from the early 1950s, of Isaias Raw, a young medical researcher based at the Universidade de São Paulo. Raw's interest in science teaching had taken shape since the late 1940s through his work as a science teacher in a São Paulo private school (conducted simultaneously with his university medical studies), where he edited a journal devoted to the teaching of science.

After getting his medical degree and a research stay in Severo Ochoa's biochemistry laboratory in New York, Raw returned to São Paulo with the idea of starting a project to change the standard paradigm of the teaching of science in Brazil. As the scientific director of IBECC's São Paulo branch, he conducted a large series of initiatives on science education and popularization, including exhibitions, clubs, fairs, talent competitions, and TV programs. Furthermore, he developed a major program for the design and production of school science equipment and experimental kits. Started as an in-house project, it soon received funding from the Conselho Nacional de Pesquisas and from several Brazilian state governments. As the project grew to industrial size, it was a major success for Raw to secure funding from the Rockefeller Foundation (1957), which already played a significant role in the funding of the new campus of the Universidade de São Paulo and especially its medical faculty.

In 1956 Raw visited the United States and became acquainted with incipient American educational projects such as the PSSC. Subsequently, Friedman was designated by the Ford Foundation to visit São Paulo, but he soon became ill and was unable to travel. However, through Raw's contacts at the Rockefeller Foundation and subsequent missions of US scientists to Brazil, it became clear that the country had an enormous potential for the development and marketing of science pedagogical packages. Thus, in 1961 a funding agreement was established with the Ford Foundation for the distribution of IBECC's experimental kits in Brazilian schools, the training of science teachers, and, last but not least, the distribution of US pedagogical materials in Brazil.[77]

The IBECC followed the progress of the PSSC project by using some of the preliminary copies of the course material and working on them between 1959 and 1960. In 1961 it published the translation of the laboratory guide and started to produce some of the PSSC equipment. A member of the IBECC attended the 1961 PSSC summer institute in Massachusetts.

A PSSC institute was held in São Paulo in January 1962, with funding from the OAS and the Ford Foundation and technical advice from the NSF. The institute staff was composed not only of Americans but also included lecturers from Chile (Darío Moreno), Costa Rica, and the IBECC (Rachel Gevertz). Participants were from Brazil (nineteen), Colombia (five), Chile (four), Paraguay (four), Argentina (three), Uruguay (three), Costa Rica (one), Nicaragua (one), Panama (one), and Peru (one). Later on that year, another PSSC summer institute was held in Brazil, this one fully developed by IBECC staff, and ran simultaneously in Costa Rica and Uruguay. By then almost all the PSSC equipment was available through local production.[78]

In this context the translation of the PSSC textbooks into Portuguese was carried forward by a team of science teachers and university physics professors at the Universidade de São Paulo, the Universidade de Minas Gerais, the Pontifícia Universidade Católica do Rio de Janeiro, and the Universidade de Brasilia, where the books were published. Between 1964 and 1971 around four hundred thousand copies of the PSSC course (split into four volumes) were sold in Brazil.[79]

Moreover, as a follow-up to its 1960 Paris conference, in 1963 the International Union of Pure and Applied Physics organized a conference on physics in general education in Rio de Janeiro with the support of UNESCO, the OAS, the Brazilian Ministry of Education and Culture, the Conselho Nacional de Pesquisas, the Centro Latinoamericano de Física, and the Centro Brasileiro de Pesquisas Físicas. The meeting gathered around 150 participants from across Latin America, Europe, the United States, and some Asian countries.[80]

This movement of teachers and physicists across countries was promoted and supported by national and international institutions. The NSF stated that its priority in relation to the science curriculum was the "development of materials potentially useful to schools across the country." However, its mission was also to cooperate with other national and private agencies specializing in foreign affairs to help circulate pedagogical materials, scientists and science teachers, and educational information, in order to fulfill "United States foreign policy goals."[81]

NSF officials confessed to being proud of the interest shown by other countries for new US curriculum materials. They were conscious of the regional importance of Latin America for the international expansion of their national projects, seeing themselves as having "special responsibilities in working with Latin American countries and the state universities in Central America." Moreover, they suggested that with regard to the sending of publications and materials, "information should be given as freely to people in other countries as to people in the United States," but since foreign relations were a complex matter,[82] discretion and cautiousness should prevail in order not to give the impression of "pushing United States materials in other countries," while helping those making requests.[83] Notwithstanding their prudence, the international circulation of PSSC materials was massive. For instance, in 1961 a copy of the PSSC Science Study Series book *Crystals and Crystal Growing* (1960) was mailed to libraries in practically every country in the world (with several copies sent to most Latin American countries).[84]

The international exposure of the PSSC project did not appear to change substantially the basic outline of the operations of ESI and the Science Teaching Center at MIT. However, in a few cases, they benefited from direct collaboration with foreign practitioners linked to countries that were developing vigorous projects of science education reform connected to the American PSSC program. Thus, in the development of its Advanced Topics program, between 1960 and 1963, ESI made use mainly of American staff and consultants. However, it also engaged some teachers from other countries, such as Sweden, Canada, Brazil, and New Zealand, who were able to attend US summer institutes and to work in Massachusetts for some time, with the help of funding from their governments or UNESCO. By 1963 the course had been used on an experimental basis in Sweden, Norway, Italy, Israel, Brazil, Uruguay, Chile, Canada, and New Zealand, and members of the Brazil team, such as Gevertz and Raw, spent long periods in Cambridge, Massachusetts. The ability of foreign physicists and teachers was valued by PSSC staff, but not always so. On some occasions foreign requests for collaboration were rejected or considered insufficiently relevant to commit to, even if they came from centers with a good record with the PSSC program such as the Universidad Nacional Autónoma de México.[85]

A different and more multilateral approach characterized the projects developed by UNESCO in Latin America. UNESCO (United Nations Educational, Scientific and Cultural Organization), established immediately after World War II, played a major role in the international development of science education programs worldwide. UNESCO's science education initiatives during the 1960s and 1970s divided the globe by coupling world regions with scientific disciplines: a program in physics teaching for Latin America, in biology for Africa, in mathematics for the Middle East, and in chemistry for Asia.[86]

UNESCO's member states represented a wide range of political approaches, from pacifist internationalism to Cold War engagement, and different priorities and ideas about how to articulate international cooperation through science, education, and culture.[87] The first decades of the organization were characterized by a growing tension between an idealized global humanism, prone to nonalignment and confident in the apolitical and universalist nature of culture, education, and science, and a pragmatic and instrumental politics, represented mainly by the United States, which sought to fight the Cold War also on the cultural and educational front and to collapse international diversity into its own national interests and outlooks.[88]

UNESCO's Division of Science Teaching was created in 1961, and its organization can be understood partly as a key element in American foreign policy aimed at placing as many US representatives as possible in relevant positions in international organizations. Its first director was Albert Baez, an American physicist trained at Stanford who had been part of the PSSC film production unit. His hiring at UNESCO was undoubtedly advantageous for the United States and for the internationalization of one version—strongly supported by the US government—of American culture expressed through science pedagogy (PSSC). It also had major consequences for the development of physics-teaching projects in Latin America.

In starting his new job in Paris, Baez prepared a pilot project for the implementation of new approaches in the teaching of science. His model was obviously the PSSC.[89] He resolved to implement such a project in Latin America, because he spoke Spanish. After presenting his project to UNESCO authorities and getting their approval, Baez constituted a team with Nahum Joel from Chile, Robert Maybury from the United States, and Alfred Wroblewski from Poland and divided the program into three sections—physics, chemistry, and biology—keeping the direction of the physics program for himself.[90]

Joel, with the help of a team including his assistant Darío Moreno, had previously developed PSSC institutes in Chile, and he had a key role in the development of UNESCO projects in Latin America. Maybury had previously met Baez at the University of Redlands in the United States, and many years later he recalled that when asked to join the UNESCO project: "Al's vision resonated with me, for as with many other professionals of that era, I was influenced by John F. Kennedy's statement: 'Ask not what your country can do for you, but ask what you can do for your country.'"[91]

At a UNESCO conference at the Paris headquarters, Baez met a member of IBECC's team, who persuaded him that his project should be developed in Brazil. Baez was invited to visit IBECC's premises in São Paulo and was convinced that Raw's initiatives had developed an advanced and extremely adequate setting for the implementation of UNESCO's project in Latin America, and that producing materials in Portuguese would not be a major obstacle to subsequently translating the materials into Spanish for the rest of Latin America. Thus, São Paulo became the headquarters for the development of the project in Latin America.[92]

UNESCO's Brazilian pilot project could rely on the network of Latin American science educationists previously developed by the IBECC and by the Chilean group led by Nahum Joel and Darío Moreno. It gathered twenty-five professors and teachers from eight Latin American countries, and in the course of a year (1963–1964) it produced five books, seven kits of inexpensive laboratory materials, eleven short films, one long film, and eight television programs. At the end of the year, another thirty-five university science teachers from seven more Latin American countries (thus, fifteen in total) attended a seminar to test and evaluate the materials. Some of them, including those from Argentina, Chile, and Venezuela, formed teams to extend the pilot project activities to their countries.[93] Baez worked as director of the Division of Science Teaching until 1967. The success of the Latin American pilot project allowed the division to develop similar projects in Africa, Asia, and the Middle East.[94]

The Transnational PSSC

The development of the national and international PSSC was subjected to different experiences that involved movement across borders and thus interaction with different national traditions all over the world, which nonetheless did not always result in the establishment of a dialogue between different cultures of science education.

For instance, the presence of foreign students at MIT and other US universities assuredly promoted the internationalization of the PSSC program through translation and use abroad. As we saw in the previous section, this was the case with students from Colombia, Mexico, Thailand, and to some extent Brazil and Chile. The agency of these students was characterized, first, by their participation in the international scene and, second, by the skills they were obliged to develop in order to understand other national cultures. They were surely transformed by their experience abroad to a greater or lesser degree. The significance of their presence is relevant to understanding not only the internationalization of PSSC and US physics at large but also the educational and research development of these US institutions. However, this issue has hitherto been rarely analyzed.[95]

PSSC's internationalization also prompted the incorporation of some physicists and foreign science teachers (Swedish and Brazilian) into some of MIT's projects, although this situation was unusual. It offered them a privileged access to MIT's scientific, educational, and cultural resources. Their experience also involved personal and cultural transformations, comparison, tensions, and partial hybridization of different national cultures.

In turn, the intensive and extensive travels of the leading PSSC staff and their exposure to different national cultures might have affected their perspectives of science and education. One would want to investigate whether this international exposure had a relevant impact on subsequent editions of the PSSC's pedagogical package. For instance, did the knowledge acquired by the PSSC leaders during their world trips to promote the project result in the subsequent adaptation of the PSSC materials to an international audience? This is difficult to say, especially because after the second edition of the course was published (1965), most of the original team disbanded because of the early deaths of some of its leaders (e.g., Friedman and Finlay), their return to physics research, or their involvement in policy (Zacharias).

Traveling abroad arguably involves personal transformation to some extent and is commonly a good antidote against nationalism; however, it does not necessarily dissolve the driving force of nationality. On the contrary, the 1960s map of the world presented a nation-state system that maintained nationality as a fundamental quality of traveling, whether as a right-of-way or as a veto-of-access (with different grades in between determined by national and international migration policies).

In this context, the international circulation of PSSC team leaders was particularly shaped by a mission that was not only educational, scientific, and commercial but also fundamentally national in nature and that linked the institutional and the personal. As in a traditional classroom, the position of the student who comes to learn and that of the teacher who comes to teach are obviously not the same, so to a large extent the PSSC staff traveled abroad to teach and to fulfill US foreign policy goals in a traditional fashion. When Morrison portrayed himself as Columbus and Friedman as "the Admiral of Cadiz," in the letter written during their 1960 world tour to publicize the PSSC, he provided an account of the PSSC international endeavor couched in a language of military and commercial conquest.[96] The implications of his metaphor are particularly powerful and inescapable, taking into account the importance of Latin America for the international PSSC and the fact that Morrison wrote these words while visiting (postcolonial) India. Anyway, without overinterpreting this metaphor, Morrison's and Friedman's correspondence during their trips to Europe, India, Pakistan, and Japan makes clear that the focus and rationale of their mission was to enlighten foreign physicists and educators through PSSC exposition and the conferring of material gifts. Although their letters also mention some sightseeing and show some interest in the cultural heritage (museums, monuments) of these countries, they say nothing about what, if anything, they learned from foreign professional counterparts with regard to other national cultures of physics teaching. In this context the world circulation of PSSC team leaders exemplifies how the international can often be driven by a purpose that is national in fundamental ways.

It is fair to say that basically the new projects developed at ESI and MIT's Science Teaching Center remained American to their core in terms of their staff and their outlook even though some of them had international ambitions. On the other hand, the extensive internationalization of the PSSC was possible thanks to a network of national, private, and international organizations, whose interests converged on this pedagogical package originally conceived at MIT. All the same, organizations are staffed and directed by humans, whose engagement in the making and practice of knowledge is a major force that does not always align completely with organizations' official statements.[97] The Division of Science Teaching, directed by Baez, defined UNESCO as a "catalyzer" and "internationalizer" but emphasized the role that individual countries, individual teachers, and specific teams of people had had in the shaping of new science-teaching materials and outlooks.[98] Opening the door to a more symmetrical interaction with other national collectives allowed in this case further effacement of national boundaries, contributing to the articulation of an international framework amenable to a potential situation of transnationalism.

In this context, there were major agents in the internationalization of the PSSC who had attributes that we can call *transnational* and that made them particularly well suited to conduct this task, while pursuing their personal agendas. Isaias Raw and Darío Moreno, for instance, were this type of actor. Here, I am going to focus on Uri Haber-Schaim and Albert Baez.

Haber-Schaim had a leading role in the practical implementation of the international PSSC; he led the production of the third edition of the

PSSC materials and was director of the PSSC project between 1961 and 1974. He was born in Berlin (Germany) in 1926, moved as a child to Palestine in 1933, and graduated from the Hebrew University in 1949, right after the creation of the State of Israel. He was part of Israel's Science Corps (HEMED), an organization of scientists connected to the army, and had a major role in the development of a defense industry in the context of growing hostilities with the surrounding nations, which made this a central project in the making of Israel. Accordingly, Haber-Schaim was sent to the University of Chicago to study nuclear physics (PhD, 1951), and he returned to the Weizmann Institute and the Israel Atomic Energy Commission (HEMED's research bases).

After repeated professional arguments with the director of the Israel Atomic Energy Commission, which led to his resignation, Haber-Schaim moved to a position in German-speaking Switzerland and shortly afterward immigrated to the United States. Haber-Schaim's clash with the Israel Atomic Energy Commission can be seen partly as the tension between a profession used to freedom (scientist) and a management based on military discipline in a war situation. This tension also suggests a clash between different political visions on how to build the Israeli nation-state. In the United States, Haber-Schaim worked at the University of Illinois (1955–1956) and as assistant professor at MIT (1957–1960). Beginning in the late 1950s he devoted himself fully to the development of the PSSC and from then on built a professional career in the field of science education, making the Boston area his home base.[99]

Albert Baez, the first director of UNESCO's Division of Science Teaching, was born in Puebla (Mexico) in 1912 but immigrated to New York with his family at the age of two. He returned for a year to Mexico when he was seven, an experience that he claimed, later in his life, had a major role in maintaining his ties with his country of birth and in preserving his ability to speak Spanish.[100] However, Baez's formal education was American. In 1933 he obtained a BA in physics and mathematics from Drew University, two years later an MA in physics from Syracuse University, and in 1950 a PhD in physics from Stanford, where he developed a research career in X-ray optics. In 1951 he obtained a UNESCO appointment in Iraq. Subsequently, he worked at the University of Redlands and again at Stanford. In 1957 he was called by Jerrold Zacharias to join the PSSC project at MIT, where he worked mainly on film production. His work for the PSSC project would shape the rest of his professional career.[101]

In his memoirs about his career at UNESCO, Baez recalled that before teaching at Redlands he worked for some time in the Cornell Aeronautical Laboratory's Operations Research Group. According to him this was an intellectually challenging job, but he became increasingly worried about being involved in scientific and technological collaboration in the war effort. He read an article in the *New York Times* about UNESCO's mission that led him to inquire about any job openings, hoping to devote his professional life to peaceful uses of science for the benefit of mankind. Later on, he was invited to collaborate in a UNESCO project to set up science laboratories at the University of Baghdad. This first mission, as well as his work for the PSSC, surely helped in his being recruited subsequently as the head of the Division of Science Teaching.

Baez's expression of humanistic ideas in his memoirs was surely genuine, although they are a later reconstruction. Baez's and Haber-Schaim's careers show a move from military-driven science to science education and in parallel from nation building to international articulation. Both had very successful careers in science education, and they found there a way to distance themselves from scientific research for military purposes. Although this might have been the goal of many of the PSSC physicists who had participated in the wartime effort,[102] the actions, careers, and language of Baez and Haber-Schaim contrast with those of Jerrold Zacharias, for instance.

However, the main argument here is that it was not by chance that individuals like Baez and Haber-Schaim were two of the major leaders in the internationalization of the PSSC. Their upbringing and life experience across several national cultures prepared them to understand and to develop the types of actions involved in internationalization, to an extent that other PSSC staff members (regardless of their competence in physics and education) were not as ready to fulfill. Beyond their language skills, their life and professional experiences were transnational, as they combined different national and cultural identities throughout their lives and used these attributes to build bridges and establish dialogues.

Baez claimed he felt linked to Mexico. In his memoirs he also confessed that in starting the UNESCO projects he originally had a rather arrogant perspective as to the superiority of American science education projects and their makers. During the development of these projects in Latin America, however, he became progressively more humble in recognizing the cleverness and capacities of Brazilian colleagues.

Haber-Schaim had an even more complex itinerary in national perspective. He, like Baez, was born in one country but grew up in another one. Moreover, he worked in the context of an emerging nation-state, a process to which he contributed through his scientific research. In the new State of Israel he worked hand in hand with scientists who shared his Israeli citizenship but were born in different nation-states. He subsequently lived in two additional countries and developed a career in the United States that was characterized by internationalism in science education research.

As transnational actors both Baez and Haber-Schaim were able to travel to foreign countries and to learn from foreign colleagues with an open mind, or at least with a mind less circumscribed by nationalist preconceptions that were rampant at the height of Cold War rivalry and the making of American hegemony—the projects that had inspired the PSSC in the first place. Their capability in this context was characterized by a relevant fluidity with regard to nationality, which allowed them to manage national allegiances with more degrees of freedom than other types of historical actors. Their UNESCO and PSSC work, respectively, shows an engagement with the communities of scientists and educators in the countries they visited. They recognized the value of "the other" for the purpose of improving science education with reference to the requirements and characteristics of each national context. In contrast, there were other historical actors holding perceptions and performing roles with a strong involvement in internationalization but having a major national bond as well, exemplified in this chapter by Zacharias, Friedman, and Morrison. The development of Baez's and Haber-Schaim's transnational agency was shaped by both the nature of their multinational upbringing and the nurture of their international experiences, shaped by their capacity to engage and communicate with different national traditions of science and education.

A question that remains is to what extent the different translations of the PSSC materials were faithful to their American originals. Were they transformed by the different national outlooks and experiences in science and education in which they were adapted? The PSSC translations were undoubtedly the product of internationalization and to some extent of transnationalism (of some of its promoters). But can the PSSC translations be seen as transnational products?

Many of the translations of the PSSC textbook were mostly literal. This was the case, for instance, for the two Spanish translations. However, the Scandinavian team that adapted the PSSC materials added new chapters to the book, and some of the translations (e.g., the one produced in Spain) divided the book into two volumes to adapt it to a two-year course.[103] In other places such as Brazil "participants utilized the text produced by the PSSC and other modern texts as a base for their own studies, but they developed and produced a set of modern learning aids, which they themselves had adapted to the local economic education needs." The Brazilian project team at IBECC thus transformed the PSSC American course mainly through a focus on methodology aimed at adapting it to local needs, but they equally focused strongly on content development. They could do so because they had an excellent starting point based on a selection of renovated curriculum contents made by the American PSSC team.[104]

With the increasing international availability of new curriculum projects in the 1960s and 1970s and the development of international teams and cross-national experiences such as UNESCO's pilot physics project at IBECC, it would be accurate to say that some of these pedagogical products not only crossed national cultures of science and education but also contributed to dissolving them. In other words, their cross-national circulation not only contributed to strengthening the action of US national science and education in a wide range of other national contexts, through direct exchanges or interactions in the field of international organizations. More important, it was arguably able to weaken the original national characteristics of these products and in addition could lead to their endowment with a somewhat lasting transnational condition. Further research based on a closer comparative analysis of several PSSC translations is still required to fully support this transnational claim.

Final Remarks

The structure of this chapter might give the impression that during the 1960s the efforts toward science curriculum reform progressively moved from the local to the national, from the national to the international, and from there to transnational science education. This has been a suitable order of presentation, chosen for conventional narrative reasons that advise following a chronological order, moving from particular to general, or going from simple to compound. However, this would be a simple linear interpretation that would hide the complexity of the world of science and education. The PSSC had local, national, international, regional, and transnational elements. All of them are relevant if we want to achieve an accurate historical understanding of this research object, and their relationships are not linear and hierarchical.

Furthermore, the coexistence of all these scales in a historical object such as the PSSC cannot be taken for granted. It is as important to know that it had each of these qualities as to understand why it had them and why we as historians confer them on the PSSC. As I showed in the first section of this chapter, nationality is a complex concept. The PSSC was American not only because almost all of its members were born in the United States or because it was developed at MIT and some other American universities and schools, but especially because it was part of a vigorous project of nation building within US territory and abroad. Nation building was performed through intranational science education reform developing and strengthening networks and communities within a country, by making comparisons with other nation-states or geopolitical regions, and by implementing large-scale programs of internationalization. Unpacking these categories is crucial to really understanding and using them appropriately and accurately.

Among the main categories discussed in this chapter, the transnational is the most elusive, since, as we saw, many international phenomena can be more akin to the national than to the transnational. However, this historical claim could be nuanced by future historiography subject to updated worldviews and cultural concerns displaying the utmost relevance of the transnational, whether in a world of nation-states or beyond it. Moreover, the distinction of the transnational from the international should play a major role in enriching the historical field with more subtle accounts integrating a wider range of objects and actors from a larger number of national cases and providing a better understanding of the phenomena that lie in the interstices of nation-states or do not succumb to the logic of the national. The examination of the production of the local, national, regional, international, and transnational PSSC presented in this chapter represents an attempt to integrate all these views and to demonstrate the importance of discussing the transnational as a vector for historiographical improvement.

The unraveling of the transitions and connections between our different scales of analysis is not a simple matter. It will require major effort by historians of science, technology, and medicine to update their tradition by overcoming the nation, which is still the most obvious site of their professional and intellectual employment.

Notes

1. R. I. Hulsizer, "The New MIT Course," *Physics Today* 20, no. 3 (1967): 55–57.

2. Anthony P. French was a British physicist seasoned at Cambridge and Los Alamos in the US-British atomic bomb project, who had returned to the United States in 1955 and had transformed his university courses at the University of South Carolina into *Principles of Modern Physics* (New York: Wiley, 1958).

3. A. P. French, "A New Introductory Course at the Massachusetts Institute of Technology," 1963, PSSC Records, MC626, box 9, folder "New Courses," Institute Archives and Special Collections, Massachusetts Institute of Technology, Cambridge, MA (hereafter MIT Archives).

4. See, e.g., U. Haber-Schaim, "Precollege: The PSSC Course," *Physics Today* 20, no. 3 (1967): 25–31.

5. Richard Gunstone, ed., *Encyclopaedia of Science Education* (Berlin: Springer, 2015).

6. S. W. Leslie and R. Kargon, "Exporting MIT: Science, Technology and Nation Building in India and MIT," *Osiris* 21 (2006): 110–130.

7. Ernest Gellner, *Nations and Nationalism* (Oxford: Basil Blackwell, 1983); Eric Hobsbawm and Terence Ranger, eds., *The Invention of Tradition* (Cambridge: Cambridge University Press, 1983).

8. A. J. Angulo, "The Polytechnic Comes to America: How French Approaches to Science Instruction Influenced Mid-Nineteenth Century American Higher Education," *History of Science* 50, no. 3 (2012): 315–38; Daniel J. Kevles, *The Physicists: The History of a Scientific Community in America* (New York: Alfred A. Knopf, 1978).

9. See, e.g., Barbara B. Clowse, *Brainpower for the Cold War: The* Sputnik *Crisis and National Defense Education Act of 1958* (Westport, CT: Greeenwood Press, 1981); John Rudolph, *Scientists in the Classroom: The Cold War Reconstruction of American Science Education* (New York: Routledge, 2002); Wayne J. Urban, *More than Science and* Sputnik*: The National Defense Education Act of 1958* (Tuscaloosa: University of Alabama Press, 2010).

10. An example of this trend can be seen in the introduction to the most recent *Companion to the History of American Science*, whose editors transform the historical observation of contemporary US world hegemony into a historiographical project per se whose goal is to demonstrate "the rapid emergence of

the United States as the global leader in science and technology in the twentieth century." Georgina M. Montgomery and Mark A. Largent, "Introduction: The History of American Science," in *A Companion to the History of American Science*, ed. Georgina M. Montgomery and Mark A. Largent (Oxford: Wiley, Blackwell, 2015), 1–5, at 4. In contrast, see Asif A. Siddiqi, "Competing Technologies, National(ist) Narratives, and Universal Claims: Toward a Global History of Space Exploration," *Technology and Culture* 51, no. 2 (2010): 425–443.

11. Pascale Casanova, *The World Republic of Letters* (Cambridge, MA: Harvard University Press, 2004).

12. Josep Simon, ed., "Cross-National Education and the Making of Science, Technology and Medicine," special issue, *History of Science* 50, pt. 3, no. 168 (2012): 251–374.

13. Ian Tyrrell, "American Exceptionalism in an Age of International History," *American Historical Review* 96, no. 4 (1991): 1031–1072; David Thelen, "Rethinking History and the Nation-State: Mexico and the United States as a Case Study; A Special Issue," *Journal of American History* 86, no. 2 (1999): 438–452; David Thelen, "The Nation and Beyond: Transnational Perspectives on United States History," *Journal of American History* 86, no. 3 (1999): 965–975.

14. Edwin H. Hall, "The Relations of Colleges to Secondary Schools in Respect to Physics," *Science* 30, no. 774 (1909): 577–586; Edwin H. Hall, "The Teaching of Elementary Physics," *Science* 32, no. 813 (1910): 129–146; C. R. Mann, "Physics Teaching in the Secondary Schools of America," *Science* 30, no. 779 (1909): 789–798; C. R. Mann, "Physics and Education," *Science* 32, no. 809 (1910): 1–5; Robert A. Millikan, "The Problem of Science Teaching in the Secondary Schools," *School Science and Mathematics* 9 (1925): 966–975; W. E. Brownson and J. J. Schwab, "American Science Textbooks and Their Authors, 1915 and 1955," *School Review* 71, no. 2 (1963): 150–180.

15. See Josep Simon, "Physics Textbooks and Textbook Physics in the Nineteenth and Twentieth Centuries," in *The Oxford Handbook of the History of Physics*, ed. Jed Z. Buchwald and Robert Fox (Oxford: Oxford University Press, 2013), 651–678; Josep Simon, "Textbooks," in *A Companion to the History of Science*, ed. Bernard Lightman (Oxford: Oxford University Press, 2016), 400–413.

16. See, e.g., F. Watson, P. Brandwein, and S. Rosen, eds., *Critical Years Ahead in Science Teaching: Report of Conference on Nation-wide Problems of Science Teaching in the Secondary Schools Held at Harvard University, Cambridge, Massachusetts, July 15 to August 12, 1953* (Cambridge, MA: Harvard University Printing Office, 1953); Rudolph, *Scientists in the Classroom*.

17. David Kaiser, "Turning Physicists into Quantum Mechanics," *Physics World*, May 2007, 28–33; Rudolph, *Scientists in the Classroom*; William C. Kelly, "Physics in the Public High Schools," *Physics Today* 8, no. 3 (1955): 12–14.

18. Jerrold Zacharias, "Memo to Dr. James Killian, Jr. Subject: Movie Aids

for Teaching Physics in High Schools, Massachusetts Institute of Technology, Department of Physics, March 15, 1956," Massachusetts Institute of Technology Oral History Program, Oral History Interviews on the Physical Science Study Committee, MC602, box 1, folder "Background Materials—PSSC," MIT Archives (hereafter MC602).

19. "Conference on the Production of Physicists," Greenbriar Hotel, WV, Mar.–Apr. 1955; "Conference on Physics in Education," New York, Aug. 1956; "Conference on Improving the Quality and Effectiveness of Introductory Physics Courses," Carleton College, MN, Sept. 1956; followed by meetings at the University of Connecticut and Wesleyan University and, in between, several AAPT annual conferences. Frank Verbrugge, "Conference on Introductory Physics Courses," *American Journal of Physics* 25, no. 2 (1957): 127–128; Walter C. Michels, "Commission on College Physics," *American Journal of Physics* 28, no. 7 (1960): 611; "Conference on Physics in Education," American Institute of Physics, Education and Manpower Division Records, 1951–1973, box 6, folder Conference on Physics Education, 1956, Niels Bohr Library and Archives, College Park, MD.

20. Walter C. Michels, "Committee on High School Teaching Materials (AIP-AAPT-NSTA), High School Physics Texts, Comments by Walter C. Michels for Meeting of 5/31–6/1, 1956," MC602, box 1.

21. There were a few exceptions, such as Sanborn C. Brown, who attended some of them.

22. Jerrold Zacharias, "A Proposal to the National Science Foundation, August 17, 1956," MC602, box 1.

23. Harry C. Kelly, NSF assistant director for scientific personnel and education, like Zacharias and many of the physicists involved early on in the PSSC project, had worked at MIT's Radiation Laboratory during the war. He would subsequently join the PSSC project board. Among the project's original members or supporters were MIT professors such as Martin Deutsch (who had worked at Los Alamos), Edwin H. Land (head of the Polaroid Company, located in Cambridge, MA, which had flourished during the war through military commissions), and Isaac Rabi (since 1940 associate director of MIT's Radiation Laboratory and recipient of the 1944 Nobel Prize in Physics), Nathaniel H. Frank, Francis Friedman, and Edward Purcell (Harvard). Both Rabi and Deutsch were Austrians of Jewish background who emigrated with their parents and were educated in the United States.

24. The Educational Testing Service was founded shortly after World War II by the American Council of Education, the Carnegie Foundation for the Advancement of Teaching, and the College Entrance Examination Board. To avoid political turmoil with regard to examination standardization at the state and the federal level and potential conflicts of interest, ETS was originally set up as a private nonprofit organization. G. Giordano, *How Testing Came to Dominate*

American Schools: The History of Educational Assessment (New York: P. Lang, 2005), 97–98.

25. Physical Science Study Committee, "Meeting of September 8, 1956," MC602, box 1.

26. Physical Science Study Committee, "Meeting of September 14, 1956," MC602, box 1.

27. Gilbert C. Finlay, "The Physical Science Study Committee," *School Review* 70, no. 1 (1962): 63–81; Educational Services Inc. (hereafter ESI), "A Partial List of Teachers Using the PSSC Course, 1962–1963, in the United States and Canada," *ESI Quarterly Report*, Winter 1962/1963, 10.

28. The role of these actors was acknowledged by Friedman in the postface of the PSSC textbook but was hidden in many of the papers advertising the project, which placed a great emphasis on MIT, as also in recent literature such as Rudolph's *Scientists in the Classroom* and in the testimonies of the MIT physicists involved in the PSSC project, such as Zacharias, collected by the PSSC Oral History Project and preserved at MIT Archives. Friedman died in 1962.

29. J. Goodlad, *School Curriculum Reform in the United States* (New York, 1964), 64.

30. ESI, *Progress Report: A Review of the Secondary School Physics Program of the Physical Science Study Committee Initiated at the Massachusetts Institute of Technology*, Watertown, 1959, 10, ESI Records, MC79, box 6, MIT Archives.

31. Although these numbers were usually extracted from the PSSC program's own records, and therefore should be submitted to a closer and more objective quantitative scrutiny, they are no doubt indicative of a rapid and large-scale expansion of the project soon after its commercialization. S. W. Daeschner, "A Review of the Physical Science Study Committee High School Physics Course" (master's thesis, Kansas State University, 1965).

32. States such as New Jersey, New York, and Ohio had around 80 PSSC teachers; Florida and Pennsylvania, approximately 100; and Maryland, California, and Illinois, almost 150. ESI Records, MC79, box 6, folder "PSSC Area Meeting Reports, 1961–1968," MIT Archives; ESI, "A Partial List of Teachers Using the PSSC Course, 1962–1963, in the United States and Canada."

33. Project Physics was directed by F. James Rutherford (New York University), Gerald Holton (Harvard), and Fletcher G. Watson (Harvard). The textbook *Introductory Physical Science* was developed by some of the people who contributed to the second edition of the PSSC textbook, but mostly not the original team who had led the project. See, e.g., Brother John Ryan, "Report of Area Physics Teacher's Meeting, February 17, 1968, Bishop David Memorial High School, Louisville, Kentucky," and "PSSC Area Meeting Report, Loyola University, New Orleans, Louisiana, March 30, 1968," ESI Records, MC79, box 6, folder "PSSC Area Reports 1968," MIT Archives.

34. Barnard College, Bryn Mawr College, Columbia University, Cornell University, Johns Hopkins University, New York University, Rutgers College, Swarthmore College, Union College, University of Pennsylvania, Vassar College, and the Woman's College of Baltimore. M. S. Schudson, "Organizing the "Meritocracy": A History of the College Entrance Examination Board," *Harvard Educational Review* 42, no. 1 (1972): 34–69.

35. CEEB, *Bulletin of Information: College Board Admissions Tests* (New York, 1966). Extracted from L. J. Karmel, *Measurement and Evaluation in the Schools* (New York: Macmillan, 1970), 299.

36. Giordano, *How Testing Came to Dominate American Schools*; C. I. Kingson, "Science Education," *Harvard Crimson*, Nov. 27, 1957.

37. ESI, "The Matter of College Boards," ESI Records, MC79, box 6, folder "Teacher lists," MIT Archives.

38. F. L. Ferris, "Testing for Physics Achievement," *American Journal of Physics* 28, no. 3 (1960): 269–278; Catherine G. Sharp, "Minutes of the Annual Meeting of the Board of Trustees of Educational Testing Services, May 3, 1960," Educational Testing Service Archives.

39. ESI, *Progress Report*, 26–28.

40. Physical Science Study Committee, "Meeting of October 13, 1956," MC602, box 1.

41. A. Calandra, "Some Observations of the Work of the PSSC," *Harvard Educational Review* 29, no. 1 (1959): 19–22; J. A. Easley Jr., "The Physical Science Study Committee and Educational Theory," *Harvard Educational Review* 29, no. 1 (1959): 4–11.

42. Namely, Harvard Project Physics; Physical Science for Nonscientists, a course given at the University of California, Berkeley; the Nuffield Science Teaching Project; lectures given by Richard Feynman; the new MIT course, Engineering Concepts Curriculum Project; the Science Courses for Baccalaureate Education; and a critical review of available college physics courses by Mark W. Zemansky, the author of an extremely successful college physics textbook (with Francis Sears).

43. L. V. Parsegian, "Baccalaureate Science," *Physics Today* 20, no. 3 (1967): 57–60.

44. G. Holton, "Harvard Project Physics," *Physics Today* 20, no. 3 (1967): 31–34; E. E. David Jr. and J. G. Truxal, "Engineering Concepts," *Physics Today* 20, no. 3 (1967): 34–40.

45. Panel on Educational Research and Development, *Innovation and Experiment in Education* (Washington, DC, 1964), 37.

46. See, for instance, a similar analogy made by CEEB's vice president for examination and research: A. S. Kendrick, "Rainy Monday," *College Entrance Examination Board*, 1967, 2.

47. John A. Douglass, "A Certain Future: *Sputnik*, American Higher Educa-

tion, and the Survival of a Nation," in *Reconsidering* Sputnik*: Forty Years since the Soviet Satellite*, ed. Roger D. Launius, John M. Logsdon, and Robert W. Smith (London: Harwood Academic, 2000), 327–362; Rudolph, *Scientists in the Classroom*; Simon, "Physics Textbooks and Textbook Physics."

48. Panel on Educational Research and Development, *Innovation and Experiment in Education*.

49. Rudolph, *Scientists in the Classroom*, 74–75.

50. These qualities are even present in Odd A. Westad, *The Global Cold War: Third World Interventions and the Making of Our Times* (Cambridge: Cambridge University Press, 2005). In contrast, see, e.g., John Krige, *American Hegemony and the Postwar Reconstruction of Science in Europe* (Cambridge, MA: MIT Press, 2006); Daniela Spenser, *The Impossible Triangle: Mexico, Soviet Russia, and the United States in the 1920s* (Durham, NC: Duke University Press, 1999).

51. An analogous long-term argument is made by Jessica Wang, "Colonial Crossings: Social Science, Social Knowledge, and American Power, 1890–1970," in *Cold War Science and the Transatlantic Circulation of Knowledge*, ed. Jeroen van Dongen (Leiden: Brill, 2015), 184–213. I thank Adriana Minor for this reference.

52. See Simon, "Cross-National Education and the Making of Science, Technology and Medicine."

53. From 1961, ESI would be involved in the creation of an institute of technology in Kanpur, India, by a consortium of eight American universities (including MIT) and funding from the US Agency for International Development and the Indian government. ESI, *1959 Progress Report*; ESI, *A Master Plan for ESI-PSSC Activities*, Watertown, 1961, 21–22 and appendix, MC79, box 6; MC602, box 2, MIT Archives.

54. ESI, *1959 Progress Report*, 17.

55. ESI, "International Interest and Problems in Connection with PSSC Physics," 1959, ESI Records, MC79, box 6, folder "Foreign Interest," MIT Archives.

56. ESI, *Physical Science Study Committee (1966): A New Physics Program for Secondary Schools*, 7, ESI Records, MC79, box 6, folder "A New Physics Program for Secondary Schools (Brochures)," MIT Archives.

57. The Swedish-Norwegian team also invited Denmark, Iceland, and Finland to join their project, but the potential expansion of the Scandinavian team met with some obstacles. The Swedish edition, though, was probably used or at least read in these countries. Letter from E. Waril to Uri Haber-Schaim, Feb. 16, 1962, Physical Science Study Records (hereafter PSSC Records), MC626, box 1, folder "PSSC International Inquiries and Reaction, 1961–1962," MIT Archives.

58. ESI, *A Master Plan for ESI-PSSC Activities*, 21–22 and appendix.

59. ESI, "International Interest and Problems in Connection with PSSC Physics."

60. This ESI program was not conceived as a course itself but as additional

chapters on more advanced topics which could be added to the teaching with PSSC materials.

61. ESI, *Quarterly Report, Summer–Fall 1964*, Watertown, 1964, MC602, box 5.

62. Letter from E. Waril to Uri Haber-Schaim, Feb. 16, 1962, PSSC Records, MC626, box 1, folder "PSSC International Inquiries and Reaction, 1961–1962," MIT Archives.

63. Before 1966 there was also a summer institute held in Colombia. ESI, *Quarterly Report, Fall–Winter 1963*, Watertown, 1963, 3–4, 24; ESI, *Quarterly Report, Summer–Fall 1964*, Watertown, 1964, MIT Archives, MC602, box 5; ESI, *Physical Science Study Committee (1966): A New Physics Program for Secondary Schools*, 7.

64. ESI, *Quarterly Report, Fall–Winter 1963*, 3–4, 24; ESI, *Quarterly Report, Winter 1962–63*, 90, Watertown, 1963.

65. ESI, "International Interest and Problems in Connection with PSSC Physics."

66. PSSC Records, MC626, box 12, folder "Correspondence 1961–1962," MIT Archives.

67. Ibid.

68. U. Haber-Schaim, "Some Guidelines on Curriculum Reform Based on the Experience of the Physical Science Study Committee," 1962, MC602, box 3.

69. ESI Records, MC79, box 6, folder "Personnel—P. Morrison Correspondence," MIT Archives.

70. An example is William Buechner, head of the MIT Physics Department, who sent PSSC materials to Korea in 1965. William Buechner Papers, 1928–1978, MC229, box 4, MIT Archives. I am grateful to Adriana Minor for this reference.

71. Letter from Philip Morrison to Edna S. Alexander, Feb. 9, 1960, 1, ESI Records, MC79, box 6, folder "Personnel—P. Morrison Correspondence," MIT Archives.

72. By the 1970s, Reverté had branches in Bogotá, Buenos Aires, Caracas, Mexico City, and Rio de Janeiro.

73. Luis Estrada, "La UNAM y yo," in *Homenaje a Luis Estrada* (Mexico City: Academia Mexicana de Ciencias, 2010), 1–6.

74. J. A. González and R. J. J. Espinosa, "Introducción al método experimental: Un nuevo curso en la Facultad de Ciencias," *Revista mexicana de física* 22 (1973): E57–E69; interview with Jorge Barojas Weber by Josep Simon, Mexico City, July 4, 2016.

75. According to the textbook's preface.

76. In English, Brazilian Institute of Education, Science, and Culture.

77. A. C. Souza de Abrantes, "Ciência, educaçao e sociedade: O caso do Instituto Brasileiro de Educaçao, Ciência e Cultura (IBECC) e da Fundaçao Brasileira de Ensino de Ciêncies (FUNBEC)" (PhD diss., Fiocruz, Rio de Janeiro,

2008); interview with Isaias Raw by Josep Simon, Butantan Institute, São Paulo, July 25, 2017.

78. Uri Haber-Schaim, "The Use of PSSC in Other Countries," *ESI Quarterly Report, Winter–Spring 1964*, MC602, box 5.

79. Souza de Abrantes, "Ciência, educaçao e sociedade."

80. Sander C. Brown, N. Clarke, and Jaime Tiomno, eds., *Why Teach Physics? Based on Discussions at the International Conference on Physics in General Education* (Cambridge, MA: MIT Press, 1964).

81. National Science Foundation, *Thirteenth Conference on Coordination of Curriculum Studies, Washington, D.C., May 13–14, 1965*, 2, 6–7, PSSC Records, MC626, box 10, folder "NSF Curriculum Conference 1966," MIT Archives; Finlay, "Physical Science Study Committee"; letter from E. Waril to Uri Haber-Schaim, Feb. 16, 1962, PSSC Records, MC626, box 1, folder "PSSC International Inquiries and Reaction, 1961–1962," MIT Archives.

82. We should note that the Educational Research Information Center was planned between 1959 and 1963 and established in 1964 by the US Office of Education.

83. National Science Foundation, *Thirteenth Conference on Coordination of Curriculum Studies*, 2, 6–7.

84. Address list appended to letter from Francis Friedman to Alan N. Holden, May 1, 1961, PSSC Records, MC626, box 11, folder "Correspondence 1961–1964," MIT Archives.

85. Letter from Robert I. Hulsizer to Jerrold R. Zacharias, Mar. 25, 1956, and letter from Augusto Moreno y Moreno to Jerrold R. Zacharias, Feb. 27, 1965, PSSC Records, MC626, box 11, MIT Archives.

86. This strict division applied to the original pilot projects but was progressively dissolved by the interaction and reciprocal feedback of the different disciplinary projects developed in each continent and the requirements of the local teachers. UNESCO, *UNESCO and Science Teaching* (Paris: UNESCO, 1966), UNESCO Archives.

87. Patrick Petitjean, "Defining UNESCO's Scientific Culture, 1945–1965," in *Sixty Years of Science at UNESCO, 1945–2005*, ed. Patrick Petitjean, V. Zharov, G. Glaser, J. Richardson, B. Padirac, and G. Archibald (Paris: UNESCO, 2006), 29–34.

88. S. E. Graham, "The (Real)politiks of Culture: U.S. Cultural Diplomacy in Unesco, 1946–1954," *Diplomatic History* 30, no. 2 (2006): 231–51.

89. Albert Baez Papers, Stanford University Archives, Stanford, CA.

90. As Baez was a specialist in x-ray optics and Joel was a crystallographer (with a PhD from London supervised by J. D. Bernal), it is likely that they had met or corresponded earlier for scientific research purposes. A few years later the team was expanded with a French physicist (Thérèse Grivet), a Belgian chemist (Robert Ganeff, who had previously led the science projects of the

OEEC), a Romanian biologist (Anne Hunwald), and an Indian biologist (Rachel John). UNESCO, *UNESCO and Science Teaching*, 6–7.

91. Robert H. Maybury, "From Model, to Colleague, to Friend: Honoring the Memory of Albert V. Baez (1913–2007)," accessed Feb. 7, 2017, http://auhighlights.blogspot.com/2007/03/from-model-to-colleague-to-friend.html.

92. Albert Baez, "The Early Days of Science Education at UNESCO," in Petitjean et al., *Sixty Years of Science at UNESCO*, 176–181. A longer version of Baez's account is preserved at UNESCO's archives.

93. Albert V. Baez, *Pilot Project in Physics Teaching* (Paris, 1964), 16–17, UNESCO Archives.

94. Baez, "The Early Days of Science Education at UNESCO."

95. Exceptions are Ross Bassett, *The Technological Indian* (Cambridge, MA: Harvard University Press, 2016); Zuoye Wang, "Transnational Science during the Cold War: The Case of Chinese/American Scientists," *Isis* 101, no. 2 (2010): 367–377.

96. Letter from Philip Morrison to Edna S. Alexander, Feb. 9, 1960, 1.

97. J. P. Sewell, *UNESCO and World Politics: Engaging in International Relations* (Princeton, NJ: Princeton University Press, 2015).

98. Division of Science Teaching—UNESCO, *Guidelines for a Massive World-wide Attack on the Problems of Science Teaching in the Developing Countries through the Use of New Approaches, Methods and Techniques* (Paris, 1965), 17, UNESCO Archives.

99. *Biographisches Handbuch der deutschsprachigen Emigration nach 1933*, vol. 2 (Munich: K. G. Saur, 1983), 565; A. Cohen, *Israel and the Bomb* (New York: Columbia University Press, 1998), 11, 36–37; M. Karpin, *The Bomb in the Basement: How Israel Went Nuclear and What That Means for the World* (New York: Simon and Schuster, 2006), 37.

100. After his UNESCO position, Baez ended up settling in California. There he was vindicated as a Hispanic figure by Hispanic science and engineering associations, and he also contributed to them. He was the president of Vivamos Mejor/USA, a philanthropic organization for the aid of the poor in Mexico.

101. F. Reimers, "Albert Vinicio Baez and the Promotion of Science Education in the Developing World, 1912–2007," *Prospects* 37 (2007): 369–381.

102. This is one of Rudolph's arguments (Rudolph, *Scientists in the Classroom*).

103. This might seem merely a minor change in form, but it is in fact a relevant one in terms of potential uses and teaching practices.

104. Division of Science Teaching—UNESCO, *Guidelines for a Massive World-wide Attack on the Problems of Science Teaching*, 17.

PART IV

The Nuclear Regime

CHAPTER TWELVE

Technical Assistance in Movement

Nuclear Knowledge Crosses Latin American Borders

Gisela Mateos and Edna Suárez-Díaz

> A man walks down the street
> It's a street in a strange world
> Maybe it's the Third World
> Maybe it's his first time around
> He doesn't speak the language
> He holds no currency
> He is a foreign man
> He is surrounded by the sound, the sound
> Cattle in the marketplace
> Scatterings and orphanages
> —"You Can Call Me Al," Paul Simon (1986)

Introduction: The Materiality of Travel

The transnational approach is synonymous with the language of movement, circulation, and flows across borders and between nodes in a network. Notwithstanding this focus on movement in this genre of historical analysis, we notice the absence of attention to *travel* itself, as if the movement from one place to another is unproblematic.[1] In this chapter we fill this lacuna. We concentrate on the materiality of movement and the intricate, different kinds of networks, contacts, and flows that make travel possible. By tracing the itinerary of a mobile radioisotope laboratory as it meandered through several Latin American countries, we highlight the challenges faced not only in crossing borders but in traveling from one town to another within any one country. These challenges were not simply bureaucratic: on the contrary, they were precipitated by an inability to imagine what travel in a "developing" country entailed, by a divergence of cultural norms and expectations between local officials and those in an international organization in Europe, and by the vagaries of nature itself, from earthquakes to floods, and their devastating effects on local infrastructures. People and things don't only move across borders: they travel. Physically crossing space involves planning, money, time, and paperwork, mundane materialities that are ignored at one's peril in a transnational approach.

The postwar period, and in particular the Cold War context, triggered an increasing number of contacts and exchanges between all sorts of agents in the new geopolitical order, particularly the newly created multilateral United Nations agencies. The growth of international markets (once colonial monopolies fell apart) and the availability of air transportation and telecommunications provided the background for scientific and technological exchanges. A sizable proportion of them involved a broad range of technical assistance programs in every area classified as a potential modernizing trigger for a newly conceptualized portion of the world: *underdeveloped* countries.[2] Agriculture, demography, infrastructure, public health, and nuclear technologies are some of the areas where it became expected *to give* and *to receive* development aid. As the large economic and sociological literature on postdevelopment studies has shown, things did not turn out as expected.[3] For historians of science and technology there are still many questions and problems that are unanswered and ignored. How, and by which economic and material means, were people—scientists and technicians—materials, and instruments moved within and across national borders? Who facilitated these movements on the ground? Which natural, political, and mundane administrative obstacles stood against this flow, and how did they affect the appropriation and adaptation of science and technology? What happened when so-called "recipient countries" were not willing to receive the supposed benefits of science and technology? Such questions are not futile or superfluous when science in movement is framed as part of technical assistance programs, an often-forgotten facet of the travel of knowledge in the second half of the twentieth century. We are fully aware that experts, both local and from neighboring countries, participated in the travel of knowledge and practices connected to radioisotope laboratories, and we will mention some of these actors in the present chapter. However, we will not deal, in depth, with scientific practices and roles involved in the interactions between international, regional, and local training experts and trainees.

By focusing on the International Atomic Energy Agency's (IAEA) Mobile Radioisotope Exhibition (MRE), we aim to address the travel of knowledge in the context of asymmetrical, nonreciprocal exchanges between countries, as embodied by the participation of multilateral agencies in technical assistance programs. To move the radioisotope techniques through Latin America, each of the national atomic commissions, the IAEA planners, and the experts and scientists involved strongly relied on the financial resources set up by the UN development machinery, through the Expanded Program of Technical Assistance (EPTA), and the expertise of its international functionaries in each country, the resident representatives of the Technical Assistance Board (UNTAB).[4] Thus, the *transnational*, in our story, is not an abstract analytic tool. It is the embodiment of the crossing of national borders through different geographic and natural accidents (like the Andes mountains or a flood), power and infrastructure asymmetries, and paper technologies specific to administrative histories. In addressing these issues, the internationalization of science and technology becomes the collective endeavor of local actors and international functionaries to move a rather rigid structure (a truck) containing a set of standardized instruments and materials through the troubled and eventful roads of *Third World* countries. It also refers to the tension between the needs and interests of local and international actors to discipline nature and technology and to the specificities and resistances of the movement within and between each country. Such an approach shows the immense adaptability and stability of the scientific practices that traveled in the context of technical assistance programs. Moreover, the *transnational*, as performed in technical assistance programs, depended on the new administrative technologies of *development planning*.[5]

The two IAEA-MRE trucks (Unit 1 and Unit 2) had been donated in 1958 by the US Atomic Energy Commission for training purposes on the several applications of radioisotopes in the context of the Atoms for Peace campaign and the ensuing creation of the IAEA in 1957.[6] Technical assistance was classified as a priority for the international agency and seen as a key instrument to shape, standardize, and control the uses of atomic energy around the world. It was also understood as a mechanism to open potential markets for the new atomic technologies and to promote the beneficial side of atomic energy to broader audiences. As such, nuclear technical assistance was modeled after the idea of stages of development, where basic radioisotope techniques (preparation, dilution, measurement) and their "everyday" applications in medicine, industry, and agriculture were seen as the first step up the nuclearization ladder, to be followed by the construction and use of research reactors, and culminating with the acquisition of power reactors.[7] The introductory technoscientific practices of radiochemistry, nevertheless, were meant to demonstrate a nation's modernity, and its possible future. We are not talking here of monumental nation-building technologies, where pride and prestige were at stake. We are talking of the mundane radioisotope techniques (dull, unremarkable, repetitive) to be used in industrial quality testing, in veterinary and dental clinical settings, and in medical therapies in middle-range hospitals. Moreover, the two trucks embodied not only science, technology, and modernization but also the deep symbolism of the "friendly atom."

In what follows we describe the real and perceived difficulties of moving the MRE Unit 2 truck through six Latin American countries (Mexico, Argentina, Uruguay, Brazil, Bolivia, and Costa Rica) as a good and localized example of the materiality of travel.[8] We focus our account on the logistics of crossing each of the national borders the International Harvester truck trespassed. In doing so, we pay attention to the actors involved in this movement (frequently made invisible in traditional accounts), and in particular the Viennese driver, Josef Obermayer, and the Argentinian physicist, Arturo E. Cairo, the acting director of Training and Exchange Programs at the IAEA.

The Travel of MRE Unit 2 across Latin America

To move the MRE around recipient countries required a sizable amount of the IAEA's United Nations Expanded Program budget, amounting to an estimated cost of US$16.00 per kilometer and a total cost of approximately US$123,900 from 1960 to 1965, or 0.8% of the total budget for technical assistance at the IAEA. To implement it, the agency relied on its bureaucracies in the Vienna headquarters and on local personnel, including two Viennese drivers, as well as trained technicians who enjoyed the agency's trust and had been given the responsibility and duty to report regularly on the trip. Starting in 1958 the IAEA also relied on the UN EPTA fund and UNTAB resident representatives in each country.[9] This mass of human and financial resources constituted a complex network that set the heterogeneous conditions for the movement of scientists, engineers, instruments, and materials.

For MRE Unit 2, crossing the borders between the United States and Mexico, Mexico and Argentina, Argentina and Uruguay, Uruguay and Brazil, and Brazil and Bolivia, to its storage depot in Costa Rica involved a number of highly specific challenges, not to mention the cancellation of the trip to Chile after the disastrous 1960 Valdivia earthquake.[10] Moreover, the itinerary itself kept changing, despite dozens of last-minute fixes and administrative arrangements and interventions. The itineraries "on paper" never seemed to consider actual times on the ground, and events and delays at one stop caused a domino effect later down the road.

The MRE trucks measured 10.5 meters long, 3.4 meters high, and 2.4 meters wide and weighed approximately thirteen tons. They had been assembled and equipped at Oak Ridge National Laboratory in Tennessee. They included a small chemistry laboratory and radiation counting room, with basic instrumentation including Geiger-Müller counters, centrifuges, and glassware. As such, they embodied the knowledge and practices of standardized radioisotope techniques, such as methods for radiation counting, dilution, and biological and medical tracing. As Nicolas Dew says, "there is no science without metrology."[11] Standardized practices are required in order to have common ground for scientific practice, but standardization relies on movement. After lengthy deliberations, the IAEA decided that a couple of drivers from the IAEA's staff needed to be commissioned for such a crucial task.

It proved to be a huge challenge to move an International Harvester truck on the Latin American roads and railroads and even onto ships and into ports.[12] A chauffeur with the type of qualifications required warrants at least a G-4 salary. The driver must have unusual driving skill to handle such a large vehicle since it is comparable to a large bus, be qualified to act as a mechanic, and have a keen sense of responsibility for the vehicle itself, including its cleanliness and day-to-day maintenance and operation. The person would also have to be willing to travel with the vehicle and be responsible for it twenty-four hours a day.[13]

Oak Ridge, Tennessee, to Nuevo Laredo, Mexico

The beginning of the trip, originally planned for December 23, 1959, was delayed a fortnight, to early January the next year. The delay was caused by the Christmas holiday season, a telling sign of the unrealistic planning elaborated by out-of-touch officials. William Pope, an electronics technician at the Oak Ridge Institute for Nuclear Science, had been hired by the IAEA for the Mexican part of the itinerary. Josef Obermayer, a bilingual (German and English) professional driver, was picked by the IAEA to take the truck through Latin America by road. He departed with Unit 2 from Oak Ridge in Tennessee to cover the almost two thousand kilometers to the Mexican border city of Nuevo Laredo. On January 5, the MRE crossed the US-Mexico border.

A young Mexican Chinese physicist, Eugenio Ley Koo, was waiting for Obermayer and Pope; his role was to act as a translator and also as the professor in charge of the radioisotope training courses. Between January and April that year, the Mexican stops included midsized cities at the center of the country—Monterrey, San Luis Potosí, Guanajuato, Guadalajara, Puebla, Mexico City, and Veracruz—where the peaceful uses of atomic energy were promoted. Arguing for the need to increase governmental funds for the development of nuclear science and technologies, a group of Mexican scientists and promoters (including Nabor Carrillo, rector of the National University of Mexico) took advantage of the IAEA exhibition and laid out a program of conferences and related shows in Mexico City that highlighted nuclear energy as a modernizing technology.

If crossing the US-Mexico border was easy (later in the trip Obermayer would recall the steak and beer he had enjoyed in Monterrey), the next part of the journey soon became a challenge. No direct shipment route existed between the Gulf of Mexico ports and Argentina (Buenos Aires). The Pan-American Highway did not go that far (and it still does not). Thus, Unit 2 of the MRE had to be taken back to New Orleans in the United States. In a letter sent from Mr. Cairo in Vienna to Mr. Adriano Garcia, he stated, "They [the Flota Argentina de Navegación de Ultramar] have suggested shipping from New Orleans, but it is also possible that they will have one of their ships deviate from its regular route to collect the mobile laboratory in Veracruz."[14]

Tampico, Mexico, to New Orleans, Louisiana, to Buenos Aires, Argentina

The truck was eventually scheduled to leave the port of Tampico for Buenos Aires on April 18. The unit was loaded on the *Lancero*, owned by the Flota Argentina de Navegación de Ultramar. The cruise took almost three weeks, passing north to New Orleans and then south to Argentina. Although the precise date is not clear, by May 10 the MRE was in Buenos Aires.

A technical problem that would come up again and again throughout the trip was the lack of a constant voltage and electric power. This fact had an impact on the itinerary itself, which was restricted to (mostly) electrified areas, and also on the instruments' performance. A telegram sent by Obermayer to his boss Cairo on June 15, 1960, reads: "Please send urgent approval to buy transformer stabilisator [*sic*] power supply big problem apr. cost 200 ds expl letter on the way. Josef Obermayer." In fact, Obermayer was confronting a problem that arises whenever one assumes that technology developed in the Western world will work "anywhere." As Joseph O'Connell has said of the US Navy, it "has found that it cannot set up an overseas base simply by sending ships, airplanes, bullets and soldiers. None of these can move freely into a new setting unless the Navy first sends the volt, the ohm, the metre, and other standards ahead to prepare the way." The IAEA, like the Department of Defense, found that scientific equipment "cannot move into new settings for long unless the setting has been prepared by rendering certain variables similar with respect to where the equipment was produced, and stable with respect to time."[15] Stable power supplies were needed for the MRE to manufacture reproducible, "universal" scientific results.

Another revealing fact about the asymmetric conditions between development planners and actors on the ground was related to Obermayer's salary and per diem (stipend) during the Argentinian journey. The recipient country had the obligation to pay half of his salary, which had been set according to the UN pay scale. However, as those in charge of the radioisotope courses in Buenos Aires claimed, there was a major disparity between Obermayer's salary, not to mention his per diem, and those of local scientists:

> Concerning the *per diem* topic . . . those received by Obermayer [1,000 Argentinian pesos per day] are of the same amount as those assigned to you,

> and to engineer Buchler, to whom I ask you to send my greetings. I want to remind you that our people do not receive the compensatory and comfortable back up of a salary in US dollars, not even one like Obermayer's, but between 6- and 8,000 [Argentinian] pesos per month. As a consequence, they don't feel their professional pride hurt, but rather their wallet.[16]

To avoid unpleasant quarrels, the IAEA decided to move the driver's salary to the UNTAB account and, more important, change his assignment to a "technical expert" post.[17]

In Argentina, the MRE traveled from Buenos Aires to Mendoza and then to Cordoba city. During this journey, the IAEA continued to push for the MRE to visit as many countries as possible, in order to optimize costs and travels. Thus, despite not being an IAEA member state, Chilean officials (Chile had no national atomic commission as yet) negotiated a MRE visit after their neighbor's. No sooner had the visit been approved than the dramatic Valdivia earthquake of May 1960 disrupted the Andean pass between Argentina to Chile:

> the road from Argentina to Chile through the Andes at the point called "Las Cuevas" cannot be utilized as it is not suitable for the vehicle. The other road connecting Argentina and Chile, at the south, is a good road but it is not possible to travel from south Chile to Santiago due to the last earthquake.[18]

Taking the truck by sea was economically unfeasible, given the other urgent priorities that the Chilean government now had. The MRE never made it to Chile: getting it there was an insurmountable hurdle.

Buenos Aires, Argentina, to Montevideo, Uruguay, to São Paulo, Brazil

Back in Buenos Aires, in November 1960, the truck was stored until its next stop in Uruguay. A new surprise was in store. At the end of this month, Dr. Hernán Durán, the UNTAB resident representative in Uruguay, wrote to Dr. Cairo declining the use of the MRE, arguing that the government was no longer interested. This change in itinerary was angrily received at the IAEA headquarters, with Cairo answering Durán's letter in an excited mood:

> I am very surprised to hear that Uruguay was no longer interested in the use of the mobile laboratory. This decision would create a most unfortunate situ-

> ation for the Agency . . . [since] all other activities have been planned in consideration of this request. Moreover, as the Technical Assistance Board has already allocated a certain amount of money for the laboratory's visit to Uruguay, it would be most unwise not to utilize the sum for this purpose.[19]

It was not until March 1961 that the University of Montevideo again showed an interest in receiving the MRE. Things had to be accelerated since, in the meantime, Brazil had committed to receive the truck on June 1. Obermayer, who at this point was back in Vienna, had to fly back to Buenos Aires, where the truck had been stored. More problems awaited him when he got there. The highway connecting Buenos Aires with Montevideo was out of order because of the season's heavy rains and the floods that had swollen the upper Paraná River. The Argentinean UNTAB representative explained the consequences:

> With the present state of the road the trip-some 200 km.- will take at least 10 to 12 days and during all this time an officer of the Argentine and Uruguayan customs have to be aboard the truck to certify that no piece has been taken out while in transit. The fees for these inspectors, plus their per diems and travel costs go heavily into money.[20]

This new obstacle was avoided, and the truck finally crossed on a boat, arriving in Uruguay, as expressed in Obermayer's letter to Cairo:

> Well, I finally made it from B. Aires to here, and not by Road but by Boat. . . . so I had to wait for a Boat since the Ferry boat to Colonia could not take a truck of this size. Anyway I get over here on the 22.4. and yesterday got the truck out of custom, which was quite a problem.[21]

For the next few weeks, from May to June 5, radioisotope courses were taught in Montevideo by Argentinean experts who had traveled to offer "technical assistance" to their more unprepared neighbors. Things ran smoothly until the MRE had to be transported to the next Latin American country. The question of how to move Unit 2 from Montevideo to Brazil posed a difficult dilemma: by sea or by road? The latter option was dropped because of the almost two thousand kilometers that Obermayer would have to drive and the high costs it entailed. After a very careful inquiry, a sea route seemed the more suitable option:

> It is possible to go by road to Rio from here, but according [to] the automobil [*sic*] club, the road is pretty bad in parts. Transport by railroad also is possible, but it seems it may take quite some time. Now the safest bet would be your guess, to go by Boat to Santos and from there by Road to Rio.[22]

As these plans were evolving, Marcello Damy Souza Santos, chairman of the National Nuclear Energy Commission (Brazil), sent a telegram to Cairo in Vienna. Souza Santos sought to cancel the MRE visit to his country, arguing that they had already established their own program on nuclear energy.[23] Once again, Cairo made a strong case to dissuade the Brazilians, who finally agreed to have the MRE for six months. On August 1, Obermayer and the MRE were on board the *Cap Palma* (fig. 12.1), which took them to the port of Santos Rio, where they arrived on August 10.

The Brazilian journey started off badly. On arriving at Santos Rio, Obermayer learned that the port had been paralyzed by a dockworkers' strike, and it took four days before the truck was offloaded. Then he had the Brazilian customs paperwork to navigate, Obermayer complain-

FIGURE 12.1. Loading MRE Unit 2 on the *Cap Palma* at Montevideo, Uruguay, 1960. *Source*: IAEA Archives, Vienna.

ing that now "something new, the Custom office wants a price on all the items of the inventory list."[24] The weather was a problem too, "[s]ince the temperature, in some of the places here is quite a lot (Santos 40°), here in R. Preto about the same, I like to thank you for the ventilador [*sic*], which helps a lot."[25] Moreover, at this point in the journey, it was not even clear *how and by whom* the MRE was to be used while in Brazil. To add to the confusion, Souza Santos was out of the country, and so "Mr. Vidal," a public relations officer from the National Nuclear Energy Commission, with no familiarity with radioisotopes whatsoever, and Obermayer were left alone to arrange the release of the truck from the customs officers.

Political problems complicated matters further. At the beginning of 1961, Jânio da Silva Quadros had been elected president of Brazil, only to resign a few months later, on August 21. The country was immersed in political turmoil, which delayed any decisions related to the MRE. More specifically, Obermayer's per diem was halted due to the closing of the banks. On August 23, the truck was finally given the green light to leave the *alfândega* ("customs" in Portuguese).

During the first weeks of October, the truck was on the road to São Paulo, even though the purpose of the exhibition was still not clear to anyone. Rather than teaching courses on radioisotopes, Unit 2 was used as propaganda for the Brazilian National Nuclear Energy Commission, to demonstrate the peaceful uses of atomic energy, and for specific research on thyroid diseases that had been explicitly requested, since the uses of radioactive iodine in diagnostics were well established. It was clear that, for Brazilian experts, the MRE had nothing new to offer. Improvisation, however, continued to rule the day. While Unit 2 was in Goiana, the rainy season arrived. Obermayer described the situation to Cairo:

> Dr. Lobo wanted to go and I told him, with the Unit now its impossible since the rain season started and the roads are very bad. So he get us an ambulance and we loaded some equipment as Scintillation Spectrometer, Zentrifugue [*sic*] and so on, in there and where [*sic*] travelling with it, via Rio, Sao Paulo, Ribeirao Preto, Araxa, Uberaba, Uberlandia, Araguari back here! Tomorrow I take the Unit to Inhumas where we stay a few days, and then we go by ambulance again to Goias Velho and a few small places around there. So you see we are cruising around a lot here.[26]

From Brazil's Alfândega *to Santa Cruz, Bolivia*

Notwithstanding his multiple adventures, nothing had prepared Obermayer on the ground, and Cairo in Vienna, for the trip from Brazil to Bolivia. After an extension of the visit in Brazil, from January to the end of March 1962, Unit 2 had been stored in Rio de Janeiro for more than a year. Meanwhile, before sending new invitations to more countries, the IAEA had asked for extra EPTA funds to cover the unexpected expenses of the truck's transportation. From Cairo's office in Vienna new invitations, offering cheaper conditions, were sent to more Latin American countries. The mobile exhibition seemed more affordable this time, and the initial response was good. However, by January 1963, the IAEA was informed that no more EPTA funds were to be provided for the mobile laboratories because of competing priorities between UN agencies. Bolivia, however, did not reconsider, and local officials agreed to pay for the transportation, half the per diem for the experts and for Obermayer, but not the experts' salaries: "As you can understand, these conditions make the IAEA's proposal practically unacceptable for countries with reduced economic media as ours."[27]

Finally, to get the MRE moving, the IAEA agreed to pay the total sum of the experts' salaries, who would travel from Argentina to teach radioisotope techniques. Different itineraries were proposed to move the MRE from Rio de Janeiro to Bolivia. All customs paperwork was to be done by the Bolivians. Radioisotope courses were scheduled for four different cities in the country: Santa Cruz, Cochabamba, Oruro, and La Paz. In an enthusiastic letter, Obermayer described the planned trip to the Vienna headquarters:

> Then I saw a Mr. TORRES the AGENTE COMERCIAL of the COMMISSÃO MISTA FERROVIARIA BRAZILEIRO-BOLIVIANA, who told me some good news! First it is possible to ship the truck via train, either from Sao Paulo or Bauru (shorter distance) via Corumbá (border) to Santa Cruz. The payment must be made at Sao Paulo up to the border and it can be arranged to pay the Bolivians share at Bolivia (Santa Cruz). The train go from Sao Paulo dayly [*sic*] to Corumbá and from there every Friday to La Paz. Wednesday next week I will get by phone the prices from Sao Paulo. The trip takes about 1 week and since it is a freight train, I believe it is advisable to fly to La Paz after loading at Bauru, to arrange for papers and payments at the other end, ok?[28]

Nice plans! But they clashed with events on the ground. One month later, Obermayer was still in Rio de Janeiro, trying to obtain the necessary documents from the *alfândega*.

> Dear Mr. Cairo!
>
> I would like to give you a short report about what happened up till now. As I wrote you last time, it is possible to ship the truck by train, and as a matter of fact about the only way at the moment, because the Rio Parana is heavy flooded and roads are closed. I have been after Mr. Vidal dayly for the necessary transit papers, but since this week there is a change in government, everything is slowed down more than usual. Also I got in touch again par telephone with the Comissão in Sao Paulo at the 19.6 and they promised to call back the exact price of shipping via UN-Office Rio, but nothing happened. After urging the UN-office to try again, we couldn't get connections with Sao Paulo, so that means wait until Monday.[29]

Days later, when Obermayer drove the truck from its storage place in Rio to Bauru, the rear axle broke and his departure was delayed once again.[30] Now quite pessimistic, the hapless driver complained about what he called "Brazilian time." This was the most challenging part of the trip, and much was to happen before he eventually arrived at Santa Cruz (fig. 12.2).

Once again Obermayer faced the problem of how to upload the truck on a train, and improvisation was the only way to get ahead:

> Well, at Bauru the railroad people did not like to take the transport at all on account of the hight [*sic*] of the truck. Just after a few hours of on going back and forth we decided to fit blocks under the frame so the truck could not sway while transported. Then we found the loading ramp to [*sic*] small and only for small cars. So around we went again and finally found a place at the Sorrocabana [*sic*] station. Then they sent a very old railcar with cracked boards where I refused to load on. So more arguments and this by 36°C. Then in order not to get stuck with the rear end of the truck, it needed a lot of boards to left the truck up, and those boards seems to be more precious as gold here. By this time I was so disgusted with all this fiddling around, I was about ready to go home! Finally in the afternoon we got the truck loaded, with a lot of scraping, yelling and confusion all around. Some characters in front yelled "vamos"

FIGURE 12.2. MRE Unit 2 on the train from Rubiacea to Corumbá, Brazil, 1963.
Source: IAEA Archives, Vienna.

> some in the rear the opposite, and after the truck was on the flatcar there where [*sic*] about 2 zentimeters? at each side missing![31]

The worst was yet to come. Obermayer flew from Rio to Corumbá, where the train was supposed to arrive on July 20, 1963. More than a week later there was still no news whatsoever of the train and its precious cargo. Our driver traveled to the train station at Bauru, where he was told the train would arrive shortly. Desperate to find the train and the truck, he traveled back to Corumbá, where the station chief made his best efforts to locate the train along the line. Then, on July 24 they received a telegram from Rubiacea, a little station a few hours from Bauru, informing them that the train had been involved in an accident. After a heated argument as to why the Corumbá station had not been told about this before, Obermayer traveled to Aracatuba in a frenzy, not without trying to locate the UNTAB resident representative at Rio, Mr. Peter. Telephone lines, however, were out of service.

Two days later, Obermayer arrived in Rubiacea, where he was able to estimate the damage to Unit 2. The direction axle was bent, in addition to there being some exterior damage. Obermayer was now stuck in

Corumbá for almost two weeks, in a place that he considered "the worst place I ever saw, with those millions moskitos you can hardly sleep . . . the telephone line is now broken down again, nobody knows for how long."[32]

Finally, the truck was released and arrived in Santa Cruz, a town on the Bolivian border, early in August. As Obermayer was trying to figure out the best way to drive the truck through Bolivia, a new surprise awaited him. When he traveled from Cochabamba in Brazil to La Paz (while still in the process of making arrangements to pay for repairs to the truck), he realized that the road to the Bolivian capital was too steep and too narrow for the truck. The length did not help either. The twisty road made it impossible for Unit 2 to reach La Paz. Unwilling to abandon the mission, in mid-August some instruments were sent to the city by bus, and a few courses were taught.

The UNTAB representantive in Bolivia, Ms. Margaret Joan Anstee, bitterly complained to IAEA official Cairo that "quite a number of the difficulties could have been avoided if our office had been brought into the picture at an earlier stage."[33] In reply, though he apologized for all the inconvenience, Cairo also argued that "they were practically helpless [in the IAEA] since most of the time we did not even know where we could reach . . . Mr. Obermayer."[34] In the event, the rest of the courses were given at the Faculty of Agronomy in Cochabamba and at the Faculty of Veterinary Science in Santa Cruz, ending on October 25.

Rio de Janeiro, Brazil, to Punta Arenas, Costa Rica

For the next two years, the MRE was stored in Rio de Janeiro, where it had returned at the end of 1963. Eventually, the IAEA donated the truck as a radioisotope laboratory to Costa Rica, as part of the UN special fund Project for the Eradication of the Mediterranean Fruit Fly in Central America. This last move was not an easy one either. The only available route from Brazil to Costa Rica was by sea. The itinerary via New Orleans was too costly. The MRE finally embarked in Rio de Janeiro on a ship specially made available by the Argentinian merchant fleet. The MRE traveled to Punta Arenas on the Pacific coast of Costa Rica, arriving on August 11, and it was unloaded in Santa Cruz port before reaching its final destination at San José.

Reflections on Technical Assistance and the Movement of Science and Technology

The portability of knowledge is a prerequisite for the standardization of science, or metrology. However, this "universalizing," or standardization, of scientific and technical practices is resisted and made visible by the multiple contingencies that constantly reconfigure knowledge itineraries. The history of MRE Unit 2's travel through six Latin American countries, though rough, most certainly was not an exception. It illustrates the many difficulties of reproducing and moving knowledge outside its curated and hygienic original locations. As Paul Simon's song states in our epigraph: "It's a street in a strange world."

The travel of knowledge and scientific practices related to radioisotope manipulations faced "obstacles" and resistances of very different kinds. Some of them had to do with practical considerations, like the size of the truck relative to the narrow roads of Latin America. International Harvester trucks had been designed for US highways and flat landscapes, where gas stations and other infrastructure were readily available. The high, tortuous, and twisty roads in the mountains between Brazil and Bolivia did not lend themselves to passage by such a gigantic truck. Its huge volume and heavy weight also posed an enormous challenge when it was loaded and unloaded from the different ships and trains used to cross national borders. No direct flights or shipping routes existed between the main urban centers and ports. For the IAEA and UN functionaries, and for Obermayer, this situation amounted to a "lumpy" obstacle course, quite at variance with the smooth itinerary they had originally planned for. This was so only because technical assistance planners had imagined an abstract landscape, where mobility was unimpeded in an imagined flat and Westernized space.[35] Latin American infrastructure was not a good fit for traveling the *American way* in those years; indeed, economic and market exchanges were locally limited, and commodities and people barely traveled, except in the very localized areas of US economic influence.[36]

A different type of resistance to the movement of knowledge lay in the lack of interconnectivity that resulted from different bureaucratic traditions (national and otherwise) embedded in customs requirements and from divergent administrative criteria, for instance, concerning the per diem payments for Obermayer and other personnel. This was, indeed, already made clear by the Uruguayans when they told Arturo Cairo of the Brazilians' inclination for excessive paperwork. There was also a disconnect between national and international agencies (typically as seen in the lack of communication and participation of the UNTAB resident representatives in each country). At a more everyday level, what the Latin American actors perceived as harmless "delays" were translated into incomprehensible obstacles, and even backwardness, by the agency's bureaucracy.[37] Even more, administrative paperwork was entangled with telephone, postal, and telegraphic communications hampered by troubled operations, which came to a standstill because of natural or political events. Time also seemed to run at a different pace: Vienna time as expressed in Obermayer's and Cairo's letters was out of sync with the rhythm and contingencies of everyday life and the countless pauses imposed by festivities, workers' strikes, and natural disasters like floods and earthquakes. For all that, Obermayer adapted and managed to overcome most material impediments. It is quite possible that the main reason the MRE kept moving was Obermayer's tenacity and inventiveness.

Still a third type of contingency, evolving from the previous two, lies in the high costs, for "recipient countries," of the transportation and maintenance of Unit 2. Technical assistance programs were meant to be cheap for donors but turned out to be very expensive for recipients, and higher than planned for by the IAEA.[38] In 1966 in the Final Report of the MRE, the agency recognized that it had to bear some of the expenses that some countries had not been able to pay. Moreover, the high cost of transporting the MRE and its precious cargo, and of paying the driver and expert personnel, competed with more pressing priorities for the Latin American countries. This was very obvious in countries like Haiti, which refused the MRE from the start, but also in Chile, which did not manage to receive it after the earthquake that shattered the country. The friendly atom seemed to be an attractive offer on paper—but not attractive enough to those who were supposed to buy/receive it. In the end, half the recipients were countries like Mexico, Argentina, and Brazil that already had nuclear science communities and facilities.

Still, the MRE exhibited interesting positive points for the countries and agents in the Latin American region. Both in Mexico and in Brazil, the exhibition's itinerary served to promote increased funding for local nuclear science and to exalt the values of modernity promoted by scientific elites. Moreover, as David Webster claims, in the polarized context of the Cold War, the UN's specialized agencies were "a relatively acceptable source of technical assistance for many governments."[39] This was certainly the case for Mexico, a country with a perceived aggressive and powerful neighbor but whose governments were equally concerned with their own nationalistic and revolutionary domestic ideologies.[40]

Concerning Latin American science, the MRE itinerary also renders visible regional asymmetries. Argentinean experts were supposed to teach the Uruguayans and Bolivians the basics of radioisotope techniques, and they did so despite all kinds of setbacks. Mexico City scientists were supposed to take their skills to the inner midsized cities. Networks, contacts, and flows are easier when linguistic and cultural barriers are lowered. In pursuing the dream of modernization, local elites were decisive in carrying out the necessary stages, both in their own countries and in those of their neighbors.

Concluding Remarks

What was left behind after the MRE truck passed was not a scientific degree or even a technical certificate on radiochemistry. The truck and the people associated with the MRE left distributed expertise, different quanta of changed local practices, and a set of conditions (a basic "atomic metrology") in which the technologies and practices of the peaceful applications of atomic energy might possibly reproduce and expand in the future. But before science, or even metrology, could possibly be established, the truck had to arrive. This, as we have shown, was by no means an easy task.

To address the actual *movement* of scientific and technological practices in an asymmetric world, we have emphasized what we call the "materialities of travel," in contrast to the model of "circulation" advanced by James Secord's article "Knowledge in Transit."[41] In previous articles, we have criticized both the metaphor of circulation—for suggesting a circular, revolving directionality and an image of downright naturalized "flow"—and Secord's conceptualization of this process as mainly concerning the practices of *communication*.[42] Though in his original article Secord mentioned the importance of "an understanding of the practices of communication, movement and translation," he soon narrowed his inquiry to the first, with crucial and conservative consequences for his whole approach.[43]

By focusing instead on travel (*movement*), and on the planned and altered itineraries that characterize most travelers' experiences, our narrative seeks to reveal the outright limits of the "circulation" of knowledge and—tangentially—to address the *translation* of transnational into local interests. Communication practices certainly are an important aspect of scientific exchanges, but in our case they are embedded in the larger question of the materialities of travel: when the telephone or the telegraph lines are out of order or when a local functionary fails to address a pressing request from the field. Nature, infrastructure, national interests, and pressing economic priorities were key when science and technology traveled as part of technical assistance and development programs during the Cold War. In Secord's account, by contrast, communication practices are salient, and even play a crucial role, because his attention is focused on exchanges playing out in a more or less symmetrical field of power during the Scientific and Industrial Revolutions.

Transnationalism not only interrogates the national frame but also subverts disciplinary boundaries. Indeed, we were often in danger of leaving aside the history of science and technology (a risk that, to be honest, is present in Secord's analysis). Our main actors include a driver and several international functionaries; they are the ones who kept the science of radioisotope applications traveling along minor roads and on ships and railroads. Still, 1,500 students and technicians took the courses offered by the MRE and its personnel on the Latin American trip, more than 130 in Mexico alone. We do not know how many of them actually used and applied the new knowledge in their everyday practices; but certainly, the MRE contributed to a basic "atomic metrology." We hope to have contributed to addressing the main question behind Secord's preoccupations, a question rightfully raised in a classic paper by Adi Ophir and Steven Shapin, "The Place of Knowledge":

> How is it, if knowledge is indeed local, that certain forms of it appear global in the domain of application? Is the global—or even the widely distributed—character of, for example, much scientific and mathematical knowledge an illusion? . . . Perhaps the days in which ideas floated free in the air are truly nearing an end. Perhaps, indeed, what we believed to be a heavenly place for knowledge we will come to see as the result of lateral movements between mundane places.[44]

As may be clear by now, our argument supports not only the relevance of mundane scientific practices and technologies but also "lateral movements between mundane places."

Acknowledgments

We thank John Krige for organizing the workshop "Writing the Transnational History of Science and Technology," held at the Georgia Institute of Technology, Atlanta (November 2–3, 2016), as well as all the participants for their generous comments on a previous version of this chapter. We also thank the generosity of Martha Riess and Leopold Kammerhofer at the IAEA Archive in Vienna. This research was made possible by UNAM-PAPIIT research grant number IN401017.

Notes

1. Examples include John Krige's account of the transnational development of ultracentrifuges and even Zuoyue Wang's treatment of Chinese students in the United States, as well as our recent book on radioisotopes in Mexico. See John Krige, "Hybrid Knowledge: The Transnational Co-production of the Gas Centrifuge for Uranium Enrichment in the 1960s," *British Journal for the History of Science* 45, no. 3 (2012): 337–357; Zuoyue Wang, "Transnational Science during the Cold War: The Case of Chinese/American Scientists," *Isis* 101, no. 2 (2010): 367–377; Gisela Mateos and Edna Suárez-Díaz, *Radioisótopos itinerantes en América Latina: Una historia de ciencia por tierra y por mar* (Mexico: CEIICH-Facultad de Ciencias, UNAM, 2015).

2. "Underdevelopment began . . . on 20 January 1949 [the day President Harry Truman took office]. On that day, 2 billion people became underdeveloped. In a real sense, from that time on, they ceased being what they were, in all their diversity, and were transmogrified into an inverted mirror of others' reality: a mirror that belittles them and sends them off to the end of the queue, a mirror that defines their identity, which is really that of a heterogeneous and diverse majority, simply in the terms of a homogenizing and narrow minority. . . . Since then, development has connoted at least one thing: to escape from the undignified condition called underdevelopment." Gustavo Esteva, "Desarrollo," in *The Development Dictionary: A Guide to Knowledge as Power*, ed. W. Sachs (London: Zed Books, 2010), 1–23, at 2.

3. Arturo Escobar, *Encountering Development: The Making and Unmaking of the Third World* (Princeton, NJ: Princeton University Press, 1994); Esteva, "Desarrollo"; Nick Cullather, "Development? It's History," *Diplomatic History* 44, no. 2 (2000): 641–653.

4. UNTAB was the 1960s successor of the UN Technical Assistance Administration, created in tandem with the United Nations. According to Webster, de-

spite the common localization of its origins in Truman's inauguration address, development programs and UN technical assistance programs were "boosted, but not created, by Truman's 'point four.' Implementation was multilateral from the start." David Webster, "Development Advisors in a Time of Cold War and Decolonization: The United Nations Technical Assistance Administration, 1950–59," *Journal of Global History* 6, no. 2 (2011): 249–272, at 252.

5. See the papers included in *Journal of Global History* 6, no. 2 (2011).

6. We should point out that these gifts had to be accepted by the IAEA, as the United States was the main contributor to its finances. Moreover, the MRE acted as a political instrument to produce obligations for the recipient countries. There is a growing literature on the different meanings and goals of the Atoms for Peace initiative, which was announced by President Eisenhower on December 8, 1953. See John Krige, "Atoms for Peace: Scientific Internationalism and Scientific Intelligence," *Osiris* 21 (2006): 161–181; Martin J. Medhurst, "Atoms for Peace and Nuclear Hegemony: The Rhetorical Structure of a Cold War Campaign," *Armed Forces and Society* 24 (1997): 571–593; Kenneth Osgood, *Total Cold War: Eisenhower's Secret Propaganda Battle at Home and Abroad* (Lawrence: University Press of Kansas, 2006).

7. On radioisotopes as instruments of US foreign policy, see Angela Creager, "Tracing the Politics of Changing Postwar Research Practices: The Export of 'American' Radioisotopes to European Biologists," *Studies in the History and Philosophy of Biology and Biomedical Sciences* 33 (2002): 367–388; Angela Creager, "Radioisotopes as Political Instruments, 1946–1953," *Dynamis* 29 (2009): 219–240; John Krige, "The Politics of Phosphorus-32: A Cold War Fable Based on Fact," *Historical Studies in the Physical and Biological Sciences* 36, no. 1 (2005): 71–91; Gisela Mateos and Edna Suárez-Díaz, "Clouds, Airplanes, Trucks and People: Carrying Radioisotopes to and across Mexico," *Dynamis* 35, no. 2 (2015): 279–305.

8. The itinerary followed by Unit 1 in Europe, Asia, and Africa will be addressed in a forthcoming article where we describe the MRE as one of the IAEA's first attempts to implement nuclear technical assistance in the context of the UN development program.

9. See Gisela Mateos and Edna Suárez-Díaz, "Atoms for Peace in Latin America," *Latin American History: Oxford Research Encyclopedias*, 2016, doi:10.1093/acrefore/9780199366439.013.317.

10. The Valdivia earthquake shattered Chile on May 22, 1960, and continues to be classified as the most powerful earthquake ever recorded.

11. Nicholas Dew, "*Vers la Ligne*: Circulating Knowledge around the French Atlantic," in *Science and Empire in the Atlantic World*, ed. James Delbourgo and Nicholas Dew (New York: Taylor and Francis, 2008), 53–72, at 56.

12. The decision-making behind the selection of two International Harvester trucks to host the MRE is not addressed in the files available to us at the IAEA

Archive. The company had several subsidiaries in Latin American countries and Asia, probably providing an argument for the availability of spare parts and mechanical services while there. But what seems to be most important is that Oak Ridge National Laboratory had organized the previous US Atoms for Peace Mobile Campaign starting in 1956, and for that, they equipped Harvester trucks too.

13. A. E. Cairo to L. Steining (chief administrator for IAEA Technical Assistance Programs), Jan. 11, 1960. The source for all letters cited in this chapter is folder SC/216-LAT-1, IAEA Archive, Vienna.

14. Cairo to Adriano García (UNTAB resident representative in Mexico), Feb. 18, 1960.

15. Joseph O'Connell, "The Creation of Universality by the Circulation of Particulars," *Social Studies of Science* 23, no. 1 (1993): 129–193, at 163.

16. Celso Papadopulos (in charge of the radioisotope courses) to Cairo, Oct. 12, 1960.

17. Bruno Leuschner (UNTAB resident representative in Argentina) to L. Steining, May 13, 1960.

18. A. M. Haymes (administrative assistance, UNTAB, Argentina) to Cairo, Oct. 18, 1960.

19. Cairo to Hernán Durán, Dec. 7, 1960.

20. Bruno Leuschner to Steiner, Apr. 17, 1961.

21. Josef Obermayer (in Montevideo) to Cairo, Apr. 30, 1961.

22. Obermayer (in Montevideo) to Cairo, June 28, 1961.

23. Telegram from Souza Santos to Cairo, May 12, 1961. On the energetic Brazilian atomic program, see Carlo Patti, "The Origins of the Brazilian Nuclear Programme, 1951–1955," *Cold War History* 15, no. 3 (2015): 353–373.

24. Obermayer (in Rio de Janeiro) to Cairo, Aug. 17, 1961.

25. Obermayer (in Riberão Preto) to Cairo, June 11, 1961.

26. Obermayer (in Goiana) to Cairo, Feb. 17, 1962.

27. Professor Ismael Escobar to Cairo, Nov. 17, 1962.

28. Obermayer to Cairo, June 13, 1963.

29. Obermayer (in Rio de Janeiro) to Cairo, June 21, 1963.

30. There is no information regarding the consequences of storage while in Latin America, though the truck ended up spending almost two years at warehouses.

31. Obermayer (in Corumbá) to Cairo, July 22, 1963.

32. Obermayer (in Corumbá) to Cairo, Aug. 2, 1963.

33. M. J. Anstee to Cairo, Sept. 6, 1963.

34. Cairo to Anstee, Oct. 2, 1963.

35. On landscapes and geographical accidents, also see Itty Abraham, "Landscape and Postcolonial Science," *Contributions to Indian Sociology* 34, no. 2 (2000): 163–187; Itty Abraham, "The Contradictory Spaces of Postcolonial Techno-Science," *Economic and Political Weekly* 41, no. 3 (2006): 210–217; Nick

Cullather, "Damming Afghanistan: Modernization in a Buffer State," *Journal of American History* 89, no. 2 (2002): 512–537; Penny Harvey and Hanna Knox, *Roads: An Anthropology of Infrastructure and Expertise* (Ithaca, NY: Cornell University Press, 2015).

36. Harvey and Knox, *Roads*; Mateos and Suárez-Díaz, *Radioisótopos itinerantes.*

37. This is so despite Cairo being an Argentinian who participated in different missions in Latin America and Asia to evaluate the state of nuclear development and the needs of "recipient countries."

38. Gisela Mateos and Edna Suárez-Díaz, "Expectativas (des)encontradas: La asistencia técnica nuclear en América Latina," in *Aproximaciones a lo local y lo global: América Latina en la historia de la ciencia contemporánea*, ed. Gisela Mateos and Edna Suárez-Díaz (Mexico City: Centro de Estudios Filosóficos, Políticos y Sociales Vicente Lombardo Toledano, 2016), 215–241.

39. Webster, "Development Advisors," 249.

40. Renata Keller, *Mexico's Cold War: Cuba, the United States and the Legacy of the Mexican Revolution* (Cambridge: Cambridge University Press, 2015); Gisela Mateos and Edna Suárez-Díaz, "We Are Not a Rich Country to Waste Our Resources on Expensive Toys: The Mexican Version of Atoms for Peace," in "Nation, Knowledge, and Imagined Futures: Science, Technology and Nation Building Post-1945," ed. John Krige and Jessica Wang, special issue, *History and Technology* 31, no. 3 (2015): 243–258.

41. James A. Secord, "Knowledge in Transit," *Isis* 95, no. 4 (2004): 654–672.

42. Mateos and Suárez-Díaz, "Expectativas (des)encontradas"; Mateos and Suárez-Díaz, *Radioisótopos itinerantes.*

43. Secord, "Knowledge in Transit," 656.

44. Adi Ophir and Steven Shapin, "The Place of Knowledge: A Methodological Survey," *Science in Context* 4, no. 1 (1991): 3–22, at 16.

CHAPTER THIRTEEN

Controlled Exchanges

Public-Private Hybridity, Transnational Networking, and Knowledge Circulation in US-China Scientific Discourse on Nuclear Arms Control

Zuoyue Wang

Beijing, May 18, 1988. Wolfgang "Pief" Panofsky, emeritus director of the Stanford Linear Accelerator Center (SLAC), had breakfast with T. D. Lee, the prominent Chinese American physicist from Columbia, before heading to the Institute for High Energy Physics (IHEP) of the Chinese Academy of Sciences. Panofsky was a pioneering particle physicist who, upon Lee's recommendation, had been appointed the official adviser to the Chinese government on its building of the Beijing Electron-Positron Collider (BEPC) at the IHEP. He had come to China to check on BEPC's progress and make recommendations to top Chinese leaders.[1]

But Panofsky also had an ulterior motive for the trip. He had participated in the Manhattan Project during World War II and had served as an experienced adviser to the US government in the postwar years. His experiences led him to become a passionate advocate for nuclear arms control, especially in his capacity as chairman of the US National Academy of Sciences (NAS) Committee on International Security and Arms Control (CISAC, pronounced "see-sak"). CISAC had been established in 1980 and had carried out fruitful back-channel face-to-face discussions with Soviet scientists in the 1980s on nuclear arms control.[2] Now

Panofsky wanted to see whether he could start a similar dialogue with scientists in China, whose government had yet to join any international nuclear arms control agreements. Frank Press, who had served as President Jimmy Carter's science adviser and was now president of the NAS, had arranged with Zhou Guangzhao, his counterpart as the president of the Chinese Academy of Sciences, for Panofsky to meet with a group of Chinese scientists for this purpose on May 23, 1988. Panofsky was uncertain about who would show up and how things would actually work out when he arrived in Beijing on May 15.

Panofsky was given a list of the potential attendees when he got to the IHEP on May 18. A sense of excitement flashed through his mind as he scanned the list and listened to Ye Minghan, IHEP director, identifying the people on it. His diary entry later that day expressed his sentiments: "This is a somewhat frightening list. The level of the attendees is extremely high; in fact it includes several people who are directly involved in atomic weapons work."[3] Heading the group was Zhu Guangya, the technical leader of China's nuclear weapons project, whom Panofsky had been anxious to meet. What may have also made a strong impression on Panofsky was the fact that several of the leading Chinese participants had been educated in the United States in the 1940s and returned to China in the 1950s. The meeting, which would take place five days later and which, as Panofsky wrote in his diary, "went extremely well" (despite the fact that Zhu did not show up), constituted one of the earliest steps taken by China and its scientists to move into international nuclear arms control. The dialogue it started may also have helped convince the authorities to sign a number of arms control agreements, including the Non-Proliferation Treaty in 1992 and the Comprehensive Test Ban Treaty in 1996.[4]

This account of Panofsky's experiences in Beijing demonstrates that scientists can indeed play an effective role in promoting international arms control discussions. That being said, a review of the participation of American and Chinese scientists in national and international nuclear arms control discussions reveals a number of tensions that might help us frame the writing of transnational histories of science and technology.[5] What was the proper role of scientists in public policy in areas such as arms control where the technical mixed intrinsically with the political? What are the potentials and limits of transnational scientific discussions and knowledge circulation in sensitive areas such as nuclear weapons? And what roles did American-educated Chinese scientists as a transnational ethnic scientific network—both those who returned to China and those who stayed in the United States during the Cold War—play in a hybrid geopolitical theater where state designs framed exchanges but successful implementation also relied on private interactions?

It is important to examine this history of US-China interactions in nuclear arms control not only for its own intrinsic value and interest but also for its policy relevance.[6] Nuclear weapons receded from public concern with the end of the Cold War in the early 1990s but have reemerged in the twenty-first century as a major public policy issue. President Barack Obama won the 2009 Nobel Peace Prize for his advocacy of nuclear disarmament, led the successful push for the 2015 international agreement, with participation by China and others, to curtail the Iranian nuclear program, but disappointed many of his supporters by approving a major program of American nuclear modernization amid a rising nuclear arms race between the United States, Russia, and China. He also failed to commit the United States to a no-first-use policy.[7] The 2016 presidential election of Donald J. Trump was even more troubling for advocates of nuclear arms control. The new US president made conflicting and, to many of his critics, irresponsible statements and took controversial policy stances on nuclear weapons (and climate change) during the campaign and early in his administration. Indeed, only days after his inauguration, alarm over Trump rose high enough for the *Bulletin of Atomic Scientists* to move its famous doomsday clock as close to the cataclysmic midnight as it had been at the height of the Cold War.[8]

The history of US-China interactions in nuclear arms control could also be valuable in providing possible lessons for seeking solutions to other global issues such as climate change. Both issues have strong scientific and technical components and pose grave threats to global security in which the United States and China are major players. Another issue of perhaps equal importance in both cases is the need for information on the Chinese national policy-making processes. I won't be able to tackle all these big questions here, but I will provide information and clues for a fuller treatment as I explore how American scientists used their transnational connections and leverages in promoting nuclear arms control discussions with China. Thus, in this chapter I will first review how American and Chinese scientists came to approach nuclear arms control before detailing and analyzing how they, under the leadership of, respectively, Panofsky and Zhu Guangya, started to engage with each other in this area.

The American Experience

If nuclear arms control can be defined as international efforts to limit the production and use of nuclear weapons, one of the earliest attempts of scientists in this direction took place in 1945 when a group of Manhattan Project scientists under James Franck drafted a report arguing against the US use of the atomic bomb on Japan. They proposed instead a demonstration in an uninhabited area as a step toward achieving international control of nuclear weapons in the postwar period. They failed in their aims, partly because of opposition by other scientists under Robert Oppenheimer who were advising Truman's policy-making Interim Committee.[9]

Yet, only a year later, Oppenheimer helped draft the Acheson-Lilienthal report, which advocated international control of nuclear weapons in order to "create deterrents to the initiation of schemes of aggression [and] even contribute to the solution of the problem of war itself." A foundation for such control is the sharing of nuclear information among all nations, the report argued, for "in the long term there can be no international control and no international cooperation which does not presuppose an international community of knowledge." The report led to the presentation of the Baruch Plan by the US government internationally at the UN and the Scientists' Movement at home to promote international control of nuclear weapons. Both of these initiatives also failed, mainly owing to the conflicting national interests of the United States and the Soviet Union.[10]

The US Atomic Energy Commission's General Advisory Committee report in October 1949 on the hydrogen bomb can be considered the next major attempt at nuclear arms control by American scientists. In it the General Advisory Committee group of scientists under Oppenheimer argued against "an all-out effort" to make the H-bomb, on the basis of both technical and moral considerations, though a division developed as to whether it should be pursued eventually. The majority, including Oppenheimer, argued that the United States should make an unqualified commitment against the making of the H-bomb, which was a possible "weapon of genocide." But a minority of two physicists, Enrico Fermi of the University of Chicago and I. I. Rabi of Columbia, who condemned the H-bomb even more strongly as "an evil thing considered in any light," nevertheless advocated only a qualified commitment against its development: "it would be appropriate to invite the nations of the world to join us in a solemn pledge not to proceed in the development or construction of weapons of this category." The implication was that if such a pledge was not accepted by other countries, especially the Soviet Union, the United States would probably have had to proceed with the H-bomb program. Once again, the scientists' argument failed, and the Truman administration launched the crash H-bomb program soon thereafter.[11]

On what basis did scientists justify their involvement in policy-making in regard to nuclear arms control? Most of them cited both their special technical familiarity with nuclear weapons and their rights as citizens to speak out on moral and political issues. Sometimes there were tensions over the balance between the two aspects among scientists themselves and among the public. In 1945 the Franck committee, for example, argued that "[w]e believe that our acquaintance with the scientific elements of the situation and prolonged preoccupation with its world-wide political implications, imposes on us the obligation" to advise the government on the atomic bomb. But the Oppenheimer panel responded that "[w]e [scientists] have, however, no claim to special competence in solving the political, social, and military problems which are presented by the advent of atomic power."[12] A year later the *New York Times* editorialized that "not even the fact that a scientist has had a share in making the atomic bomb qualifies him to map national policy or read the future."[13] A group of scientists from the Manhattan Project's laboratory in Oak Ridge, Tennessee, answered that not only did they have the right of ordinary citizens to speak out on political issues, but like the Franck group, they felt that their familiarity with the technical issues obligated them to get involved: "Since we are best equipped with this knowledge, we have assumed the responsibility of aiding in the education of those who are not aware of the revolutionary nature of atomic power. The scientists appear as strangers in the public eye only because they have never before seen a development with such far-reaching implications and they feel compelled to step out of their laboratories and warn the people of impending dangers to civilization."[14]

Why not leave international affairs to the experts? The Oak Ridge scientists questioned whether the latter existed—"Who are the experts, and where is the evidence of their handiwork?"—and concluded that "perhaps the time is ripe for some logical reasoning to be injected into the art of international diplomacy."[15]

Given these debates, it was no surprise that when the General Advisory Committee scientists made their recommendation against the H-bomb, critics attacked not only their conclusion but also their competence and appropriateness in reaching it. Edward Teller, the politically conservative nuclear physicist, for example, argued that "it is *not* the scientists' job to determine whether a hydrogen bomb should be constructed, whether it should be used, or how it should be used. This responsibility rests with the American people and their chosen representatives."[16] His argument, however, did not prevent him from becoming the most forceful advocate for the making of the H-bomb. Later, in the government's case justifying the removal of Oppenheimer's security clearance, Oppenheimer's and other scientists' suitability to play a role in policy once again came under official attack.[17] Everyone tried to patrol the boundary between the technical and the political and reshape it to fit their own largely predetermined policy positions.

If strengthening national security was only implied in the above examples of scientists' advocacy of arms control, it became a central argument in 1958 when the President's Science Advisory Committee (PSAC) suggested that the Eisenhower administration move aggressively to pursue a nuclear test ban with the Soviet Union to freeze the then American superiority in nuclear weapons design. Eisenhower had established the PSAC group of moderate scientists in the aftermath of the Soviet launch of the *Sputnik* satellite in 1957, as he became increasingly concerned over the dangers of nuclear war and the threat posed by the growing military-industrial complex. He accepted PSAC's recommendation for a test ban against opposition by critics such as Teller and thus started the US government on the road that eventually led to the signing of the Limited Test Ban Treaty between the United States, the Soviet Union, and the United Kingdom in 1963.[18]

Even though Eisenhower's trust and the public shock over *Sputnik* elevated the status and self-confidence of scientists advocating arms control, they continued to face the tension between the technical and the political, even within PSAC. When the committee met in 1958 to vote on its resolution that a nuclear test ban was in the overall interests of the United States, one member, Herbert York, a physicist and former director of the Livermore nuclear weapons laboratory then at the Pentagon's Advanced Research Projects Agency, objected on the basis that such a matter was beyond the competence of the scientists. Jerome Wiesner of MIT, speaking for the majority, told York in a private conversation that he was right that PSAC members were no experts on arms control but neither was anyone else; in addition, the president could ask for advice from anyone on any subject. Rabi went even further in arguing for scientists to lead a broader program of arms control.[19]

PSAC's advocacy of arms control helped make possible the Geneva Conferences on a nuclear test ban in 1958, which, ironically, set a precedent of a split between the technical and the political in arms control: specialists from negotiating parties tackled technical issues first before diplomats tried to reach political agreements. Scientists from the United States, the Soviet Union, and the United Kingdom did reach technical agreements on how to police a test ban, leading many to praise science and scientists for opening a new era of scientific diplomacy, but new nuclear test data and research soon dashed hopes for a prompt political accord when they exposed loopholes in the original scheme. Doubts came back to haunt scientists as to the wisdom of integrating science and scientists into international politics. Even some of the scientist-participants themselves developed skepticism toward the dual-track technical and political negotiations. For example, Panofsky, then a PSAC consultant and soon-to-be member, headed the US delegation in the technical discussions on detecting nuclear testing in space in Geneva in 1959. The experiences convinced him that "[t]he social experiment of separating scientific and political considerations was essentially a failure. One clear symptom of that failure was that whenever disagreements arose between the U.S. and Soviet scientific delegates, the positions were the same: the Soviets argued that verification would be technically easier and more effective than U.S. specialists believed. This polarized disagreement corresponded to the political interests of the two parties involved."[20]

Such experiences transformed PSAC's thinking on arms control. Its early hope of devising a technical solution changed to what I call "technological skepticism," a conviction that technology by itself—in the form of either better weapons or better arms control systems—would never solve the problem of the nuclear arms race. At its meeting with Eisenhower on July 12, 1960, PSAC members told the president that "the United States will have to make a purely political decision" on the desirability of a nuclear test ban.[21]

Indeed, historians generally agree that the 1963 Limited Test Ban Treaty was possible thanks to President Kennedy and the Soviet leader Nikita Khrushchev rethinking the Cold War after the Cuban Missile Crisis of 1962 rather than to any new technological breakthroughs. Kennedy was further motivated by a hope, in vain as it turned out, to use the test ban to somehow prevent China from becoming a nuclear power.[22] Yet, to say that the limited test ban was a political achievement does not mean that scientists did not contribute to its success: international scientific discussions did help clarify many technical issues involved in monitoring a test ban. According to Panofsky, the eventual treaty's prohibition of nuclear testing in outer space rested "heavily" on the 1959 Geneva negotiations in which he was involved.[23] Perhaps most important, science advising based on international discussions helped to remove what Kennedy called the "vague fears" concerning the risks involved in a test ban and arms control measure, which were often exaggerated by opponents of arms control.[24]

American Scientists, China, and Arms Control

Once China tested its first atomic bomb in 1964, it moved from the background to center stage of the American debate over nuclear weapons and arms control policies. In the American national debate over whether the United States should launch a massive antiballistic-missile (ABM) system in the late 1960s, the Johnson administration conceded that any ABM system technically feasible at the time would not be adequate in dealing with the Soviet nuclear offensive power. But in 1967 it decided to go ahead with the modest Sentinel ABM on the grounds that it was both feasible and necessary to counter the limited but unpredictable Chinese nuclear threat.[25]

Moderate scientists opposed both Sentinel and, later, Safeguard (the name given by the Nixon administration to the repackaged ABM) as a destabilizing development in the nuclear arms race. Frustrated by such internal dissent, Nixon followed earlier critics of scientists' involvement in policy, announcing that there should be a new understanding: "political people stay out of science *and* science people stay out of politics."[26] Following the reasoning to its logical end, Nixon, in 1973, after his reelection, dissolved PSAC and nearly the entire presidential science advising system that Eisenhower had established in the late 1950s.[27]

Now in exile from the federal government, moderate scientists interested in arms control sought to influence policy in two ways. One was by having several of them serve as informal advisers to Henry Kissinger—mainly Panofsky, Richard Garwin of IBM, Sydney Drell, also of the

SLAC, and Paul Doty of Harvard, who was a friend of Kissinger's and acted as leader of the group. Kissinger was involved in arms control policy first as Nixon's national security adviser and then as the secretary of state under both Nixon and Gerald Ford. The activities of the Doty group, all members of PSAC at one time or another, had started before PSAC's dissolution. They now gained added significance as Kissinger negotiated the ABM treaty and other arms control measures with the Soviet Union since PSAC itself was banished from the White House.[28] The other approach was by opening new, nongovernmental channels, such as the NAS, the Federation of American Scientists, and the Natural Resources Defense Council, to engage in nuclear arms control activities with Soviet scientists. Chinese scientists were involved as well after Nixon's historical trip to Beijing in 1972.[29] Under the sponsorship of the American Academy of Arts and Sciences, Doty led another group, with some overlap with those advising Kissinger, that also entered into informal discussions with Soviet scientists on arms control in the early 1970s.[30] Nongovernmental scientific groups and individuals, of course, had been involved in arms control since the beginning of the nuclear arms race, and some of the PSAC insiders had actually participated in some of them themselves. Wiesner, for example, was involved in the Pugwash conferences on policy issues, which were attended by scientists from both the East and the West, before becoming John F. Kennedy's science adviser and PSAC chairman in 1961. But now that PSAC was gone, these outside channels gained added utility.[31]

Between Nixon's 1972 trip and the formal reestablishment of US-China diplomatic relations in early 1979, the Chinese government actually preferred to deal with nongovernmental organizations and private individuals rather than federal agencies in bilateral interactions; official interactions should wait, it argued, until the United States withdrew its recognition of Taiwan and established diplomatic relations with the mainland. In this connection, left-leaning organizations such as Science for the People and even the moderate Federation of American Scientists were given favorable receptions in China. Chinese American scientists such as T. D. Lee led the way for American scientists visiting China but non-ethnic-Chinese scientists also found their paths to the long-closed Middle Kingdom. Luckily for the NAS, in the mid-1960s it had established, in cooperation with the Social Science Research Council and the American Council of Learned Societies, a group that became known as the Committee on Scholarly Communication with the People's Republic of China (CSCPRC). The CSCPRC played the most active role in sponsoring US-China scientific exchanges in the 1970s.[32]

When leading American scientists, especially those who had been actively involved in nuclear weapons and arms control issues in the PSAC system, visited China, they sought to engage Chinese scientists in nuclear arms control via a strategy characterized by connections and leverages. Despite the cessation of official relations for nearly a quarter of a century, connections between American and Chinese scientists existed and survived via private, personal networking. Many leading Chinese scientists had trained in the United States, especially in the 1940s, returning home in the 1950s. Following Nixon's 1972 trip, a number of them were able to quickly reestablish ties with their American classmates, professors, and friends, especially those fellow Chinese students who had decided to stay in the United States.[33] All these multifaceted ties helped pave the way for US-China scientific dialogue on nuclear arms control. Thus, when Wiesner, then back at MIT after serving as Kennedy's and very briefly as Johnson's science adviser, visited China in 1974, he was able to reestablish connections with Zhou Peiyuan, a leading Chinese physicist who had received his PhD from Caltech in 1928 and who had met Wiesner at several Pugwash conferences in the 1950s and in 1960 in Moscow.[34] As will be detailed later, Panofsky's own efforts to promote arms control discussion with Chinese scientists would also depend on personal connections between him and T. D. Lee and between Lee and Zhu Guangya.[35]

Leverage was another key factor in promoting US-China scientific discussions on arms control: visits of American scientists to China were highly valued by the Chinese government and by Chinese scientists for the introduction of cutting-edge science and technology into a country that was a decade or more behind international developments during and after the destructive Cultural Revolution (1966–1976). Such was the motivation for the invitation from the Chinese Electronics Society to Richard Garwin to visit China in 1974 and for the invitations from the Chinese Academy of Sciences to Panofsky starting in 1976. During their visits, they delivered lectures on a wide variety of technical topics ranging from computers and low-temperature physics in Garwin's case to high-energy physics by Panofsky. Yet, both leveraged their scientific and technological prominence in China to promote their interest in nuclear arms control by engaging Chinese scientists in discussions and planning future ones.[36]

Panofsky's efforts are also illuminating in showing how China actually started its processes and institutions to engage in technical discussions on nuclear arms control. When Panofsky and the NAS first proposed initiating discussions on arms control with the NAS's Chinese counterpart, the immediate reaction of the Chinese Academy of Sciences was that it, as a formally civilian institution, was not the proper partner for this endeavor. So it was through trial and error and connections, including his friendship with T. D. Lee, that in 1988 Panofsky connected up with Zhu Guangya, who was able to organize the group of Chinese bomb physicists to engage in discussions on arms control with Panofsky and later the NAS's CISAC, as described earlier and as will be detailed later.

Chinese Scientists and Nuclear Arms Control

At this point, one may ask what motivated the Chinese scientists to engage in discussions on nuclear arms control? Here I will focus on Zhu Guangya as a leader of both the bomb project and arms control discussion in China and perhaps one of the most low-key and understudied figures in modern science in China. Zhu studied physics at the famed Southwest Associated University during the War of Resistance against Japan. In 1946 the Chinese government, then still under the Nationalist leader Jiang Jieshi (Chiang Kai-shek), sent him, along with about half a dozen other talented students, including T. D. Lee, to the United States specifically to study how to make atomic bombs. Rebuffed of course by the US security restrictions, Zhu ended up studying and receiving a PhD in nuclear physics at the University of Michigan in 1950. With an older brother being an underground Chinese communist still serving in the Nationalist government, Zhu decided to return to mainland China, now under Communist control, in the same year. Once the bomb project started in the late 1950s, he was appointed its chief overall organizer responsible for making technical and organizational recommendations in meetings with top leaders such as Premier Zhou Enlai and Marshal Nie Rongzhen.[37] When Panofsky visited Beijing in 1988, Zhu was still a top leader in charge of the nuclear weapons complex in his position as vice chairman of the Scientific and Technological Committee of the Commission on Science, Technology, and Industry for National Defense (COSTIND).

Zhu's first foray into nuclear arms control came in 1963 when Zhou Enlai asked him to draft a report in response to the signing of the Limited Test Ban Treaty that year. His report, titled "The Ban on Nuclear Tests Is a Big Scam," pointed out that the United States and the Soviet Union had already conducted just about all the tests they needed, and whatever they still needed they could now get from underground tests, which were allowed under the treaty. China still needed to make its initial tests in the atmosphere, tests that were banned under the treaty, leading Zhu to conclude that the main purpose of the treaty was to try to prevent China from succeeding in its nuclear weapons program. His main recommendation was that China should accelerate its mastery of underground nuclear testing for its advantages in terms of secrecy, reduction of radioactive fallout, and acquisition of data not possible from testing in the air. He was given the job of organizing such underground tests while testing in the air continued.[38]

Arms control attracted Zhu's attention again in 1986 when Chinese leader Deng Xiaoping—under international pressure and overruling those in charge of the nuclear weapons program—decided that China would stop nuclear testing in the atmosphere.[39] Learning from this experience, Zhu and other Chinese bomb scientists recognized that international politics could suddenly change China's nuclear weapons policy. They foresaw that the United States, having approached the theoretical limit of nuclear weapons design, would soon push for a comprehensive nuclear test ban that would eventually affect all Chinese tests.[40]

The above sequence of events indicates that Chinese scientists' initiation into arms control was mainly a reaction to outside developments, both international and within China, that were external to the nuclear weapons system. This may explain in part Zhu Guangya's and other weapon scientists' initial motivation in engaging with Panofsky and other American scientists in arms control discussions: they needed to know what might be coming in international arms control developments and expected that such developments would eventually affect their work on nuclear weapons through a change in the Chinese government's own nuclear policy (in addition, possibly, to their own independent desires to promote international nuclear arms control).

The Beginning of US-China Scientific Discussions on Arms Control

Thus, the stage was set for Panofsky's first meeting with Chinese scientists on nuclear arms control on behalf of CISAC on May 23, 1988, at the IHEP. Aware of the significance of the moment and the role of the meeting atmosphere in trust building and facilitating communication, Panofsky was grateful, as he noted later in his diary, that "the staff had done a magnificent job setting up a square table to accommodate the 30 or so participants."[41] Besides Panofsky, the only other American present was Oren Schlein, an undergraduate student then studying international relations at Nanjing University whom Panofsky had recruited to take notes. The striking theatrics of Panofsky alone facing dozens of Chinese nuclear weaponeers or analysts in a room might actually have worked to his advantage: the latter might have felt more comfortable in a setting like this and more willing to exchange views with someone who had already gained widespread respect among Chinese scientists for his work on the BEPC, which would bode well for his proposal for continued dialogue.

As Panofsky scanned the room, he found that two prominent Chinese scientists on the list had not shown up, as he noted later in his diary: "We had anticipated that Dr. Zhu Guangya, the Vice Minister and Head of the Commission of Science, Technology and Industry for National Defense, would probably not come and he didn't. To our disappointment Zhou Peiyuan also did not come although he was expected." Zhu's absence was understandable because of his high position with the Chinese nuclear weapons complex and the sensitivity of the subject of the meeting. But Zhou's no-show was puzzling to Panofsky because Zhou had been an early participant in the Pugwash conferences of international scientific discussions on arms control, dating back to the 1950s, and he had also met with several American scientists, including Panofsky, who had visited China and promoted arms control. His absence may have indicated that there was an insider-outsider division of labor in China, as in the United States, between insiders like Zhu's associates and outsiders like Zhou who represented the public, "activist" face of China's interest in arms control.[42]

The meeting, apparently conducted mostly in English, was chaired by Zhou Guangzhao. Before becoming president of the Chinese Academy of Sciences, Zhou had worked on Chinese nuclear weapons and had, as a distinguished theoretical physicist, been a visiting scholar at Virginia Tech in the mid-1980s. He had also just been elected a foreign member of the US NAS in 1987. The Chinese scientists in attendance included Chen Nengkuan, a metallurgical physicist who had returned from the United States in 1955 and worked on the detonation of the Chinese atomic bombs; Yu Min, the chief designer of the Chinese hydrogen bomb; Cheng Kaijia, a nuclear physicist who had returned to China in 1950 after receiving his PhD two years earlier working under Max Born at the University of Edinburgh and a chief architect of the Chinese nuclear tests; He Zuoxiu, another theoretical physicist who had worked on the bombs and was then at the Chinese Academy of Sciences Institute of Theoretical Physics; and Hu Side, a leader of the Institute of Applied Physics and Computational Mathematics of the Chinese Academy of Engineering Physics, the institution mainly responsible for Chinese nuclear weapons design. Hu would soon become a leader on the Chinese side in dialogues with CISAC. Also present were two participants who were not nuclear physicists but defense analysts, Liu Huaqiu of the COSTIND's China Scientific and Technological Information Center for National Defense and Zou Yunhua, an assistant to Zhu Guangya and one of the few women in the room. Two attendees without any connection to nuclear weapons systems were Ye Minghan and Xie Jialin from the IHEP itself, with whom Panofsky had worked closely on the BEPC for several years.[43]

The meeting opened with Zhou Guangzhao's "somewhat flattering introduction" of Panofsky, followed by Panofsky expressing his gratitude in turn for "the presence of so many distinguished members of the academy and other scientists interested in military affairs." The world was in a "very difficult condition," he said, because it was "complex and difficult" to steer a middle course between "unfettered arms competition" and unilateral disarmament. In solving this dilemma he believed that scientists had a special role to play: "They are first citizens of their country, but they are also an international community that can communicate somewhat easier than officials of government. They are also a resource that can be used by government whenever opinions on difficult subjects are necessary before final decisions are taken."[44] He credited Doty's American Academy of Arts and Sciences group, which had engaged with Soviet scientists in the 1970s, for helping persuade the Soviets that "in nuclear strategic matters offense and defense are very much inter-related." This in turn had led the latter to reverse its earlier position that "there should be no control of any kind on defense but only on offense," and so to reach agreement on the ABM treaty with the United States in 1972.[45] Bringing this discussion on arms control home to the Chinese scientists—and probably trying to convince them of the merits of continued dialogue with CISAC—Panofsky pointed out, at this juncture, that the ABM treaty was beneficial to all countries but especially to China, whose modest nuclear deterrent "can be continued without concern of a ballistic missile defense being generated."[46]

Panofsky then described CISAC's membership of sixteen with brief comments on each member. What must have stood out to his Chinese audience, mostly bomb and missile makers then contemplating a move into arms control, was the fluidity in the identities of CISAC members, especially across the military/civilian and nuclear arms design/control boundaries. For example, Panofsky described Lew Allen as director of the Jet Propulsion Laboratory, a largely civilian institution managed by the California Institute of Technology (Caltech) for the National Aeronautics and Space Administration, but he also added that "earlier he was chief of staff of the U.S. Air Force."[47] He did not mention but presumably the Chinese side would have learned from public sources that Allen, with a PhD in nuclear physics from the University of Illinois at Urbana-Champaign, had worked on designing nuclear warheads at Los Alamos National Laboratory and served as the deputy to the director of central intelligence and director of the National Security Agency before becoming the highest-ranked uniformed officer in the US Air Force.[48] Similarly, Panofsky mentioned Richard Garwin as "a special science advisor to IBM" who had also been "involved in nuclear weapons designs" (he had helped miniaturize and weaponize the hydrogen bomb in the 1950s); Alexander Flax as the home secretary of the NAS who had been "director of the Institute of Defense Analysis and was involved in major activity in engineering of missiles"; Marshall Rosenbluth as plasma physicist at MIT who had "dealt with fluid dynamics problems of nuclear weapons."[49]

As Panofsky read off the names of other CISAC members to the Chinese scientists in attendance, they might have recognized many of them from their prominence in American nuclear or science policy in the past or from their involvement in US-China scientific exchanges under the sponsorship of other institutions. Indeed, the continuity between PSAC and CISAC under Panofsky was striking, ensuring a sense of continuity in American scientists' advocacy of nuclear arms control. Besides Panofsky and Garwin, other former PSAC members in CISAC in 1988 included Doty; Marvin Goldberg, then director of the Institute for Advanced Study in Princeton; Charles Townes, then a professor of physics at the University of California, Berkeley, who had been another former director of the Institute of Defense Analysis; and Wiesner. Yet another CISAC member, Spurgeon Keeny, had served as a staff member for PSAC. Chinese scientists would also have been familiar with Michael May, a nuclear weapons designer who was then associate director at large of the Lawrence Livermore National Laboratory after having served as its director.[50]

In his introduction Panofsky highlighted for the Chinese scientists the public-private hybrid nature, as well as the broad purview, of CISAC's activities: "The committee's objectives are to study and report on scientific and technical issues germane to international security and arms control; engage in discussions with similar organizations in other countries; develop recommendations, statements, conclusions and other initiatives for presentation to both public and private audiences; to respond to requests from the executive and legislative branches of the U.S. government; and to expand the interest of U.S. scientists and engineers in international security and arms control."[51] Calling CISAC's dialogues with its counterpart in the Soviet Academy of Sciences its "principal current activity," Panofsky further emphasized the public-private duality of the process: "Although these meetings have no official status, appropriate officials of the U.S. Government have been kept fully informed on the plans for and the proceedings of these meetings. In order to encourage frank discussion, it has been agreed that the meetings should be private without communiqués, joint statements, or public reports."[52]

In a way, the low-key manner of these transnational scientific exchanges helped to create a hybrid public-private space in which personal networking and direct contacts on sensitive technogeopolitical issues generated information and understanding not only in keeping with the advocacy of arms control by the scientist-participants themselves but also of value to the nation-states involved, which in turn not only acquiesced to but sometimes even encouraged such undertakings. Indeed, as Panofsky reported to the Chinese scientists, CISAC's arms control discussions had expanded to several Western European academies of sciences, a fact that he probably hoped would help convince his Chinese counterpart to follow suit. "The purpose of our committee is to explore and study solutions to these problems on which eventually our survival depends," Panofsky declared, "but to do so in a problem solving rather than a negotiating or argumentative spirit." "Therefore we do so in a manner which is private and in no way serves the purpose of public relations or public pressure," he added.[53]

CISAC's public-private duality clearly caught Chinese scientists' attention, perhaps as they themselves pondered a similar institutional setup for their country. They asked Panofsky questions about how CISAC operated, especially how it pulled off its public-private mode of operation, how American scientists more broadly participated in public policy, and specific technical and policy issues related to nuclear arms control. "Where does your budget come from?" was the first question for Panofsky after his introductory remarks on CISAC. "It comes entirely from private foundations and general funds accumulated over the years," Panofsky answered. Responding to another question about what other venues of US-Soviet discussions existed besides CISAC, Panofsky mentioned Pugwash, the Dartmouth conferences on US-Soviet relations, and the Federation of American Scientists, but he distinguished their public activism from CISAC's own back-channel approach: "These differ from our group in that they influence public opinion and increase the sensitivity of people in the world to the problems of arms control. That means that they give not only the opportunity for technical discussions but they also have a public relations purpose. Our discussions must not influence public opinion, but in the interest of having the frankest possible discussion they must be totally private, but with the understanding that our discussions are an open channel to the governments."[54] "Do your conclusions or advice influence government decisions?" a Chinese participant then asked. Panofsky acknowledged that his committee did studies at the request of the government and reported discussions to it, "but we can't be certain if our advice has influenced government decisions."[55] Nevertheless, his account of the existence of such varied institutions and approaches on arms control in the nongovernment sector must have made a deep impression on the Chinese audience.

In the spirit of problem solving and probably as a model for possible future discussions, Panofsky then made a presentation entitled "The Prospects for Deep Cuts in Nuclear Armaments: The Role of China." In it he reviewed the then-current US-Soviet negotiations on reducing nuclear weapons on both sides and expressed his hope that Chinese scientists would join CISAC for "informal but substantial" discussions on

China's entry into nuclear arms control, which would become necessary if the United States and the Soviet Union were to undertake deep cuts in their strategic nuclear weaponry, for example, by more than 75 percent, as China and others had called for and as the superpowers were contemplating.[56]

Much of the discussion following Panofsky's presentation stayed at the general level without much controversy, perhaps naturally, given the sensitivity of the topics and the novelty of the format for the Chinese scientists. Chinese participants asked and Panofsky answered questions on, for example, the American military's attitude toward strategic nuclear reductions, possible American and Soviet nuclear structures after these cuts, technical implementation of such cuts, effects of nuclear test bans on weapon improvements, and the nuclear winter phenomenon.

But at least one contentious technical issue did emerge when discussing the issue of the possible development of nuclear weapons by Japan, a topic that was brought up by He Zuoxiu: whether it was a good idea to reprocess nuclear fuel. Panofsky argued against the technology: "I am concerned with a technical matter about Japan: the reprocessing of nuclear fuel. Reprocessing is a step to acquiring nuclear weapons, and in my view the economic justification for reprocessing is very weak for civilian nuclear power. This is because reprocessing only extends the amount of nuclear fuel by a relatively small amount at a very large capital cost. Therefore, I am suspicious of any nation who wishes to acquire reprocessing capacity because in my view the economic motivation for doing that is not very good."[57] While the transcript of the meeting recorded no further discussion on reprocessing nuclear fuel, He Zuoxiu apparently did not agree with Panofsky's analysis. He claimed to have found an easy way to accomplish reprocessing using an accelerator and attributed Panofsky's continued opposition to his approach to a desire not only to prevent nuclear proliferation but also to protect the market for American nuclear fuel.[58] Panofsky, for his part, was critical toward He Zuoxiu as someone who resisted his and CISAC's efforts to persuade China to give up peaceful nuclear explosions (PNEs) by citing American and Soviet failures in this area.[59] He Zuoxiu later did change his mind on PNEs but he insisted that this was due not to Panofsky's argument but to his own recognition that such uses would create unacceptable nuclear pollution. Without fuller access to Chinese archives, it is difficult to determine whether Panofsky's argument or He's switch of positions had any effect on Chinese policy, but it is possible that they did. China gave up its insistence on the right to carry out PNEs when it joined the Nonproliferation Treaty in 1992 and the Comprehensive Test Ban Treaty in 1996.[60]

To return to the May 1988 meeting in Beijing, the question of institutional asymmetry also became a major subject of discussion and concern on the part of the Chinese scientists. Everyone, including Panofsky, recognized that CISAC, a private group with support from philanthropic foundations, had no counterpart in China. "There exists no such organization in China to support this kind of research," as one of the Chinese participants pointed out, while expressing his agreement with Panofsky that unofficial exchanges of views between scientists were "very necessary."[61] Even if such an organization were to be established in China, it would by default be a governmental organization with all the restrictions that would come with it. The nonofficial status of Panofsky and CISAC gave them the freedom and independence that might not be available to their Chinese counterparts. At one point, for example, Panofsky expressed his "criticism of the present U.S. doctrine, which has been responsible for driving the numbers [of required nuclear weapons] up to high levels." He and his CISAC colleagues believed that a reduction of nuclear weapons by a factor of 4 would "not change matters much."[62] At another point Panofsky expressed his hope that if the Soviets agreed to reduce asymmetrically its conventional forces in Europe, the United States would not need to pursue the so-called "extended deterrence" of threatening to use nuclear weapons to fend off a conventional Soviet invasion of its allies.[63] In contrast, several Chinese speakers explained China's long-standing pledge of "no first use" of nuclear weapons in any conflict and its insistence on the precondition of deep nuclear reductions by the superpowers before it would enter into a nuclear arms control agreement, but no one questioned directly any aspects of Chinese policy.

Nevertheless, the focus on technical discussions and the promise of confidentiality offered a possible way to solve the problem of asymmetry, as Chinese scientists could also claim to speak as individual specialists and not government representatives. In addition, Panofsky was confident that CISAC's long experience of working with its counterpart in the Soviet Union would help alleviate such problems with China, as he noted in his diary:

> The Chinese expressed concern about the problem of getting financial support on their side in case bilateral meetings were instituted. An interesting

> remark: one speaker said if we accept government support for our bilateral negotiations, then we will not be speaking as independent scientists and you wouldn't want that, would you? I replied that we are fully aware of the fact that in discussions with China or the Soviet Union we were not in a symmetrical situation, that an organization strictly analogous to the National Academy of Science simply did not exist; this would not detract from the value of having informal discussions which then can be briefed to governments.[64]

Indeed, it was with this understanding of informality and confidentiality that Chinese scientists, led by Chen Nengkuan and He Zuoxiu, engaged in an extended discussion with Panofsky on a wide range of issues. At one point, one Chinese participant expressed the view that for China a quota test ban, in which each nuclear state was allowed a limited number of tests, was preferable to a threshold test ban (only underground tests below a certain threshold were allowed) or a comprehensive test ban because China needed the tests to verify the reliability of its nuclear weapons.[65] In general, both sides agreed on the inseparability of reduction in nuclear weapons from conventional weapons and defensive systems in space, the importance of the survivability of the Chinese nuclear deterrent, and opposition to the possible development of nuclear weapons by Taiwan and Japan. Getting back to his criticism of US nuclear doctrine, Panofsky said that he hoped that once the United States abandoned extended deterrence it could also adopt a no-first-use nuclear policy and then "our strategy can become similar."[66]

At the end of the meeting, Panofsky's central objective of setting up a mechanism for CISAC to continue dialogue on arms control with a Chinese counterpart remained uncertain, even though most of the Chinese participants had expressed support for such an idea. "One young member made an eloquent speech about how much he had learned from these discussions," Panofsky noted in his diary.[67] Zhou Guangzhao, who participated actively in the discussion, ended the seminar by thanking Panofsky and by calling for "continuing such discussions, either as an occasional gathering or a more continuous series of meetings."[68] Both he and Panofsky knew that the decision was not Zhou's but would require approval at a higher level in the national security system. Panofsky intuited that Zhu Guangya, as a leader of COSTIND, was likely the pivotal figure in this process. As mentioned earlier, Zhu was on the invitation list as the highest ranking of the Chinese scientists, but he, as Panofsky had expected, did not show up.[69] His absence was likely a reflection of the caution and sensitivity with which he and other leaders of the Chinese nuclear weapons program approached Panofsky and the NAS's overture.

At this critical juncture Panofsky's friend T. D. Lee came to his rescue.[70] Having signed a statement by American Nobel laureates (Lee had shared the Nobel Prize in Physics in 1957) endorsing the Limited Test Ban Treaty in 1963, Lee supported Panofsky's efforts in arms control in China. The latter happily recorded in his diary that on May 23, 1988, "T. D. Lee arranged for me to have lunch tomorrow with Zhu Guangya who is really the key person who will make the decision on the future arms control discussions."[71] At the same time he had confirmation that he would also meet with Fang Yi, the Chinese vice premier, to report on the BEPC, and with Winston Lord, the US ambassador to China, to debrief him and his staff on his activities, especially in regard to the arms control discussions (when he did, Lord and his staff encouraged his arms control efforts). Not without some excitement and satisfaction, he wrote in his diary on May 23, "so tomorrow is again going to be a day when I will be interacting with three different dignitaries."[72]

Yet, even as he looked forward to a direct contact with Zhu, Panofsky realized that this forthcoming meeting with Zhu but without the presence of Zhou Guangzhao presented him with "a slight diplomatic problem": whether he should "respect Zhou Guangzhou's final indefinite decision" or negotiate directly with Zhu, which amounted to going "over Zhou Guangzhao's head."[73] Not surprisingly he chose the latter and used the lunch with Zhu, with Lee and his wife present, to talk about arms control: "Zhu pointed out that there was some 'political sensitivity' in setting up a similar committee of scientists in China but the conversation ended by [his] saying 'I will do my best.' Nothing better could be expected at this point."[74]

Only years later did we learn, from the recollections of Zou Yunhua, who not only was Zhu's assistant on arms control but would also spend time at SLAC working with Panosky, that Lee did more than just introduce Panofsky to Zhu Guangya for this crucial meeting. According to Zou, on May 23, 1988, the day of Panofsky's seminar at the IHEP, Lee had tried to invite Zhu to attend the meeting in the afternoon after learning that he did not show up in the morning; Zhu declined "due to a busy schedule." Later, after arranging the Panofsky-Zhu lunch meeting, Lee wrote not only to Zhu to convey Panofsky's appreciation of Zhu's "scholarly style" but also to Nie Rongzhen, the Chinese marshal still influential in Chinese nuclear and defense policy-making in the 1980s, to vouch for Panofsky's goodwill and seek Nie's approval for Zhu's participation in CISAC dialogues: "Professor Panofsky is an internationally well-known physicist with great achievements who has enthusiastically assisted with the building of the Beijing Electron-Positron Collider for many years. Four years ago he even came to work in Beijing only three months after a heart surgery, earning widespread praise from Chinese leaders and scientists. . . . He will be discussing problems of nuclear disarmament with relevant Chinese experts, and he would very much like to see Mr. Zhu Guangya participating in these discussions as a formal member of the Chinese side."[75] It is not clear that Lee's letter to Nie worked, but it is very likely that it helped to make it easier for Zhu to get more involved in arms control in the future.

Indeed, as Zhu promised and as Panofsky hoped, what became known as the Chinese Scientists' Group on Arms Control (CSGAC), with Zhu as chairman, was set up in 1991, initially under the sponsorship of Zhu's COSTIND and later of the ostensibly nongovernmental Chinese People's Association for Peace and Disarmament, as a counterpart to CISAC in arms control dialogue.[76] This was an important step in reaching institutional symmetry in the CISAC dialogue, but important differences remained. For example, there was a disparity between the positions and responsibilities of the leaders on each side: Zhu, as a high-ranking government official, carried active responsibility for China's nuclear weapons program, while Panofsky did not do so with regard to the American program even though, as he made clear to the Chinese scientists, he and his committee maintained close communication with the US government. Similarly, even though some CISAC members were quasi–government employees and the committee maintained close ties with the US State Department, including the US embassy in Beijing, it was a nongovernmental organization, whereas all members of the CSGAC, even though it operated nominally under the nongovernmental Chinese People's Association for Peace and Disarmament, were Chinese government employees with close and often direct ties to the national security system. Perhaps sensitive to this fact, Zhu declined to seek funding from foreign private foundations for Chinese arms control research.[77] Nevertheless, because of the informal nature of the discussions, such institutional differences did not seem to impede the interactions between CISAC and the CSGAC.

Even before the CSGAC was formally established, the COSTIND, as Zhu Guangya had promised Panofsky, arranged for the visit to Beijing of a delegation from CISAC in October 1988 and sponsored CISAC's first formal, two-day meeting with Chinese scientists on arms control.[78] The timing was designed to coincide with a major meeting of the CSCPRC, of which Panofsky was also a member.[79] Besides Panofsky, who headed the group, the CISAC delegation also included the aforementioned Allen, Garwin, May, and Townes as well as CISAC member John Steinbruner, an influential political scientist and policy analyst then at the Brookings Institute in Washington, DC, and CISAC staff director Lynn Rusten.[80]

As evidence of what Panofsky had described as CISAC's practice of keeping the US government "fully informed," CISAC delegation members met in September 1988 for a daylong pre-trip conference at the NAS in Washington, DC, which included meetings with representatives of the US State Department. The latter not only briefed them on China's security and arms control policy-making processes but also expressed interest in their initiative.[81] On arriving in Beijing CISAC delegation members attended a reception given by the CSCPRC's Beijing office, where they met Ambassador Lord. Lord encouraged CISAC's efforts, as he did in May to Panofsky, calling the forthcoming meeting "the first indication of Chinese scientific interest in arms control at a quasi-official level."[82] Lord also cautioned Panofsky about potential problems in the meeting next day and offered some advice: "Ambassador Lord warned me that the Chinese may be uncomfortable with the arms control meeting and the general tactic would be to have formal presentations take the entire period so that there would be relatively little time for discussions. He suggested one should chair the meeting in such a way that that won't happen."[83] Once again matters of sensitivity and communication took central place in the process of transnational exchanges on nuclear arms control. This perceived reluctance and caution on the part of the Chinese participants in transnational arms control discussions may have also reflected internal debates in China over the propriety of scientists' involvement in policy that mirrored those in the United States, as mentioned earlier.

Meanwhile, the scale and composition of Chinese participants differed from those in the May 1988 meeting, perhaps as a result of a desire for symmetry. Instead of the "frightening list" of around thirty, it was now reduced to six people plus an interpreter. These included He Zuoxiu, Hu Side, Liu Huaqiu, and Zou Yunhua, who all had attended the May meeting, with the addition of Du Xiangwan, Hu Side's chief lieutenant at the Institute of Applied Physics and Computational Mathematics and later a leader of the CSGAC, and Huang Zuwei, a missile specialist from the Ministry of Aerospace Industry.[84] These Chinese participants were in general younger than the leaders of the Chinese nuclear weapons projects who had attended the May meeting, and their selection probably also reflected Zhu Guangya's determination to professionalize nuclear arms control as a new field of study in China. These participants would form the core of CSGAC later on.

All these activities and developments built up excitement and anxiety for the formal opening of US-China scientific communication on the sensitive topic of nuclear arms control on October 7 and 8, 1988, in Beijing. The site for the conference, the hall of the Union of Chinese Students Who Have Returned from Study Abroad in Europe and America (Oumei Tongxue Hui), was somewhat unusual but fitting in many ways. It not only was located near the Beijing Hotel, where CISAC delegation members had stayed, but also represented one of the few nongovernmental (at least nominally) organizations in China at the time, with a strong transnational symbolism.[85] Thus, the choice of site may have signified the desire on the Chinese side to seek symmetry in terms of the public-private institutional hybridity that marked CISAC's operations.

Compared with Panofsky's May meeting, the October meeting went into more technical depth, at least during the first day, October 7. After opening remarks by Hu Side and Panofsky, May was the first to speak, on the topic of "arms control in space." May gave "a summary update of the major technical problems and of some strategic issues associated with arms control in space . . . intended to be a conversation opener between our two groups on the subject." In fact, he also made arguments that he said represented the views of some or most of the other members of CISAC. For example, he endorsed the Outer Space Treaty of 1967, which China joined in the early 1980s, because it banned stationing weapons of mass destruction in space and was beneficial "both in the arms race sense and in the sense of crisis stability"; he believed that the ABM treaty then in force between the United States and the Soviet Union did "ban the development, test and deployment of space-based ABM systems and components." Perhaps out of sensitivity about his own position as a scientist in a government lab, May did not mention explicitly the Strategic Defense Initiative (SDI) system recently proposed by President Reagan, but it was clear that he believed that carrying it out would have violated the ABM treaty. Finally, May concluded that for arms control measures such as the ABM treaty to work it was necessary (and possible) to ban antisatellite systems as well.[86] He was followed by two Chinese scientists, Du Xiangwan and Huang Zuwei, who responded to May with their own papers on the subject.[87]

As the workshop got under way, covering not only arms control in space but also nuclear test limitations, drastic US-Soviet nuclear reductions, the role of China and other nuclear states in such scenarios, and the impact of nuclear weapons on regional stability in Asia, it became clear that both sides agreed on the general desirability of arms control but they diverged on technical and political feasibilities. For example, in Du and Huang's responses to May, they advocated nuclear arms control in space in general even more passionately than May but placed most of the burden on the two superpowers. Du ended his presentation with the following plea: "It's a pity that in today's world the force for arms control is still not much stronger than the force for arms race, so scientists should give full play to their knowledge and conscience to promote international justice, peace and development."[88] Huang presented several measures to achieve "the goals of non-weaponization in space," including "bilateral negotiations [that] should be held between two super powers on banning of space weapons."[89] May in turn responded that he "personally agreed with some of the measures discussed in the Chinese papers" but there were strong advocates for both offensive and defensive nuclear weapons systems in the United States who argued that arms control had to be verifiable and "compatible with the standards of present policy." "This sort of reply to the fairly idealistic Chinese proposals, that is, going back to what might be practical in the light of existing policies, was fairly typical of most American replies over the next two days."[90] Despite such problems, CISAC scientists took these exchanges in their stride, according to May: "In general, the Chinese were not prepared at this meeting to go into any quantitative or specific consequences of their views. This was not a surprise to us. Typically, such discussions come after a few preliminary meetings."[91]

It should be noted that despite the overall general nature of the discussions, technical knowledge exchanges did enter into the conversations and played a part in enhancing mutual understanding. At one point, for example, Hu Side, speaking for the Chinese side, explained his skepticism toward a threshold test ban treaty, which had been a topic of discussion at the May meeting. According to May's notes:

> While he [Hu] thought there was little possibility for a CTB (comprehensive test ban) in the near term, he also thought there was little value in intermediate threshold test bans. He thought the 10 kt limit had little military meaning because "the major problem is the primary which is usually below 10 kt," and even the 1 kt limit would permit experiments. He thought that the U.S. and the Soviet Union should instead lead in drastically cutting nuclear tests, to 1 or 2 or 3 a year, in order to restrain the arms race.[92]

May did not record the reactions of the American scientists in the room but one can imagine that they would have understood Du's technical arguments perfectly and would likely have found them helpful in understanding the Chinese position.

As the leader of both CISAC and its delegation in Beijing who had worked hard to bring about this dialogue, Panofsky was greatly encouraged by what happened in the seminar, especially in regard to the main purpose of trust building. During the first day he noted that the "atmosphere at the meeting was good and Chinese had prepared quite a few papers." "In fact," he continued, "the main problem was to keep a discussion going, rather than simply focusing on the papers." This could have been due in part to what Lord had warned about—the Chinese tendency to avoid discussion—but Panofsky took it as a positive sign of Chinese seriousness. At the end of the second day Panofsky again recorded in his diary his feeling that "the sessions went quite well." The second day of meetings was "less technical" than the first day, but included an important session on plans for future dialogues, which concluded, to Panosky's delight, with a "definite agreement that the process should continue."[93] Once again, Panofsky's sense of progress derived from not only the formal meetings but also informal contacts. On October 8, 1988, the last day of the seminar, for example, Panofsky and his colleagues were told "in several private discussions" that the Chinese position had "softened significantly as to the preconditions for participating in the arms control talks from their official position of 'deep cuts and the three stops,'" that is, the superpowers agreeing to stop production, stop deployment, and stop testing of nuclear weapons. And on the important question of the continuation of the dialogue, Panofsky was encouraged once more by his personal contacts with Zhu Guangya himself, who again was absent from the formal discussions but showed up at the official banquet marking the end of the seminar on October 8, seated next to Panofsky.

FIGURE 13.1. Chinese American physicist T. D. Lee brought together Americans and Chinese in scientific discussions on nuclear arms control in 1988 in Beijing. From left: Pan Zhenqiang, Zou Yunhua, Xu Huijun (Mrs. Zhu), Zhu Guangya, Wolfgang Panofsky, T. D. Lee, Du Xiangwan, Jeanette Lee (Mrs. Lee).
Source: Courtesy of the Panofsky family and SLAC Archives.

As Panofsky recorded in his diary: "I gave him a brief report [on] what had been happening in the talks. He did not seem to be interested in any detail, and we switched to small talk and physics. Then at the end he came back to the subject; he expressed the view that the discussion should continue with larger representation on their part, including some social scientists."[94]

Only years later did some of the Chinese decision-making processes become known in memorial articles written after Zhu's death in 2011. Hu Side, who became head of the Chinese Academy of Engineering Physics and a leader of the CSGAC seminars, recalled in a 2012 paper that the cautious Zhu Guangya had to receive clearance from the highest level—the Central Military Commission—to open the dialogue with Panofsky's CISAC in 1988. Then during the October 1988 seminar: "He at first asked us to talk with them. We reported to Chairman Zhu every night on what happened, and he would suggest to us what questions to ask and what opinions to express."[95] No wonder Zhu Guangya was not interested in Panofsky's telling him about what had happened at the seminar!

The scheduled CISAC-CSGAC meetings in the United States in 1989 or 1990 apparently did not take place, most likely owing to the hiatus in NAS exchanges with China resulting from the Chinese government's crackdown on the prodemocracy student protest in Tiananmen Square in Beijing in June 1989. The seminar resumed, however, in October 1991 at Irvine, California, which gave Zhu a chance to revisit the United States for the first time since he left in 1950, as he led a delegation of CSGAC members to discuss arms control with CISAC. And it appears that the earlier meetings with American scientists had prompted Zhu and other Chinese scientists to strengthen Chinese efforts and institutions pertaining to technical studies on nuclear arms control. On September 14, 1990, most likely in anticipation of the forthcoming Irvine trip, Zhu had held a meeting within the COSTIND to expand personnel on arms control studies. He also helped make arms control physics a new branch of applied physics and started the training of students in this field in China.[96]

At the Irvine meetings in 1991 Zhu did participate in the seminar actively and spoke about his own understanding of the role of nuclear weapons in the world:

> There seems to be a consensus that due to their enormous destructiveness, nuclear weapons play a major role in strategic deterrence. But if one reflects more deeply, one should reach the conclusion that the military significance of nuclear weapons has been exaggerated. . . . We also appreciate this argument in your report, which is that all nuclear powers should reach a political consensus that nuclear weapons would not be used for any other purpose except for deterring others from using them, and that gradually all nuclear powers should solemnly declare their pledge to no first use of nuclear weapons.[97]

From someone who had devoted his entire career to making nuclear weapons for China, it was a remarkable statement on the limits of nuclear weapons as a technological solution to international political conflicts, one that PSAC and Panofsky would endorse. Zhu's transformation from a bomb maker to an advocate of arms control and scientists' social responsibilities also mirrored that of PSAC members and Zhu's American counterparts in the CISAC-CSGAC dialogues such as Panofsky, Garwin, and May. The evident consensus by both sides at the Irvine conference that the only purpose of nuclear weapons should be the prevention of other states from using them was a remarkable development and came close to China's official position on no first use of nuclear weapons.[98]

Meanwhile, Zhu and his colleagues also used their knowledge of developments in international nuclear arms control to help inform their advice to the Chinese government on both the production and the control of nuclear weapons. In 1986 they had convinced the Chinese government to accelerate its own underground test schedule in anticipation of a possible US proposal for a comprehensive nuclear test ban.[99] When the United States, as expected, made the proposal for a comprehensive test ban in 1992, Zhu and others persuaded the Chinese government to launch a second accelerated test series. All these tests were designed to perfect China's neutron bomb and the miniaturization and weaponization of some of the new warheads. On July 29, 1996, hours after the last shot of the second series took place, the Chinese government announced that it would observe a moratorium on testing.[100] Just as in the United States, national security and arms control were integrated in China.

But to say that Chinese scientists became engaged in arms control out of institutional self-interest does not mean that Chinese scientists did not share the objectives of their American counterparts in trying to reduce and limit the scale of nuclear weapons programs in the world, as Zhu's 1991 Irvine speech and the continued Chinese participation in the CISAC-CSGAC meetings and in international arms control regimes indicate. In a 1992 coauthored paper on the need for technical evaluations of weapons systems, Zhu expressed a view that was close to the kind of technological skepticism to which Panofsky and other PSAC scientists had subscribed in their advocacy for nuclear arms control: "the introduction of a new kind of weaponry usually leads to a new round of arms race. Thus, full evaluations of the effects of weapons could make those decision-makers in charge of the development of such weapons aware that, owing to the existence of countermeasures, the development of such weapons is not very meaningful. In addition, a weapon tends to be glorified or mythologized during its early stage of development; evaluations by scientists could reveal the true functions and practical capabilities."[101] Equally important, the institutional foundation and the training of several generations of nuclear arms control specialists in China under Zhu's leadership, facilitated by the CISAC-CSGAC process, ensured that China would be gradually integrated into the international discourse on arms control and that there would be expertise and a vested interest for arms control within China.

The year 2018 marked the thirtieth anniversary of the beginning of the CISAC-CSGAC seminars. Without access to all the relevant information, owing in large part to widespread archival restrictions in China, the sensitivity of the subject matter, and the confidential character of these discussions, it has been difficult to gauge accurately the impact of these interactions. But it is clear that both sides valued the dialogue enough to have continued them, not only during Panofsky's chairmanship of CISAC until 1993 but also through those of John Holdren of Harvard (1993–2004), who would later become President Barack Obama's science adviser, and of Raymond Jeanloz of UC Berkeley (2004–). The CISAC-CSGAC dialogues, which were modeled after CISAC's interactions with the Soviet Academy of Sciences, were in turn replicated in its exchanges with India. Panofsky believed that the clarification of the cost-benefits of PNEs in the CISAC-CSGAC discussions probably helped persuade the Chinese government to withdraw its objection to the prohibition of PNEs and ratify the Non-Proliferation Treaty in 1992, which has been an important step in China's involvement in nuclear arms control.[102] More recently, a direct achievement of the CISAC-CSGAC dialogues, unusually well publicized in view of the traditional reticence associated with such exchanges, is the making and publication of the *English-Chinese Chinese-English Nuclear Security Glossary* in 2008.[103] Its publication and positive reception again indicated the centrality of matters of communication and language in transnational scientific interactions on sensitive subjects.

Garwin, the longest-serving member of CISAC, agreed with Panofsky's positive assessment of the process. As he reflected on the CISAC-CSGAC meetings in 2014, "important achievements of these interactions, which take place without publicity and with no open reports, include a deep understanding of the attitudes on the two sides and, in particular, the Chinese government's signing the Comprehensive Nuclear-Test-Ban Treaty."[104] He also noted that the CISAC-CSGAC mechanism "convened the first meeting between Chinese and US nuclear laboratory and forensic experts," which helped, at least indirectly, initiate the fruitful and extensive US-China Lab-to-Lab Program (CLL), in which American and Chinese nuclear experts worked together on nuclear arms control and nonproliferation.[105]

Remarkably, the CLL followed the CISAC-CSGAC model of high-level but "unofficial" interactions to build trust and enhance mutual technical understanding. In many ways, however, it went beyond talks and into actual technical collaborations. For example, the program's flagship MPC&A Project in the late 1990s aimed to develop "a joint demonstration of technologies for nuclear material protection, control, and accounting (MPC&A)." For this project US nuclear weapons labs provided sensors and computer hardware and software and used them to work with Chinese nuclear institutions to help strengthen China's technical ability to control and secure its nuclear materials and reduce proliferation risks. The CLL ended in 1999 during controversies over the congressional Cox Report and the Wen Ho Lee case and accusations that China stole US nuclear secrets, but its legacy can be seen in renewed US-China collaboration on nonproliferation under Obama, which led to a joint project to remove enriched nuclear fuel from Ghana in 2017.[106] However, as *Science* reported in 2017, the Chinese government insisted on an apology from the US government for the Cox Report and an acknowledgment of "past cooperation," including presumably not only the CLL but also the CISAC-CSGAC dialogues, as "legal and mutually beneficial," before it would be willing to move ahead with broader collaboration in nuclear arms control. The US Congress, dominated by the Republicans, which had directed the making of the Cox Report in the first place, not only refused "to send the letter" but also opposed "such cooperation with an assertive China."[107]

Conclusion

What lessons can we draw from this discussion of US-China scientific interactions in nuclear arms control? First, it appears that American scientist-activists, through nongovernmental channels, took the lead in approaching Chinese scientists on the significance of nuclear arms control. But once the Chinese scientists studied the matter and realized its importance, they actively organized studies, joined international efforts, and persuaded their government to take necessary actions, a pattern that would be repeated in the case of climate change.[108] The CISAC-CSGAC case indicates the importance of transnational face-to-face scientific interactions, especially in sensitive areas such as nuclear weapons, building personal trust and understanding among policy-influencing scientists, which in turn helped their respective nation-states to seek mutually acceptable compromises and common ground. As Zhu's assistant Zou Yunhua observed, the close personal connections that T. D. Lee helped Panofsky and Zhu to develop between each other paved the way for the successful continuation of the CISAC-CSGAC dialogue: "From my perspective, what made it possible for CSGAC under Chairman Zhu and NAS-CISAC under Professor Panofsky to carry out bilateral academic exchanges was that the bridge maker between them was Chairman Zhu's good friend Professor T. D. Lee. Furthermore, Professor Panofsky loved China and made momentous contributions to the development of China's science and technology, which Chairman Zhu appreciated very much."[109] Thus, nation-states framed the context for nuclear dialogue but private or semiprivate transnational networking and scientific interactions played an important, even critical, role in moving the process forward.

Such movements of knowledge and circulation of people across national borders appeared to be effective in bringing China into the international system of nuclear arms control in the past and likely was a key element in dealing with other nuclear cases. For example, in the negotiations over the Iranian nuclear program during the Obama administration, public-private hybridity also played a role as personal and professional connections between the US secretary of energy Ernest J. Moniz and the Iranian nuclear official Ali Akbar Salehi helped reach the international agreement.[110] Former and current members of CISAC made their contributions: in the 2000s Panofsky helped initiate a dialogue between a small group of American scientists and their Iranian counterparts on Iran's nuclear program, and then Richard Garwin, the PSAC veteran who served as a major lieutenant to Panofsky in CISAC's discussions with Russia and China, organized influential public campaigns of American scientists in support of the deal under both Obama and Trump.[111]

But the US-China case also reveals tensions over scientists' identity as both experts and citizens, their choices to be outsider-activists or insiders working within the system, and their understanding of the potentials and limits of technological solutions to social and political problems. In this case, as so often in transnational scientific interactions, trust is a central issue. As John Krige and others have argued, the increasingly stringent national security regimes in the United States and elsewhere have posed growing obstacles to international scientific communication, especially in sensitive areas such as nuclear weapons.[112] Geopolitical disagreements, suspicion, and secrecy certainly often disrupted US-China scientific discussions on nuclear arms control, as happened in the aftermath of the 1989 Tiananmen tragedy and again in the publication of the congressional Cox Report and Wen Ho Lee case in 1999–2000, leading to charges of Chinese theft of US nuclear secrets. It was amply clear that the CISAC-CSGAC exchanges were conducted under the tight framing of geopolitical interests on both sides both before and after the end of the Cold War. Yet, at least in the beginning of the CISAC-CSGAC contacts on arms control in 1988, such difficulties did not seem to pose impassable obstacles to the start of dialogue. Aside from the Panofsky–He Zuoxiu technical dispute over nuclear fuel reprocessing, it appeared that secrecy and security requirements did not prevent the two sides from carrying out discussions at both technical and policy levels. Perhaps therein lies the value of person-to-person communication, in which a certain level of personal trust can be established and can help to overcome or at least lessen institutional and national differences. The public-private hybridity in CISAC and, to a lesser degree, in the CSGAC also probably helped create room for flexibility and maneuverability. By acting as a private organization but keeping the US government "fully informed," CISAC maintained a degree of independence while also alleviating problems with security and export controls.

It should also be noted that in this case scientists with transnational connections and leverages played a low-key but quite effective role in the policy process. As the political scientist Matthew Evangelista noted in his study of US-Soviet nuclear relations in the 1980s, transnational scientific organizations such as the Federation of American Scientists played influential roles, especially when the international and domestic circumstances were favorable.[113] In the beginning of US-China nuclear dialogue in the late 1980s and early 1990s, it was American transnational scientist-activists such as Panofsky who used their international scientific prominence to help promote policy initiatives in arms control; and it was well-placed Chinese scientists like Zhu Guangya who used their own scientific prominence and contributions to China's nuclear weapons program to mobilize efforts in promoting nuclear arms control in China and internationally. Trust that was first built in the area of high-energy physics and then reinforced through face-to-face interactions helped to overcome considerable resistance to the arms control agenda advocated by Panofsky and other American scientists in China. And finally one should not underestimate the critical roles of other, less visible transnational connectors such as T. D. Lee and other members of ethnic scientific networks who helped bring these scientist-activists together to advocate both science and nuclear arms control. A similar case could be made for the importance of complex personal interactions under state sponsorship in climate change and other areas of international policy discussions. The rich layering in transnational scientific interactions deserves close historical examination.

Acknowledgments

I sincerely thank Jeremi Suri for inviting me to present an earlier version of this paper at the workshop "Nuclear Arms Control and Climate Change Negotiations: Shared Lessons and Possibilities," held at the University of Texas, Austin, in 2014. I am grateful to him, Michael Gordin, Perrin Selcer, and other participants at that workshop for feedback. I thank John Krige for inviting me to present another version of the paper at the workshop "Writing the Transnational History of Science and Technology," held at the Georgia Institute of Technology in 2016, and for the feedback I received from him and other participants at this workshop. I thank Gordon Barrett, Jean Deken, Richard Garwin, He Zuoxiu, Laura O'Hara, Maxine Trost, Peter Westwick, and Xiong Weimin for assistance with materials, interviews, or feedback; and, finally, two anonymous reviewers for insightful feedback.

Notes

1. Wolfgang K. H. Panofsky, "Diary, May 13–25, 1988," Wolfgang Panofsky Papers, ser. V, box 51, folder 1, SLAC Archives, Stanford, CA.

2. For more on CISAC, see information and links on its official website (accessed Jan. 2018): http://sites.nationalacademies.org/PGA/cisac/index.htm.

3. Panofsky, "Diary, May 13–25, 1988," 16 (entry for May 18). In this chapter, names of Chinese in China are rendered in Pinyin, with family names (e.g., Ye) first and given names (e.g., Minghan) second.

4. Ibid., 31 (entry for May 23); Wolfgang K. H. Panofsky, *Panofsky on Physics, Politics, and Peace: Pief Remembers* (New York: Springer, 2007), 153–154.

5. For a recent review, see John Krige and Kai-Henrik Barth, eds., *Global Power Knowledge: Science and Technology in International Affairs*, Osiris, vol. 21 (Chicago: University of Chicago Press, 2006).

6. For a recent history of Chinese arms control efforts, see Evan S. Medeiros, *Reluctant Restraint: The Evolution of China's Nonproliferation and Practices, 1980–2004* (Singapore: NUS Press, 2009), which mentions CISAC's positive role in the training of Chinese arms control experts (230).

7. Steven Erlanger and Sheryl Gay Stolberg, "Surprise Nobel for Obama Stirs Praise and Doubts," *New York Times*, Oct. 10, 2009, A1; William J. Broad and David E. Sanger, "Race for Latest Class of Nuclear Arms Threatens to Revive Cold War," *New York Times*, Apr. 17, 2016, A1; David E. Sanger and William J. Broad, "Obama Unlikely to Vow No First Use of Nuclear Weapons," *New York Times*, Sept. 6, 2016, A1.

8. Science and Security Board of the *Bulletin of the Atomic Scientists*, "It Is Two and a Half Minutes to Midnight," Jan. 26, 2017, accessed Feb. 2017, http://thebulletin.org/sites/default/files/Final%202017%20Clock%20Statement.pdf. See also Jonah Engel Bromwich, "Doomsday Clock Moves Closer to Midnight, Signaling Concern among Scientists," *New York Times*, Jan. 27, 2017, A17.

9. Richard G. Hewlett and Oscar E. Anderson Jr., *A History of the United States Atomic Energy Commission*, vol. 1, *The New World: 1939–1946*, pbk. ed. (Berkeley: University of California Press, 1990). The full text of the report (accessed Dec. 2013) can be found at http://www.dannen.com/decision/franck.html.

10. Richard G. Hewlett and Francis Duncan, *A History of the United States Atomic Energy Commission*, vol. 2, *Atomic Shield: 1947–1952*, pbk. ed. (Berkeley: University of California Press, 1990); Alice Kimball Smith, *A Peril and a Hope: The Scientists' Movement in America, 1945–47* (repr., Cambridge, MA: MIT Press, 1970). The Acheson-Lilienthal report (accessed Dec. 2013) can be found at http://www.learnworld.com/ZNW/LWText.Acheson-Lilienthal.html.

11. Herbert F. York, *The Advisors: Oppenheimer, Teller, and the Superbomb* (Stanford: Stanford University Press, 1976). The full text of the General Advisory Committee report (accessed Dec. 2013) can be found at http://www.pbs.org/wgbh/amex/bomb/filmmore/reference/primary/extractsofgeneral.html.

12. For a discussion of this issue, see, e.g., Zuoyue Wang, *In* Sputnik*'s Shadow: The President's Science Advisory Committee and Cold War America* (New Brunswick, NJ: Rutgers University Press, 2008), 19–20.

13. "Topics of the Times," *New York Times*, Oct. 21, 1945, E8.

14. "Scientists Not Different," *New York Times*, Nov. 3, 1945, 11. See also Smith, *A Peril and a Hope*, 175–177; Wang, *In* Sputnik*'s Shadow*, 336.

15. "Scientists Not Different," 11.

16. See Wang, *In* Sputnik*'s Shadow*, 29.

17. Ibid., 46. Notably, such questioning of scientists' proper role in policymaking recurred in the debates over climate change as well. In 2004, for example, James Hansen, director of NASA's Goddard Institute and one of the most outspoken scientists advocating actions on global warming, was pressured by the George W. Bush administration to refrain from making public statements on cli-

mate policy. See Andrew C. Revkin, "Climate Expert Says NASA Tried to Silence Him," *New York Times*, Jan. 29, 2006, 1, 20. In response, Hansen, echoing the Oak Ridge scientists, argued that "I don't think my opinion about policies has any more weight than that of anybody else, but I shouldn't be prevented from saying it and I shouldn't be prevented from connecting the dots." Transcript of *Hot Politics*, PBS Frontline program aired in 2007, accessed Dec. 2013, http://www.pbs.org/wgbh/pages/frontline/hotpolitics/etc/script.html.

18. Wang, *In* Sputnik*'s Shadow*, 121–215.

19. Ibid., 125.

20. Wolfgang K. H. Panofsky, "Response to the CTBT Discussion," in *Scientific Cooperation, State Conflict: The Role of Scientists in Mitigating International Discord*, ed. Allison L. C. de Cerreño and Alexander Keynan, Annals of the New York Academy of Sciences 866 (New York: New York Academy of Sciences, 1998), 262.

21. Wang, *In* Sputnik*'s Shadow*, 141.

22. Glenn T. Seaborg with Benjamin S. Loeb, *Kennedy, Khrushchev, and the Test Ban* (Berkeley: University of California Press, 1981); Gordon H. Chang, "JFK, China, and the Bomb," *Journal of American History* 74, no. 4 (Mar. 1988): 1287–1310.

23. Panofsky, *Panofsky on Physics, Politics, and Peace*, 66.

24. Wang, *In* Sputnik*'s Shadow*, 225, 345.

25. Ibid., 274–280.

26. Ibid., 293.

27. Ibid., 305–308.

28. Ibid., 295. For example, Kissinger consulted with some members of the Doty group on August 15, 1975, before meeting with Soviet ambassador Anatoly Dobrynin on arms control. See John H. Kelly, "Memorandum of Conversation," Aug. 15, 1975, accessed Oct. 2016, https://fordlibrarymuseum.gov/library/document/0332/033200495.pdf.

29. On the Federation of American Scientists, see, e.g., Jeremy Stone, *"Every-man Should Try": Adventure of a Public Interest Activist* (New York: Public Affairs, 1999). On the Natural Resources Defense Council, see Kai-Henrik Barth, "Catalysts of Change: Scientists as Transnational Arms Control Advocates in the 1980s," *Osiris* 21 (2006): 182–206.

30. Richard L. Garwin, "Pief's Contributions to Arms Control and Nuclear Disarmament," First Panofsky Lecture, XVII Amaldi Conference, Hamburg, Germany, Mar. 14, 2008, accessed Oct. 2016, http://fas.org/rlg/PIIS10a.pdf.

31. Walter A. Rosenblith, ed., *Jerry Wiesner: Scientist, Statesman, Humanist; Memories and Memoirs* (Cambridge, MA: MIT Press, 2003), 259–263, 277–278. The physicist Leo Szilard was a one-man international institution for nuclear arms control during the Cold War. See, e.g., Barton J. Bernstein, "Leo Szilard: Giving Peace a Chance in the Nuclear Age," *Physics Today* 40, no. 9 (Sept. 1987): 40–47.

32. Zuoyue Wang, "US-China Scientific Exchange: A Case Study of State-Sponsored Scientific Internationalism during the Cold War and Beyond," *Historical Studies in the Physical and Biological Sciences* 30, pt. 1 (1999): 249–277; Wang, *In* Sputnik*'s Shadow*, 303. See also Kathlin Smith, "The Role of Scientists in Normalizing U.S.-China Relations, 1965–1979," in de Cerreño and Keynan, *Scientific Cooperation, State Conflict*, 114–136; Richard P. Suttmeier, "Scientific Cooperation and Conflict Management in U.S.-China Relations from 1978 to the Present," in de Cerreño and Keynan, *Scientific Cooperation, State Conflict*, 137–164.

33. Zuoyue Wang, "The Cold War and the Reshaping of Transnational Science in China," in *Science and Technology in the Global Cold War*, ed. Naomi Oreskes and John Krige (Cambridge, MA: MIT Press, 2014), 343–369.

34. Rosenblith, *Jerry Wiesner*, 259–263.

35. Panofsky, "Diary, May 13–25, 1988."

36. Richard L. Garwin, "China Trip: Transcribed Notes of a Trip to the Chinese People's Republic, March 18 to April 17, 1974," courtesy of Dr. Garwin; Wolfgang Panofsky, "Trip Report: Trip to People's Republic of China, October 5–22, 1976," Panofsky Papers, ser. V, subser. P, box 50, folder 3.

37. Gu Xiaoying and Zhu Mingyuan, *Women de fuqin Zhu Guangya* [Our father Zhu Guangya] (Beijing: People's Press, 2009).

38. Qian Wei, "Zhu Guangya," *Zhongguo xinwen zhoukan* [China Newsweek], Mar. 7, 2011, 71–73, at 73.

39. Gu and Zhu, *Women de fuqin Zhu Guangya*, 136; Qian, "Zhu Guangya," 73.

40. Xi Qixin, *Zhu Guangya zhuan* [A biography of Zhu Guangya] (Beijing: People's Press, 2015), 564–566.

41. Panofsky, "Diary, May 13–25, 1988," 30 (entry for May 23).

42. Ibid. On Zhou Peiyuan and Pugwash, see Gordon Barrett, "China's 'People's Diplomacy' and the Pugwash Conferences, 1957–1964," *Journal of Cold War Studies* 20, no. 1 (Winter 2018): 140–169.

43. Oren Schlein, "[Transcript of] Meeting to Explore Possible Future Discussions between CISAC of the U.S. National Academy of Sciences and an Appropriate Committee of PRC Scientists," Institute of High Energy Physics, Beijing, China, May 23, 1988, and attached invitation list, Panofsky Papers, ser. V, box 51, folder 1. On Panofsky and Xie, see Zuoyue Wang, "China, *Sputnik*, and American Science," *APS [American Physical Society] News* 20, no. 10 (Nov. 2011): 4, 7.

44. Schlein, "[Transcript of] Meeting," 2. See also Panofsky, "Diary, May 13–25, 1988," 31.

45. Schlein, "[Transcript of] Meeting," 2.

46. Ibid., 2–3.

47. Ibid.

48. Douglas Martin, "Gen. Lew Allen, Who Lifted Veil on Security Agency, Is Dead at 84," *New York Times*, Jan. 9, 2010, D8; oral history interviews with Lew Allen Jr. by Heidi Aspaturian, 1991 and 1994, accessed Jan. 2018, http://oralhistories.library.caltech.edu/203/1/Allen%2C_L._OHO.pdf. Allen was the first National Security Agency director to publicly testify in the US Congress in 1975. He was elected a member of the National Academy of Engineering in 1978.

49. Schlein, "[Transcript of] Meeting," 2. For more on Garwin, see Wang, *In* Sputnik*'s Shadow*, esp. 187, 247, 297–298, 303; Joel Shurkin, *True Genius: The Life and Work of Richard Garwin* (Amherst, NY: Prometheus Books, 2017).

50. Schlein, "[Transcript of] Meeting," 3.

51. Ibid., 5.

52. Ibid., 6.

53. Ibid.

54. Ibid., 9.

55. Ibid., 10.

56. Wolfgang Panofsky, "The Prospects for Deep Cuts in Nuclear Armaments: The Role of China," May 23, 1988, Panofsky Papers, ser. V, box 51, folder 1.

57. Schlein, "[Transcript of] Meeting," 21.

58. Interview with He Zuoxiu by Xiong Weimin, July 30, 2014, Beijing, transcript available in Documentation Center of the Research Office on the History of the Chinese Academy of Sciences, the University of the Chinese Academy of Sciences, Beijing; interview with He Zuoxiu by Zuoyue Wang, July 15, 2016, Beijing. On He Zuoxiu's roles in the Chinese politics of science, see also H. Lyman Miller, *Science and Dissent in Post-Mao China: The Politics of Knowledge* (Seattle: University of Washington Press, 1996); H. Lyman Miller, "Xu Liangying and He Zuoxiu: Divergent Responses to Physics and Politics in the Post-Mao Period," *Historical Studies in the Physical and Biological Sciences* 30, no. 1 (1999): 89–114. In the interview with me he did acknowledge Panofsky's contributions to the development of China's high-energy physics and to international arms control efforts.

59. Interview with Wolfgang Panofsky by Zuoyue Wang, Jan. 23, 1998, Berkeley, CA.

60. Interview with He Zuoxiu by Xiong Weimin, July 30, 2014, Beijing; interview with He Zuoxiu by Zuoyue Wang, July 15, 2016, Beijing; Panofsky, *Panofsky on Physics, Politics, and Peace*, 154. In the 2000s and 2010s, He Zuoxiu actually became very critical of rapid Chinese expansion in civilian nuclear power. See He Zuoxiu, "Bixu tingzhi heneng fazhan de 'dayuejin'" [The "Great Leap Forward" in nuclear energy development must be stopped], *Huanqiu shibao* [Global times], Feb. 24, 2012, accessed Jan. 2018, http://opinion.huanqiu.com/1152/2012-02/2466570.html.

61. Schlein, "[Transcript of] Meeting," 28.

62. Ibid., 11.

63. Ibid., 21.

64. Panofsky, "Diary, May 13–25, 1988," 33 (entry for May 23).

65. Ibid.

66. Schlein, "[Transcript of] Meeting," 16–22.

67. Panofsky, "Diary, May 13–25, 1988," 33 (entry for May 23).

68. Schlein, "[Transcript of] Meeting," 30.

69. Panofsky, "Diary, May 13–25, 1988," 30–34 (entry for May 23).

70. On Lee and the test ban, see Wang, *In* Sputnik*'s Shadow*, 229.

71. Panofsky, "Diary, May 13–25, 1988," 31 (entry for May 23).

72. Ibid., 34. As an example of the benefits for internal communication from transnational interactions, Panofsky recorded (35) that his mention of the Chinese interest in a quota test ban caught the attention of Lord and his staff, who had actually never heard of such a proposal before.

73. Ibid., 34.

74. Ibid.

75. Zou Yunhua, "Shenqie huainian enshi yiyou Zhu Guangya" [Cherishing deeply the memory of my mentor and friend Zhu Guangya], in Bianjizu [Editorial group], *Fengfan changcun tiandijian: Zhu Guangya tongzhi shishi yizhounian jinian wenji* [Lasting legacy in the world: Memorial essays one year after the passing of Comrade Zhu Guangya] (Beijing: People's Press, 2012), 228–237, at 233.

76. Xi, *Zhu Guangya zhuan*, 566.

77. Zou Yunhua, "Shenqie huainian," 231. Among CISAC members in this period, Panofsky was director emeritus of SLAC, Michael May was associate director at large of the Lawrence Livermore National Laboratory, and Lew Allen Jr. was director of the Jet Propulsion Laboratory. All three laboratories were owned by the US government but managed by Stanford University, the University of California, and the California Institute of Technology, respectively.

78. Wolfgang Panofsky, "Trip Report: People's Republic of China, October 3–9, 1988," and Wolfgang Panofsky, "Diary, October 3–9, 1988," Panofsky Papers, ser. V, subser. P, box 51, folder 3.

79. Panofsky, "Diary, May 13–25, 1988," 33 (entry for May 23).

80. On Steinbruner, see Bart Barnes and David Hoffman, "John D. Steinbruner, Scholar of Foreign and Defense Policy, Dies at 73," *Washington Post*, Apr. 25, 2015, accessed Jan. 2018, https://goo.gl/DoQ3az.

81. "CISAC China Workshop, September 13, 1988, Participants," Sept. 14, 1988, box 5, folder 4, "Trip Report China and Russia," and "CISAC China Workshop . . . September 13, 1988 Agenda," box 5, folder 5, "Travel China NAS-

CISAC 1988," both in Michael May Papers, Lawrence Livermore National Laboratory Archives, Livermore, CA.

82. Michael M. May, "Foreign Trip Report," 1, Oct. 24, 1988, box 5, folder 4, "Trip Report China and Russia," May Papers.

83. Panofsky, "Diary, October 3–9, 1988," 9 (entry for Oct. 6).

84. "Chinese Participants," undated, box 5, folder 5, "Travel China NAS-CISAC 1988," May Papers. Because of his involvement in discussions with CISAC, Huang Zuwei helped start studies on arms control among China's aerospace experts. See Medeiros, *Reluctant Restraint*, 230.

85. Most leading Chinese scientists in the twentieth century had overseas-study background, and the union in question also included those who had studied in the Soviet Union and Russia. On the transnational features of modern Chinese science, see Wang, "The Cold War and the Reshaping of Transnational Science in China." In 2016 the Chinese party-state made the union much more a part of the government in an effort to attract Chinese scientists and students currently studying abroad to return to China to aid its drive to become a "strong power in science and technology [*keji qiangguo*]." See Xinhua Press Agency, "Zhonggong zhongyang bangongting yinfa 'guanyu jiaqiang Oumei Tongxue Hui (liuxuerenyuan lianyihui) jianshe de yijian'" [The Executive Office of the Central Committee of the Chinese Communist Party prints and distributes "Suggestions on strengthening the Union of Chinese Students Who Have Returned from Study Abroad in Europe and America (Association of Those Who Have Returned from Study Abroad)"], *People's Daily*, Aug. 3, 2016, accessed Jan. 2018, http://politics.people.com.cn/n1/2016/0803/c1001-28608880.html.

86. Michael May, "Arms Control in Space," Sept. 23, 1988, box 5, folder 4, "Trip Report China and Russia," May Papers.

87. Du Xiangwan, "Relations among Space Arms Control, Nuclear Disarmament and Nuclear Test Ban—the Necessity and Possibility for a Convention on the Prohibition of Outer Space Weapons," and Huang Zuwei, "Non-weaponization of Outer Space," Panofsky Papers, ser. V, subser. I, box 138, folder 4, "China Trip—CISAC/PRC Arms Control Discussion."

88. Du, "Relations," 4.

89. Huang, "Non-weaponization," 7.

90. May, "Foreign Trip Report," 2.

91. Ibid.

92. Ibid., 3.

93. Panofsky, "Diary, October 3–9, 1988," 10 (entry for Oct. 8).

94. Ibid.

95. Hu Side, "Zhu zhuren wei women zhiming fazhan fangxiang" [Chairman Zhu pointing the way for our development], in Bianjizu [Editorial group], *Fengfan changcun tiandijian*, 116–117.

96. Gu and Zhu, *Women de fuqin Zhu Guangya*, 199–200.

97. Zhu is quoted in ibid., 200–201.

98. Ibid., 199–202; Panofsky, *Panofsky on Physics, Politics, and Peace*, 163–164. Trump's statements on nuclear weapons, widely viewed as reckless, during his presidential campaign and his early presidency led two Democratic lawmakers to propose a bill in January 2017 to commit the United States to a no-first-use nuclear policy, but it had little hope of passing because of expected opposition by the Republican majority in Congress. See Emily Tamkin, "Lawmakers Introduce Bill Restricting First Use of Nuclear Weapons," *Foreign Policy*, Jan. 24, 2017, accessed Feb. 2017, http://foreignpolicy.com/2017/01/24/senator-and-congressman-introduce-restricting-first-use-of-nuclear-weapons-act-trump/.

99. Gu and Zhu, *Women de fuqin Zhu Guangya*, 130.

100. Qian, "Zhu Guangya," 73.

101. Zhu is quoted in Xi, *Zhu Guangya zhuan*, 570.

102. Panofsky, *Panofsky on Physics, Politics, and Peace*, 154. On recent developments, see Bates Gill, "China and Nuclear Arms Control: Current Positions and Future Policies," *SIPRI Insight on Peace and Security*, Apr. 2010, accessed Dec. 2013, http://books.sipri.org/product_info?c_product_id=406.

103. See the CISAC's website (accessed Oct. 2016), http://sites.nationalacademies.org/pga/cisac/index.htm. The glossary is available there for free downloading.

104. Richard Garwin, "Maintaining International Dialogue," *Proceedings of the National Academy of Sciences* 111, suppl. 2 (June 24, 2014): 9333–9334, at 9334.

105. Ibid. See also Nancy Prindle, "The U.S.-China Lab-to-Lab Technical Exchange Program," *Nonproliferation Review*, Spring–Summer 1998, 111–118.

106. For the Cox Report, see *Report of the Select Committee on U.S. National Security and Military/Commercial Concerns with the People's Republic of China*, House Report 105-851, https://www.gpo.gov/fdsys/pkg/GPO-CRPT-105hrpt851/pdf/GPO-CRPT-105hrpt851.pdf.

107. Richard Stone, "Atomic Bonding," *Science* 357, no. 6354 (Sept. 1, 2017): 862–865.

108. See Zuoyue Wang, "Scientists and Arms Control: The US-China Case and Comparisons with Climate Change," paper presented at the conference "Nuclear Arms Control and Climate Change Negotiations: Shared Lessons and Possibilities," University of Texas, Austin, Jan. 2014.

109. Zou, "Shenqie huainian," 232.

110. David E. Sanger, "No. 2 Negotiators in Iran Talks Argue Physics beyond Politics," *New York Times*, Mar. 29, 2015, A1.

111. Spurgeon M. Keeny Jr., "In Memoriam: Wolfgang K. H. Panofsky," *Arms Control Today* 37, no. 9 (Nov. 2007): 51–52; William J. Broad, "29 U.S. Sci-

entists Praise Iran Nuclear Deal in Letter to Obama," *New York Times*, Aug. 9, 2015, A8; William J. Broad, "Top Scientists Urge Trump to Abide by Iran Nuclear Deal," *New York Times*, Jan. 3, 2017, A13.

112. John Krige, "Elements for a Transnational History of Knowledge Circulation in the Cold War," paper presented at workshop "Writing the Transnational History of Science and Technology," Georgia Institute of Technology, Atlanta, Nov. 2016. See also Barth, "Catalysts of Change."

113. Matthew Evangelista, *Unarmed Forces: The Transnational Movement to End the Cold War* (Ithaca, NY: Cornell University Press, 1999).

Afterword: Reflections on Writing the Transnational History of Science and Technology

Michael J. Barany and John Krige

As scholars of the past, historians are informed by current events and debates but do not routinely have to confront the sense that they are being outrun by the present. We write now at what feels like a precipitous time for the transnational formations whose histories and implications are the focus of this volume. One of us (JK) has already written a good deal about the history of international and transnational science and technology. Hoping to foster collective reflection on such histories, he began planning this volume and its associated workshop early in 2016, at a time when migrants from war-torn countries in Europe were being met by new waves of racism and xenophobia. The other (MB), a recent PhD in the field, learned of the project through an announcement at the quadrennial joint meeting of the British, American, and Canadian professional organizations for historians of science, in the summer of 2016 in Edmonton. He was dining at a Thai restaurant on the Canadian prairie with a mix of European and North American historians based in Scotland, Canada, and the United States as returns from the United Kingdom's "Brexit" poll began rolling in, and it gradually dawned on the group that we would wake up the next morning in a rather less open world. The project workshop convened in Atlanta, Georgia, on the eve of the 2016 presidential election in the United States, many of whose citizens had already cast their votes in what would become, in that state and the nation alike, a narrow but decisive victory for Donald Trump. We revised and assembled our papers during a tumultuous political transition, amid denunciations, refugee crises, travel bans, detentions, talk of border walls, and other developments that bore directly on the questions treated herein and that demanded formal responses from our professional societies.

With nationality, nationalism, xenophobia, and division regularly in the news, it may seem quaint to make transnational science and technology our focus. Some scholars have, indeed, suggested that transnational history can never have the popular appeal or vital relevance of national histories that give readers a sense of place, of belonging, of identity.[1] Recent events and the scholarship in this volume alike suggest the opposite view, that states and localities and their concerns, prerogatives, tensions, and conditions come most clearly and potently into view from a transnational perspective, which requires intensifying rather than avoiding attention to local and national scales. The transnational historian must take quite literally the old exhortation to think globally and act locally, constantly embedding situated productions and their archival records in border-crossing patterns of movement and action. Global thinking gives purchase to otherwise unaccountable local phenomena by embedding them in longer and wider genealogies and drawing out the tensions they expose between (often global-seeming) ideals and (inevitably local) practices.

These wider views of narrower practices give us a foothold for grappling with the single most important phenomenon that has us feeling overtaken by the present moment: the US political leadership's retreat from superpower hegemony in matters economic, geopolitical, environmental, and moral. The outsized presence of American actors and institutions in some of the histories in this volume derives in large measure, by direct and indirect means, from the predominant role that the United States has played in the twentieth century as a global power and its mobilization of knowledge as an instrument of legibility and of rule. The transnational history represented here is a recent and geopolitically specific phenomenon. It was born in the United States in the 1990s in the wake of the Cold War, in deliberate methodological and ideological opposition to American exceptionalism, a sense of American distinctiveness, and the associated tendency to conceptualize the United States as "in many ways the clearest embodiment of the idea of the self-sufficient sovereign state," as Akira Iriye puts it.[2] The chapters in this volume do not propound American exceptionalism or self-sufficiency, but neither could they avoid the thoroughgoing presence of the United States as a crucial geopolitical node in the systems by which knowledge moves or fails to move across borders. Indeed, the critique of American exceptionalism can be an uncomfortable one for historians of contemporary science and technology precisely to the extent that the US research system does, as matter of fact, seem to be exceptional, qualitatively and quantitatively, and to differ from other national systems—not only as regards the extent of federal and corporate investment in research and development ($344 billion out of the world total R&D expenditure of $962 billion in 2007, for instance) but also as regards aspiration.[3] The pursuit of American scientific and technological preeminence on an international stage has provided an overarching rationale for a US national research system which has, of necessity, been built along transnational dimensions. Technoscientific leadership has dovetailed with political and military might, each reinforcing and reconfiguring the other. As Marilyn Young reminds us: "America may not be exceptional, but it is exceptionally powerful, and can only be decentered so far without obscuring that power."[4] If the nation and state take on particular salience in transnational history, the American nation and state in their changing geopolitical moments across the twentieth century and into the present do so all the more in the transnational histories of science and technology assembled here.

The significant presence of the United States as actor in many of the chapters in this project was also one of the project's strengths. Intellectually and pragmatically, it furnished common conceptual and historical backdrops and points of reference for scholars with substantially varied methodological approaches and subject expertise. These points of coherence show the geographical and historical specificity of the volume's accounts while opening it up as a comparative benchmark and methodological guide. As Josep Simon observes, this collection can "provide models or exemplars that other scholars could apply to other cases built from a geopolitical centre different to the US or even to case studies displaying a more clear multipolarity with regard to knowledge exchange."[5] Such work does exist, particularly in transnational studies of European integration, though that region has its own "center" and "periphery" too, as Kostas Gavroglu and his colleagues in the outer ring of European countries remind us.[6] The time has come to put these various local worlds into conversation with each other.

This volume's chapters emphasize how quickly the state apparatus springs into view when we take knowledge as the transnational object, even if our focus is not the regulatory state. The long-standing importance of scientific and technological knowledge to the exercise of state power and to the construction of national prestige requires that we reckon with the state—its actions, effects, and limits—whenever we study the production and movement of knowledge across borders. This does not subvert the goal of transnational history to break the bounds of the national container and to situate the state in a web of interconnections and interdependencies that are less visible from a national approach. On the contrary, in many of the chapters presented in this volume the state emerges as an actor *precisely because* its borders are being crossed. Seeing like a state involves being aware of where borders lie and policing them in the name of national sovereignty and other prerogatives.

It bears emphasizing that states controlled the movement of knowledge and knowledgeable bodies centuries before the transnational moment explored here. Indeed, they have attempted to do so wherever policing knowledge's borders gave state actors a competitive edge over rival powers. Consider, for example, Spanish control of maritime knowledge in the sixteenth century. Alison Sandman has described the complex body of laws governing pilots, charts, and nautical instruments that officials of the Spanish Casa de la Contratación developed in the early sixteenth century.[7] Protecting knowledge of Spanish sea routes from rival powers who might attack their ships and infringe their colonial monopolies required policing people, records, and instruments. Chart makers were forbidden to sell their charts to foreigners. Naturalized experts found their loyalty questioned. Pilots, who could easily flee the country with their precious knowledge of sea routes, could be licensed only if they were native or naturalized Castilians, and they had to guarantee loyalty by having significant family or property ties to the kingdom. Cosmographers, who taught pilots the use of instruments to establish latitude and longitude, were obliged to swear oaths not to share their knowledge with foreigners.

The strong parallels between such early modern forms of transnational knowledge control and the modern ones studied here lead us to ask what, precisely, is modern about the latter. The modern state constructed "rational," systematic, bureaucratic apparatuses that regulated political, economic, and military potentialities as part of the process of nation building. As Daniels and Krige show in the first two chapters, the striking feature of America after 1945 was the institutionalization and expansion of regimes of control developed for wartime contingencies or invoked temporarily in response to a specific crisis. Unstable moments of contingency—in the late 1940s, in the 1990s and early 2000s, or in 2017—leave lasting imprints on the structures of transnational modernity as the state seeks to maintain control over the changing geopolitical economy of knowledge production and circulation. Borders that began in artificial and in contingent, limited operations of geopolitics become aggressively affirmed, as refugees crossing into or out of Austria, Hungary, Mexico, and the United States know well. Xenophobia's modern transnational configuration resounds in policing regimes that are justified in the name of protecting the "nation" from being undermined by "foreigners" within and without.

This pronounced and asymmetrical salience for borders and state control leads us to emphasize terms like "travel" and "transit" rather than "circulation," "flow," or similar language that risks minimizing the problems of crossing in transnational undertakings. By stressing the labor and contingency engaged in moving knowledge, contributors focus attention on practice and expand the range of "actors" to include the material infrastructure and equipment needed to produce knowledge at different sites. Asif Siddiqi specifically draws our attention to the dispersed sites where transnational knowledge is negotiated in tandem with borders and geopolitics. The sitedness of such transnational labor gives Siddiqi a potent framework for dissecting how transnational actors assemble the resources required to construct a network of interrelated, interdependent people, ideas, and things that together constitute the transnational network. The deep history of specific sites can explain their attraction as poles in a transnational network of knowledge makers, weaving together multiple geographical and temporal scales that condition transnational endeavors.

Reading transnational history through the lens of knowledge easily—but not necessarily—marginalizes those who have no direct say in, or control over, the social transformations brought about by the transnational moment. They have neither the expertise nor the financial or social capital to be enrolled in the knowledge networks that are put in place to satisfy the aspirations of transnational elites. The lives of peasant farmers in India (Kumar) or of native villagers in Malindi (Siddiqi) are disrupted, sometimes massively, by the transformation of their local practices brought about by external actors and projects. Without a forceful voice of their own in the historical record, they often appear reduced to "complications," or to compliant resources to be drawn on as and when the need arises. Foregrounding sites and contingencies can effectively recover the agency of non-elites *as transnational actors*, an effort realized as well in Mateos and Suárez-Díaz's attention to the marginalized-yet-central personnel who literally drove the International Atomic Energy Agency's transnational projects. A richer transnational agenda would deliberately investigate the limits and possibilities of "creole" or métis knowledge in the construction of transnational networks that engage social strata other than the elite.

The portability of transnational knowledge/power subverts efforts to draw stark national divisions. Those the state sees as knowledgeable individuals are not treated in the same way as refugees or members of diasporic communities. What people know makes them both desirable *and* dangerous from the state's point of view, and state controls to regulate their movements open some avenues of identity and assertion while foreclosing others. Knowledgeable individuals navigate "hybrid" identities that combine a sense of self as a knowledge producer with a sense of national and political allegiance, with different contexts calling for the performance of different selves—a phenomenon most explicitly developed in Adriana Minor's study of physicist Manuel Sandoval Vallarta. When knowledgeable bodies move across borders, the potential contradictions between these selves can be the source of considerable anxiety to them and their colleagues and of suspicion to immigration officials.

The historical pattern of English linguistic hegemony means that anglophone sources grant current transnational historians access to large swaths of the technoscientific past. Indeed, specific resources with their own anglophone priorities and biases, such as Google Books and other massively searchable online repositories, make an already preponderantly anglophone corpus appear even more so, threatening to drown out nonanglophone voices. It is all too easy to ignore nonanglophone sources altogether or to assume (usually erroneously) that the easiest-to-access anglophone materials are representative of the far-flung multilingual body of potentially available documentation as a whole. Transnational historians are limited not just by what is in "the archives" but by which of those archives and which of their respective contents are geographically, financially, and linguistically accessible. By exploiting economies of representation that have historically concentrated anglophone sources in central nodes of exchange, we risk uncritically reinforcing those nodes as naturalized fixtures in transnational networks.

Contributors to this volume confronted these tensions in a variety of ways, writing and presenting in English while drawing on both anglophone sources and those in local languages from non-US archives at sites of interest in their analyses. Our own border crossings open up both archival sources and experiential sensitivities to the transformations, selective accommodations, and contestations that arise when the transnational meets the local, when new knowledge and the social relations in which it is embedded traverse networks at various scales. Like the subjects we study, we interacted predominantly in English but admitted numerous interstices filled with Spanish, Portuguese, and other languages, especially over coffee or beer as we digested and reframed our findings. The workshop and volume's subsidiary focus on North-South interactions in the Western Hemisphere created countervailing coherences in idioms and reference points that weighed, in some ways, against American anglophone domination. Language operated through the workshop as a means of authentication, projection, consensus, and exchange. We wore our many nations of origin and travel, and the cosmopolitan foundations and aspirations of our research, in our accents, our gestures, and our circumlocutions.

Beyond the importance of working in archives in several countries, and in several languages, historians of science and technology require sources and literatures for transnational arguments that differ from those for studies where borders are less at issue—a metaphorical disciplinary border crossing to match the geopolitical borders just discussed. Relationships between people who circulate across borders often engage very different actors and institutions from those operating on the national scale and are studied in academic fields—like international relations, diplomacy, law, and comparative cultural studies—that can be quite different from those that are used for writing a national history. The situation is all the more challenging because there is already a rather weak coupling between historians of science and technology and historians writing social and cultural history, though certain fields like environmental history or the history of capitalism are creating inroads into these academic divisions. A transnational history of science and technology must draw on insights from other disciplines to analyze the individuals and institutions responsible for promoting and negotiating border crossing, perforce imposing a heavy intellectual burden on its practitioners.

Beyond its methodological contributions, it must be said that this project derives from and, we believe, advances distinctively cosmopolitan political convictions rooted in the challenges of our moment. While transnational scholarship need not critically confront chauvinism and its associated geopolitical barriers, such an approach does offer powerful resources to understand and subvert nationalistic discourses that other and exclude. The desire to engage with the world, to gain from and nurture fruitful interactions across borders, can be a powerful motivation for the kind of research and collaboration required of transnational history. Such research and collaboration can challenge facile exceptionalisms and underscore hidden dependencies, while situating both with respect to the operation of state power. If history is to be a virtuous resource for present politics, it may ultimately have to be transnational history.

Notes

1. E.g., Michael McGerr, "The Price of the 'New Transnational History,'" *American Historical Review* 96, no. 4 (1991): 1056–1067, at 1066; Ann Curthoys and Marilyn Lake, eds., *Connected Worlds: History in Transnational Perspective* (Canberra: Australian National University Press, 2005), 14.

2. Akira Iriye, "Internationalizing International History," in *Rethinking American History in a Global Age*, ed. Thomas Bender (Berkeley: University of California Press, 2002), 47–62, at 51.

3. For the R&D expenditures, see http://www.aaas.org/sites/default/files/migrate/uploads/Chart.-Shares-of-Total-World-RD-2007.pdf.

4. Marilyn B. Young, "The Age of Global Power," in Bender, *Rethinking American History in a Global Age*, 274–294, at 291.

5. Private communication with John Krige, Mar. 25, 2017

6. Kostas Gavroglu et al., "Science and Technology in the European Periphery: Some Historiographical Reflections," *History of Science* 46 (2008): 153–175; Martin Kohlrausch and Helmut Trischler, *Building Europe on Expertise: Innovators, Organizers, Networkers* (New York: Palgrave Macmillan, 2104); Erik van der Vleuten and Arne Kaijser, eds., *Networking Europe: Transnational Infrastructures and the Shaping of Europe, 1850—2000* (Sagamore Beach, MA: Watson Publishing International, 2006).

7. Alison Sandman, "Controlling Knowledge: Navigation, Cartography, and Secrecy in the Early Modern Spanish Atlantic," in *Science and Empire in the Atlantic World*, ed. James Delbourgo and Nicholas Dew (New York: Routledge, 2008), 31–51.

Contributors

Michael J. Barany
Society of Fellows
Dartmouth College
Hanover, NH 03755
USA

Mario Daniels
BMW Center for German and European Studies
Georgetown University
Washington, DC 20057
USA

Olival Freire Jr.
Instituto de Física
Universidade Federal da Bahia
Salvador, BA 40210-340
Brazil

Miriam Kingsberg Kadia
234 UCB
University of Colorado Boulder
Boulder, CO 80309
USA

John Krige
Kranzberg Professor
School of History and Sociology
Georgia Institute of Technology
Atlanta, GA 30332
USA

Prakash Kumar
Department of History
Pennsylvania State University
University Park, PA 16802
USA

Neil M. Maher
Federated History Department
New Jersey Institute of Technology—Rutgers University
Newark, NJ 07102
USA

Gisela Mateos
Centro de Investigaciones Interdisciplinarias en Ciencias y Humanidades
Universidad Nacional Autónoma de México
Mexico City
Mexico

Adriana Minor
Instituto de Física—Laboratório Ciência Como Cultura
Universidade Federal da Bahia
Salvador, BA 40210-340
Brazil

Tiago Saraiva
Department of History
Drexel University
Philadelphia, PA 19104
USA

Asif Siddiqi
Department of History
Fordham University
Bronx, NY 10458
USA

Indianara Silva
Departamento de Física
Universidade Estadual de Feira de Santana
Feira de Santana, BA 44036-900
Brazil

Josep Simon
Escuela de Medicina y Ciencias de la Salud
Universidad del Rosario
Bogotá 31 111221
Colombia

Edna Suárez-Díaz
Estudios de la Ciencia y la Tecnología
Facultad de Ciencias
Universidad Nacional Autónoma de México
Mexico City
Mexico

Zuoyue Wang
Department of History
California State Polytechnic University
Pomona, CA 91768
USA

Index

Page numbers followed by "f" indicate figures.